T0360809

HOW TO MEASURE THE INFINITE
Mathematics with Infinite and Infinitesimal Numbers

HOW TO MEASURE THE INFINITE
Mathematics with Infinite and Infinitesimal Numbers

Vieri Benci

Università di Pisa, Italy

Mauro Di Nasso

Università di Pisa, Italy

 **World Scientific**

NEW JERSEY · LONDON · SINGAPORE · BEIJING · SHANGHAI · HONG KONG · TAIPEI · CHENNAI · TOKYO

Published by

World Scientific Publishing Co. Pte. Ltd.

5 Toh Tuck Link, Singapore 596224

USA office: 27 Warren Street, Suite 401-402, Hackensack, NJ 07601

UK office: 57 Shelton Street, Covent Garden, London WC2H 9HE

Library of Congress Cataloging-in-Publication Data

Names: Benci, V. (Vieri), author. | Di Nasso, Mauro, 1963–   author.
Title: How to measure the infinite : mathematics with infinite and infinitesimal numbers /
   by Vieri Benci (Università di Pisa, Italy), Mauro Di Nasso (Università di Pisa, Italy).
Description: New Jersey : World Scientific, 2018. | Includes bibliographical references and index.
Identifiers: LCCN 2018024555 | ISBN 9789812836373 (hardcover : alk. paper)
Subjects: LCSH: Nonstandard mathematical analysis. | Calculus. | Series, Infinite.
Classification: LCC QA299.82 .B46 2018 | DDC 511.3--dc23
LC record available at https://lccn.loc.gov/2018024555

British Library Cataloguing-in-Publication Data

A catalogue record for this book is available from the British Library.

For any available supplementary material, please visit
https://www.worldscientific.com/worldscibooks/10.1142/7081#t=suppl

Printed in Singapore

Preface

A belief in the infinitely small does not triumph easily. Yet when one thinks boldly and freely, the initial distrust will soon mellow into a pleasant certainty ... A majority of educated people will admit an infinite in space and time, and not just an "unboundedly large." But they will only with difficulty believe in the infinitely small, despite the fact that the infinitely small has the same right to existence as the infinitely large.

(Paul du Bois-Reymond, *On the paradoxes of the infinitary calculus*, 1877)[1]

Non-Archimedean mathematics is a branch of mathematics based on the use of *infinitesimal* and *infinite* numbers. We believe that it is a very interesting and rich field and – in many circumstances – non-Archimedean mathematics allows to construct models of the physical world in a more elegant and simple way, with respect to the traditional Archimedean mathematics grounded on the real numbers.

Throughout the history of human thought, the nature of infinite and infinitesimal quantities has been investigated by the best minds, and their actual use in calculus led to the modern development of mathematics. However, in the first years of the last century, the infinitesimals were banned from the world of mathematics, as their use was considered inappropriate for a rigorous work. The following words of

---

[1] As translated by J.J. O'Connor and E.F. Robertson in *"Paul David Gustav du Bois-Reymond"*, http://www-groups.dcs.st-and.ac.uk/history/Biographies/Du_Bois-Reymond.html.

Bertrand Russell describe the prevailing opinion in the scientific community: "Infinitesimals as explaining continuity must be regarded as unnecessary, erroneous, and self-contradictory."[2]

As a result, non-Archimedean mathematics has been forgotten until the 1960s when Abraham Robinson introduced his *Nonstandard Analysis* (NSA). Nonstandard Analysis represents a milestone in non-Archimedean mathematics, since it grounded the use of infinitesimal numbers on firm foundations, and provided powerful tools, most notably the *Transfer Principle*.

This book is an introduction to *Alpha-Theory*, which can be considered as a variant of NSA based on the existence of a new "infinite" number "Alpha" and on the notion of "Alpha-limit" of any sequence. There are relevant differences between Alpha-theory and the usual approaches to NSA, both about the methods and about the underlying philosophy.

The Alpha-limit of a sequence of mathematical entities is a new object that preserves most of the properties of the "approximating entities." Such new objects are named *internal* and – in general – they are not present in the usual mathematics. The fundamental example is given by the non-Archimedean field of *hyperreal numbers*, obtained as the set of Alpha-limits of real sequences; indeed, such a field include infinitesimal and infinite numbers. A nice feature of Alpha-limits is that they allow to recover most of the language, tools and results of (standard) analysis in a straightforward way. We remark that by following this approach, one can introduce non-Archimedean mathematics avoiding the technicalities of formal logic, which are used only at a later stage when treating more advanced questions. Another difference with respect to NSA is that one works in a single universe, instead of assuming the existence of two distinct mathematical universes, namely the standard universe and the nonstandard universe.[3]

Let us now examine the differences between Alpha-Theory and NSA in the "philosophy." Usually, non-Archimedean mathematics follows the traditional point of view of Archimedes and Leibniz, in that it is considered as a methodology that takes advantage of infinitesimal and infinite numbers as mere devices to simplify the proofs of theorems. On the contrary, our ideas are closer to the viewpoint of Paul

---

[2] B. Russell, §324 of *"Principles of Mathematics,"* 1903.

[3] This idea of a single universe is shared with E. Nelson's approach [**77**] to NSA, called *Internal Set Theory* (IST). However, our Alpha-Theory and IST are quite distinct (this is commented in §3.4 of Chapter 13).

Du Bois-Reymond, Giuseppe Veronese, Tullio Levi-Civita and Federigo Enriques: we think that infinitesimals have the same status as the others mathematical objects, and therefore they can be used to build models. In our opinion, the advantages of a theory which includes infinitesimal and infinite numbers rely more on the possibility of constructing new models, than in the techniques used in the proofs. Actually, Chapter 10 on the *Brownian motion* and the whole Part 5 on the *numerosity theory* in this book, are fully inspired by this principle.

Contents

Plan of the Book

This book is organized in five parts.

The *Alpha-Calculus* presented in Part 1 is a new elementary axiomatics for developing a calculus grounded on the use of infinitesimal and infinite numbers. Basically, it is a simplified version of *nonstandard analysis*, that still allows for a complete and rigorous treatment of all the basics of calculus. An important feature with respect to the usual approaches to nonstandard analysis, is that Alpha-Calculus is presented in a truly elementary way, and it needs no more prerequisites than traditional calculus. Alpha-Calculus is then generalized to *Alpha-Theory* in Part 2, where the fundamental principles of nonstandard methods, namely the *transfer principle* and the *saturation property*, are introduced and proved. A selection of applications is given in Part 3 to illustrate the use of Alpha-Theory in different mathematical contexts. Part 4 is devoted to the foundations of the presented material. In particular, it is shown that Alpha-Theory incorporates the full strength of the methods of nonstandard analysis, and can be formalized as a general "nonstandard set theory" that is a conservative extension of the usual Zermelo-Fraenkel set theory. The final Part 5 is entirely dedicated to the theory of *numerosity* and its close relationships with Alpha-Theory.

Acknowledgements. The authors thank Marco Forti for many valuable discussions; and thank Jerry Keisler for useful remarks on a preliminary version of this book.

Notation

- $\mathbb{N} = \{1, 2, 3, \ldots\}$ is the set of natural numbers.
- $\mathbb{N}_0 = \{0, 1, 2, 3, \ldots\}$ is the set of non-negative integers.
- $\mathbb{Z}$ is the set of integer numbers.
- $\mathbb{Q}$ is the set of rational numbers.
- $\mathbb{R}$ is the set of real numbers.
- If R in an ordered ring, then
  - $R^+ = \{x \in R \mid x > 0\}$
  - $R_0^+ = \{r \in R \mid x \geq 0\}$.
  - $R^- = \{x \in R \mid x < 0\}$
  - $R_0^- = \{r \in R \mid x \leq 0\}$.
- $\mathbb{N}^*, \mathbb{Z}^*, \mathbb{Q}^*, \mathbb{R}^*$ are the sets of hypernatural, hyperinteger, hyperrational and hyperreal numbers, respectively.
- $|A|$ is the cardinality of a set A.
- $\mathfrak{c} = |\mathbb{R}|$ is the cardinality of the *continuum*.
- $\mathcal{P}(A) = \{B \mid B \subseteq A\}$ is the powerset of a set A.
- $A^c = X \setminus A$ is the complement of a subset $A \subseteq X$.
- $\mathfrak{Fin}(A)$ is the set of finite parts of a set A.
- if $(X, <)$ is a linearly ordered set, and $a, b \in X$, then
  - $[a, b]_X = \{x \in X \mid a \leq x \leq b\}$;
  - $(a, b)_X = \{x \in X \mid a < x < b\}$.
- $\mathfrak{Fun}(A, B) = \{f : A \to B\}$ is the set of functions from A to B.
- $\mathrm{dom}(f)$ is the domain of a function f.
- $\mathrm{ran}(f)$ is the range of a function f.
- $f_{\restriction A}$ is the restriction of the function f to the set A.
- $\lim_{n \uparrow \alpha} \varphi(n)$ or $\lim_\alpha \varphi$ is the Alpha-limit of a sequence φ.
- $\imath : \mathbb{N} \to \mathbb{N}$ is the identity map.
- $c_a : \mathbb{N} \to \{a\}$ is the constant sequence with value a.

- $\xi \sim \zeta$ means that the numbers ξ and ζ are infinitely close, that is, $\xi - \zeta$ is infinitesimal.
- $\xi \simeq \zeta$ means that the positive hyperreal numbers $\xi, \zeta \in \mathbb{R}^{*+}$ have the same order of magnitude, i.e. $a < \frac{\xi}{\zeta} < b$ for suitable $a, b \in \mathbb{R}^{+}$.
- $\xi \ll \zeta$ or $(\zeta \gg \xi)$ means that $\xi \in \mathbb{R}^{*+}$ has a smaller order of magnitude than $\zeta \in \mathbb{R}^{*+}$, i.e. $\frac{\xi}{\zeta} \sim 0$. We agree that $0 \ll \xi$ for all $\xi \in \mathbb{R}^{*+}$.
- $\xi \approx \zeta$ means that $\xi, \zeta \in X^*$ are infinitely close with respect to the gauge $\approx$.
- $\xi \sim_f \zeta$ means that $\xi \in \mathbb{R}^*$ is finitely close to $\zeta \in \mathbb{R}^*$, i.e. $\xi - \zeta$ is finite.
- The monad of x is the set of numbers that have infinitesimal distance from x:

$$\mathsf{mon}(x) = \{\xi \mid \xi \sim x\}.$$

So, $\mathsf{mon}(0) = \{\xi \mid \xi \sim 0\}$ is the set of infinitesimal numbers.
- The galaxy of x is the set of numbers that are finitely close to x:

$$\mathsf{gal}(x) = \{\xi \mid \xi \sim_f x\}.$$

So, $\mathsf{gal}(0)$ is the set of finite numbers.
- $\mathbb{Q}^*_{\mathrm{fin}}$ is the set of all finite hyperrational numbers.
- $\mathbb{R}^*_{\mathrm{fin}}$ is the set of all finite hyperreal numbers.
- $\mathbb{N}_\infty = \mathbb{N}^* \setminus \mathbb{N}$ is the set of all infinite hypernatural numbers.
- $\mathbb{R}_\infty = \mathbb{R}^* \setminus \mathbb{R}^*_{\mathrm{fin}}$ is the set of all infinite hyperreal numbers.
- $\alpha = \lim_{n \uparrow \alpha} n$ is the Alpha-limit of the identity sequence.
- $\eta = \frac{1}{\alpha}$ is the reciprocal of α.
- $\mathcal{Q} = \{A \subseteq \mathbb{N} \mid \alpha \in A^*\}$ is the family of qualified sets.
- $\mu_\alpha : \mathbb{N} \to \{0, 1\}$ is the Alpha-measure.
- $\mathscr{C}(E)$ is the set of continuous real-valued functions defined on the topological space E.
- $\mathscr{C}_c(\Omega)$ is the set of continuous real-valued functions defined on Ω with compact support.
- $\mathscr{C}^k(\Omega)$ is the set of real-valued functions defined on the set $\Omega \subseteq \mathbb{R}^n$ which have continuous derivatives up to the order k.
- $\mathscr{D}(\Omega)$ is the set of the infinitely differentiable real-valued functions with compact support defined on the set $\Omega \subseteq \mathbb{R}^n$.
- $B_r(x) = \{y \in \mathbb{R}^n \mid |y - x| \le r\}$ is the closed ball in $\mathbb{R}^n$ with center x and radius r.
- $\|A\|$ is the internal cardinality of a hyperfinite set A.
- $\mathbb{H}(n) = \{\pm \frac{i}{n} \mid i = 0, 1, \ldots, n^2\}$ is the grid of ratio $\frac{1}{n}$.

- $\mathbb{H}(\boldsymbol{\alpha}) = \lim_{n \uparrow \alpha} \mathbb{H}(n)$.
- "iff" is an abbreviation for "if and only if".
- "*a.e.*" is an abbreviation for "μ_α-almost everywhere".
- By abusing notation, sometimes we write $\{a_n\}$ to mean the sequence $\langle a_n \mid n \in \mathbb{N} \rangle$.[4]

---

[4] Notice that $\{a_n\}$ would denote the *range* of the sequence.

Historical Introduction

1. Ancient times

Already at the dawn of the ancient Greek and Hellenistic civilization, the problem of the existence of infinitesimal quantities – as closely related to the notions of continuity and of motion – was a central theme of the philosophical and mathematical discussion.

Historically, the first explicit discussion goes back to the celebrated paradoxes of the philosopher Zeno (5th century BC), including the famous "Achilles and the Tortoise" conundrum. In his *Elements*, dated around 300 BC, Euclid seems to be well aware of the problem of infinitesimals. In the part covering the theory of ratio and proportion, one reads the following definition:

> *Magnitudes are said to have a ratio with respect to one another which, being multiplied, are capable of exceeding one another.*
>
> $\big($Book 5, Definition 4 of Euclid's *Elements*$\big)$[5]

Notice that homogeneous magnitudes that *do not* have "a ratio with respect to one another," are precisely magnitudes A and B that are one *infinitesimal* with respect to the other, that is, such that for every natural number m:

$$\underbrace{A + \ldots + A}_{m \text{ times}} < B.$$

Nowadays, the assumption that all magnitudes have a ratio to one another, is known as *Archimedean property*. The use of the adjective "Archimedean" is of recent origin (it goes back to the 19th century),

---

[5] As translated by R. Fitzpatrick, in *"Euclid's Elements of Geometry,"* 2007.

and it is founded on the repeated use that Archimedes made of that property, which he credited to Eudoxus of Cnidus.

Archimedes (3th century BC), the greatest mathematician of ancient times, systematically used infinitesimals to "guess" the right computation of surfaces or volumes. Although Archimedes fully accepted infinitesimals as a powerful heuristic tool, nevertheless he did not accept their use in rigorous proofs. To this end, he always supplied proofs by the so-called *method of exhaustion*. From the point of view of a contemporary mathematician, the method of exhaustion can be seen as a rigorous technique of analysis, that can be fully developed within the axiomatic framework of Euclid's Elements. Archimedes' procedure is highlighted in his fundamental work *"The Method."*

> *Now the fact here stated is not actually demonstrated by the argument used; but the argument has given a sort of indication that the conclusion is true. Seeing then that the theorem is not demonstrated, but at the same time suspecting that the conclusion is true, we shall recourse to the geometrical demonstration which I myself discovered.*

(Archimedes, *The Method*)[6]

The level of awareness and of mathematical rigor that emerge from these words is striking.[7] It is a relevant fact that what is now considered by many as one of the major works of ancient mathematics, was discovered in modern times only in 1906, and in a completely unexpected way. Indeed, behind the script of a 13th century prayer book that the philologist J.L. Heibergh was examining, the faint writing of an older manuscript appeared: it was the only survived copy of "The Method".[8] The story of that finding hints the possibility that many fundamental works of ancient mathematics are lost forever.

---

[6] As translated by T.L. Heath, in *"The Method of Archimedes"*, Cambridge University Press, 1912.

[7] About the (underestimated) achievements of the scientific thought in the Hellenistic age, we warmly suggest the book by L. Russo, suggestively entitled: *"The Forgotten Revolution: How Science Was Born in 300 BC and Why it Had to Be Reborn,"* Springer, 2004.

[8] The reader interested in the adventurous story of this lost manuscript, can find plenty of details in the popular book *"The Archimedes Codex"* by R. Netz and W. Noel, Da Capo Press, 2007.

2. The rise of calculus

Following a long period of almost complete oblivion, the works of Euclid and Archimedes were recovered and studied by mathematicians in the 16th and 17th centuries, laying the groundwork for integral and differential calculus. Among the most relevant mathematicians of that age, let us recall Simon Stevin (1548–1620), Luca Valerio (1552–1628), Johannes Kepler (1571–1630), Bonaventura Cavalieri (1598–1647), Pierre de Fermat (1601–1665), Evangelista Torricelli (1608–1647), René Descartes (1596–1650), John Wallis (1616–1703), and Blaise Pascal (1623–1662).

In the years preceding the discovery of calculus, the idea of an infinitesimal quantity did not play a big role in mathematics; two relevant exceptions are the work by John Wallis (see his book *"Treatise on the Conic Sections,"*, 1655) and the "indivisibles" of Bonaventura Cavalieri (see his book *"Geometria indivisibilibus continuorum nova quadam ratione promota,"* 1653).

Modern calculus started with the methods elaborated by Gottfried Leibniz (1646–1716) and by Isaac Newton (1643–1727). With respect to this, it is worth mentioning that in the first years of the 18th century a great controversy arose about the authorship of the invention of calculus between Leibniz and Newton, fomented by the competition between the English and the European mathematical schools. Most of the recent historiography concluded that – although Newton's studies preceded Leibniz' by a few years – the two scientists discovered calculus independently of each other.

Leibniz made a direct use of infinitesimal numbers; for example, he introduced the by now familiar notation $\frac{df}{dx}$ as ratio of infinitesimal increments. His philosophical position about the status of infinitesimal quantities is not clearly and unequivocally expressed in his writings, and is still the object of studies and disputes. According to authoritative interpretations, Leibniz saw the infinitesimal numbers as "ideal" elements that are only introduced for convenience, in order to simplify proofs and to facilitate the discovery of new results. He believed that their use would be ruled by the same laws of real numbers and that would lead to exact results, but he never provided a fully clear and satisfying justification to such an assumption. This vision would later inspire the formulation of the so-called *Leibniz transfer principle*, which was proposed by Abraham Robinson in the 1960s as the foundation of *nonstandard analysis*.

Many of the fundamental theorems of calculus were discovered between the late 17th and early 18th centuries, and were justified by explicitly using infinitesimal quantities; among them: Rolle's Theorem (1691), L'Hospital's rule (1696), Taylor's formula (1715). It is just the case to note that, from the point of view of contemporary mathematics, such original proofs can hardly be considered as correct.

The first comprehensive monograph on the new infinitesimal analysis it is traditionally considered *"Analyse des infiniments petits"* of Guillaume de l'Hospital (1661–1704). The book opens with some definitions and postulates. We include one, to give an idea of how the notion of infinitesimal was used in those early times of calculus.

> *We suppose that a curved line may be considered as an assemblage of infinitely many straight lines, each one being infinitely small, or (what amounts to the same thing) as a polygon with an infinite number of sides, each being infinitely small, which determine the curvature of the line by the angles formed amongst themselves.*

(G.F. L'Hôpital, Postulate II of *Analyse des infiniments petits*, 1696)[9]

Alongside the enthusiasm generated by the extraordinary results which were reached, soon it began to spread the concern for the very dubious logical bases of that new mathematics. The philosopher George Berkeley (1685–1753), made explicit that criticism on the foundations of infinitesimal analysis. In the famous book *"The analyst"* addressed to "an infidel mathematician," Berkeley supported the thesis that the objects and methods of the new calculus did not have more justification than "religious mysteries and points of faith."

> *And what are these fluxions? The velocities of evanescent increments? And what are these same evanescent increments? They are neither finite quantities nor quantities infinitely small, nor yet nothing. May we not call them the ghosts of departed quantities?*

(G. Berkeley, *The analyst*, 1734)

Beyond the wit of the phrases listed here, the criticism contained in *"The Analyst"* were well argued, and hit a sore spot of infinitesimal calculus, as it was practiced at the time. In particular, Berkeley highlighted the contradictions in the use of the infinitesimal increments

---

[9] As translated by R.E. Bradley, S.J. Petrilli, and C. E. Sandifer in *"L'Hôpital's Analyse des infiniments petits,"* Birkhäuser, 2015.

in the calculation of derivatives. To understand how mathematicians used to thinking at the time, we quote Leonhard Euler (1707–1783), who is considered one of the greatest mathematicians of history.[10]

> [Differentials] since they are without quantity, they are also said to be infinitely small. Hence, by their nature they are to be so interpreted as absolutely nothing, or they are considered to be equal to nothing. Thus, if the quantity x is given an increment ω, so that it becomes $x + \omega$, its square x^2 becomes $x^2 + 2x\omega + \omega^2$, and it takes the increment $2x\omega + \omega^2$. Hence, the increment of x itself, which is ω, has the ratio to the increment of the square, which is $2x\omega + \omega^2$, as 1 to $2x + \omega$. This ratio reduces to 1 to $2x$, at least when ω vanishes. Let $\omega = 0$, and the ratio of these vanishing increments, which is the main concern of differential calculus, is as 1 to $2x$. On the other hand, this ratio would not be true unless that increment ω vanishes and becomes absolutely equal to zero.

(L. Euler, Institutiones calculis differentialis, 1755)[11]

In formulas, if f is the "square" function, x is a fixed point, and ω denotes an infinitesimal increase, the above argument can be summarized as follows.[12]

$$f'(x) \;=\; \frac{(x + \omega)^2 - x^2}{\omega} \;=\; \frac{x^2 + 2x\omega + \omega^2 - x^2}{\omega}$$

$$=\; \frac{\omega(2x + \omega)}{\omega} \;=\; 2x + \omega \;=\; 2x.$$

Berkeley polemically pointed fingers on the contradictions of such a kind of reasoning where, at first, it is supposed that $\omega \neq 0$ (otherwise the difference quotient is not defined), and later, ω is ignored by equating it with zero. So, the infinitesimal increment ω is considered at the same time both different and equal to zero, and therefore it deserves to be called "a ghost of departed quantities."

---

[10] For historical accuracy, it should be pointed out that Euler's quotation dates a few years later than Berkeley's criticism. However, it is in no way influenced, and reproduces in an exemplary way a type of argument that was very common throughout the 18th century.

[11] As translated by J.D. Blanton, *Foundations of Differential Calculus*, Springer, 2000.

[12] Compare with the nonstandard proof of the derivative of $f(x) = x^2$ given later on in Example 3.13.

3. The ban of infinitesimals

Despite the discomfort and embarrassment caused by Berkeley's criticism, the direct use of infinitesimals remained a long-standing common practice. Mathematicians, scientists, and engineers continued to use infinitesimals, and discovered new correct results. Indeed, the enthusiasm for the many new important results that were being discovered did not help with foundational studies. Among the most prominent mathematicians of this era of great developments, let us mention, besides Leonard Euler, Colin MacLaurin (1698-1746), Jean D'Alembert (1717-1783), and Joseph Fourier (1768-1830).

After the important contributions of Lazare Carnot (1753-1823) and Louis Lagrange (1736-1813), a relevant step towards greater rigor was done by Louis Cauchy (1789-1857), whose *"Cours d'analyse de l'École Polytechnique"* is traditionally considered the first systematic treatise of modern analysis. Introducing the contemporary point of view, Cauchy grounded all fundamental concepts of calculus (namely continuity, derivation, integration) on the theory of limits.

Only in the second half of the 19th century, the foundational research culminated with the so-called "ϵ-δ formalization," due to Karl Weierstrass, the one currently adopted. Finally, calculus was put on firm foundations, but the price to be paid was the ban of infinitesimal and infinite numbers. Along with the contributions of Richard Dedekind and Georg Cantor, Weierstrass's work closed an amazing era for the foundations of mathematics. The enthusiasm of contemporaries for the great achievements is well represented by the following words by Bertrand Russell:

> *Zeno was concerned, as a matter of fact, with three problems ... These are the problems of the infinitesimal, the infinite, and continuity ... From him to our own day, the finest intellects of each generation in turn attacked the problems, but achieved, broadly speaking, nothing. In our own time, however, three men – Weierstrass, Dedekind, and Cantor – have not merely advanced the three problems, but have completely solved them. The solutions, for those acquainted with mathematics, are so clear as to leave no longer the slightest doubt or difficulty.*

> (B. Russell, *Mysticism and Logic*, 1901)

Alongside a further impetus of research, an intense work started to put previous research on sound basis within the new foundational framework.

Inevitably, mathematicians started to share the common view that, in the words of B. Russell:

> *Infinitesimals as explaining continuity must be regarded as unnecessary, erroneous, and self-contradictory.*

(B. Russell, *The principles of Mathematics*, 1903).

However, the elimination of the infinitesimal numbers from mathematical practice was neither immediate nor painless, as the following quotations testify.

> *The idea of an infinitesimal involves no contradictions ...*
> *As a mathematician, I prefer the method of infinitesimals to that of limits, as far easier and less infested with snares.*

(C.S. Pierce, *The Law of Mind*, 1891.)

> *I have at last become fully satisfied that the language and idea of infinitesimals should be used in the most elementary instruction – under all safeguards of course.*

(A. de Morgan, *Grave's Life of W.R. Hamilton*, 1882–1889.)

4. Non-Archimedean mathematics

Although out of the mainstream, the study of numerical systems containing infinitesimals continued throughout the late 19th and the beginning of the 20th centuries, thanks to the work of Giuseppe Veronese [**93, 94**], Tullio Levi-Civita [**69**], Paul du Bois-Reymond [**41**], and others.[13] However, soon their research was almost completely forgotten, and it was not until the 1960s that non-Archimedean mathematics retrieved a role in mathematics, with the advent of A. Robinson's *nonstandard analysis* [**83**]. Since then, the nonstandard methods have been successfully applied and have led to new results in such diverse fields of mathematics as functional analysis, measure and probability theory, stochastic analysis, hydromechanics, combinatorial number theory, *etc.*

While Veronese, Levi-Civita and Bois-Reymond considered the infinitesimals as legitimate objects of mathematics to be used to construct models, with nonstandard analysis you come back to the concept of Archimedes and Leibniz, who (roughly speaking) viewed infinitesimals as convenient tools for calculation, but not suitable to model the physical world. Our *Alpha-Theory* presented in this book, can be seen as a

---

[13] For a detailed report and discussion on the history of non-Archimedean mathematics, the reader is warmly recommended to consult P. Ehrlich's works [**43, 44**].

simplified approach to nonstandard analysis where "infinitesimals are taken seriously," in the spirit of the afore mentioned researchers.

Part 1

Alpha-Calculus

The *Alpha-Calculus* is a new elementary method for developing a calculus grounded on the use of infinitesimal and infinite numbers. Basically, it is a simplified version of *nonstandard analysis*, that still allows for a complete and rigorous treatment of all the basics of calculus. Most notably, *Alpha-Calculus* is expressed in the everyday language of mathematics and only assumes a few basic notions from algebra.

In the first chapter, we study the class of the ordered fields that properly extend the real line. The *Archimedean property* necessarily fails, and one can distinguish between infinitely small, finite, and infinitely large numbers. A few basic algebraic properties of numbers according to their "size" is then established, and the fundamental notion of *standard part* is introduced.

After "warming-up" with the general properties of fields that extend the real line, in the second chapter we introduce the axioms of Alpha-Calculus that rule the properties of the *hyperreal line* $\mathbb{R}^*$. The hyperreal numbers are an ordered field that extends the real line, and that satisfies remarkable additional properties. Most notably, every subset $A \subseteq \mathbb{R}$ and every real function $f : A \to B$ can be canonically extended to a subset $A^* \subseteq \mathbb{R}^*$ (called the *hyperextension* of A), and to a function $f^* : A^* \to B^*$ (the *hyperextension* of f), respectively. This is done in such a way that all "elementary properties" are preserved under the hyper-extensions.[14]

In the third chapter, the main basic notions and results of calculus are developed within the Alpha-Theory, and a selection of relevant classic results – such as the *Extreme Value* and the *Intermediate Value* theorems, and the *Fundamental Theorem of Calculus* – are proved. In particular, a notion of *grid integral* is introduced, which generalizes the usual Riemann integral and makes sense for *all* real functions.

According to the pedagogical purposes of this part, in the exposition the *transfer principle* is never mentioned. Rather, on a case by case basis, we prove directly only those instances that are actually needed to obtain the desired results.

---

[14] The notion of "elementary property" will be given a precise meaning in Chapter 5 by using the formalism of first-order logic.

CHAPTER 1

Extending the Real Line

In this chapter we study the basic properties of ordered fields that properly extend the real line. In such fields the Archimedean property necessarily fails and in consequence, formalizations of the intuitive notions of a "small" number and of a "large" number can be naturally given.

1. Ordered fields

DEFINITION 1.1. *An ordered field is a couple* $(\mathbb{F}, \mathbb{F}^+)$ *where* $\mathbb{F}$ *is a field, and the subset* $\mathbb{F}^+ \subset \mathbb{F}$ *of positive elements satisfies the following:*

- *If* $x, y \in \mathbb{F}^+$, *then* $x + y \in \mathbb{F}^+$ *and* $x \cdot y \in \mathbb{F}^+$;
- $\mathbb{F}$ *is the disjoint union* $\mathbb{F}^- \cup \{0\} \cup \mathbb{F}^+$, *where* $\mathbb{F}^- = \{-x \mid x \in \mathbb{F}^+\}$ *is the subset of negative elements.*

The ordering on $\mathbb{F}$ *is defined by setting:*

$$x < y \iff y - x \in \mathbb{F}^+.$$

It is easily verified that the relation $x < y$ as defined above is a *linear* ordering.

Following the common practice, we identify each $n \in \mathbb{N}$ with the corresponding iterated sum of the neutral element $1 \in \mathbb{F}$, *i.e.*:

$$n = \underbrace{1 + \ldots + 1}_{n \text{ times}}.$$

Thus, by considering opposites and reciprocals, we will directly assume that $\mathbb{Q} \subseteq \mathbb{F}$. Accordingly, by $n \cdot x$ we mean the iterated sum $\underbrace{x + \ldots + x}_{n \text{ times}}$.

In the ancient Greek geometry, the following property (attributed to Eudoxus) was considered:

> *Magnitudes are said to have a ratio to one another which are capable, when multiplied, of exceeding one another.*
>
> (Euclid's *Elements*, Book V, Definition IV).[1]

In modern mathematics, that property is referred to as the *Archimedean property*, and it is naturally formulated in the context of ordered fields.

DEFINITION 1.2. *An ordered field* $\mathbb{F}$ *is* Archimedean *if the iterated sums of any positive element* y *surpass any given number* x. *In formulas:*

$$\forall x > 0 \; \forall y > 0 \; \exists n \in \mathbb{N} \quad n \cdot y \geq x.$$

Equivalently,

$$\forall z > 0 \; \exists n \in \mathbb{N} \quad n \geq z.$$

Clearly, the latter property is obtained from the former by taking $x = z$ and $y = 1$. Conversely, the former follows from the latter by considering the ratio $z = x/y$.

2. Infinitesimal numbers

Another notion that has been repeatedly considered in the history of mathematics is that of an "infinitely small" or "infinitesimal" quantity. Again, thanks to the language of ordered field, a precise formalization is possible.

DEFINITION 1.3. *Let* $\mathbb{F}$ *be an ordered field. An element* $\varepsilon \in \mathbb{F}$ *is called* infinitesimal *if* $-\frac{1}{n} < \varepsilon < \frac{1}{n}$ *for all* $n \in \mathbb{N}$.

---

[1] Euclid's *Elements*, written approximately 300 BC, have been used as a textbook by virtually all educated people in western history until the beginning of the 20th century. In that volume, whose importance in the historical development of mathematical and scientific thought could hardly be over-estimated, substantial parts of geometry and number theory are developed, and the basis of the modern notion of proof were founded. (See [**46**] for a modern commented edition.)

Trivially, the neutral element 0 is infinitesimal. Notice that in the real field $\mathbb{R}$, the number 0 is the unique infinitesimal element. The next result directly follows from the definitions.

PROPOSITION 1.4. *An ordered field is* Archimedean *if and only if its only infinitesimal number is 0.*

PROOF. By contradiction, let $\varepsilon \neq 0$ be an infinitesimal number. Without loss of generality, let us assume that $\varepsilon > 0$. The number $z = 1/\varepsilon$ provides a counter-example to the Archimedean property, because $n < 1/\varepsilon$ for all $n \in \mathbb{N}$.

Conversely, let $z > 0$ be a counter-example to the Archimedean property, *i.e.* $n < z$ for all $n \in \mathbb{N}$. Then its reciprocal $\varepsilon = 1/z$ is a non-zero infinitesimal number, because $0 < \varepsilon < 1/n$ for all $n \in \mathbb{N}$. □

From this point on, we will assume that a non-Archimedean field $\mathbb{F}$ has been fixed. In this section, we investigate the algebraic properties of numbers according to their "size". Besides infinitesimal numbers, we will also consider the analogous notion of an infinite number. To simplify things, let us adopt the usual *absolute value* notation:

$$|\xi| = \begin{cases} \xi & \text{if } \xi \geq 0\,; \\ -\xi & \text{if } \xi < 0\,. \end{cases}$$

DEFINITION 1.5. *Let $\xi \in \mathbb{F}$. We say that:*
- *ξ is* infinitesimal *if $|\xi| < \frac{1}{k}$ for all $k \in \mathbb{N}$;*
- *ξ is* finite *(or* bounded*) if there exists $k \in \mathbb{N}$ with $|\xi| < k$;*
- *ξ is* infinite *(or* unbounded*) if $|\xi| > k$ for all $k \in \mathbb{N}$ (equivalently, if ξ is not finite).*

The inverse of an infinite number is infinitesimal. Conversely, the inverse of a non-zero infinitesimal number is infinite. Clearly, all infinitesimal numbers are finite. We remark that in any non-Archimedean field there are plenty of finite numbers that are neither infinitesimal nor rational; indeed, for every non-zero $q \in \mathbb{Q}$ and for every non-zero infinitesimal ε, the number $q + \varepsilon \notin \mathbb{Q}$ is finite and non-infinitesimal.

In the next proposition, we itemize a list of simple algebraic properties that match the intuition of "small", "finite", and "infinite" quantities.

PROPOSITION 1.6.
(1) *If ξ and ζ are finite, then $\xi + \zeta$ and $\xi \cdot \zeta$ are finite;*

(2) *If ξ and ζ are infinitesimal, then $\xi + \zeta$ is infinitesimal;*

(3) *If ξ is infinitesimal and ζ is finite, then $\xi \cdot \zeta$ is infinitesimal;*

(4) *If ξ is infinite and ζ is not infinitesimal, then $\xi \cdot \zeta$ is infinite;*

(5) *If $\xi \neq 0$ is infinitesimal and ζ is not infinitesimal, then $\frac{\xi}{\zeta}$ is infinite;*

(6) *If ξ is infinitesimal and ζ is finite, then $\frac{\zeta}{\xi}$ is infinitesimal.*

PROOF. All proofs follow from the definitions in a straightforward manner. As an example, let us prove (4). Since ζ is not infinitesimal, there exists $n \in \mathbb{N}$ such that $|\zeta| > 1/n$. Now let $k \in \mathbb{N}$. As ξ is infinite, certainly $|\xi| > n \cdot k$, and so $|\xi \cdot \zeta| = |\xi| \cdot |\zeta| > (n \cdot k) \cdot \frac{1}{n} = k$. As this is true for all given $k \in \mathbb{N}$, we conclude that $\xi \cdot \zeta$ is infinite.   □

PROPOSITION 1.7. *Non-Archimedean fields are not complete.*

PROOF. Just notice that the set of all infinitesimal numbers is upper bounded, but it has no least upper bound. In fact, if x is larger than all infinitesimals then so is $x/2 < x$.   □

3. The smallest non-Archimedean field

In this section we show that one can easily construct a non-Archimedean field by simply adding a new "infinite" number α to the set of rational numbers. The resulting $\mathbb{Q}(\alpha)$ can be seen as a canonical structure, because it is the "smallest" non-Archimedean field.

Let us denote by $\mathbb{Q}[\alpha] = \{a_0 + a_1\alpha + \ldots + a_n\alpha^n \mid a_i \in \mathbb{R}, a_n \neq 0\}$ the ring of all polynomial expressions in the indeterminate α. Notice that $\mathbb{Q}[\alpha]$ is an integral domain.

DEFINITION 1.8. $\mathbb{Q}(\alpha)$ *is the* fraction field *generated by* $\mathbb{Q}[\alpha]$, *that is:*

$$\mathbb{Q}(\alpha) = \left\{ \frac{P(\alpha)}{Q(\alpha)} \mid P(\alpha), Q(\alpha) \in \mathbb{Q}[\alpha], \ Q(\alpha) \neq 0 \right\}.$$

The set $\mathbb{Q}(\alpha)^+$ *of positive elements of* $\mathbb{Q}(\alpha)$ *is defined by setting*

$$\frac{a_0 + a_1\alpha + \ldots + a_n\alpha^n}{b_0 + b_1\alpha + \ldots + b_m\alpha^m} \in \mathbb{Q}(\alpha)^+ \iff \frac{a_n}{b_m} > 0.$$

We remark that the above definition is coherent with the intuition of α as an "infinite" positive number. In fact, it is easily verified that the following property holds:

$$\frac{P(\alpha)}{Q(\alpha)} \in \mathbb{Q}(\alpha)^+ \iff \exists M \text{ such that } \frac{P(x)}{Q(x)} > 0 \text{ for all } x > M$$

So, the positive elements of $\mathbb{Q}(\alpha)$ are obtained as ratios of polynomials that are positive from some point on.[2]

PROPOSITION 1.9. *The pair* $(\mathbb{Q}(\alpha), \mathbb{Q}(\alpha)^+)$ *is an ordered field.*

PROOF. As pointed out above, if $\frac{P(\alpha)}{Q(\alpha)}, \frac{R(\alpha)}{S(\alpha)} \in \mathbb{Q}(\alpha)^+$, then there exist M, N such that

- $\frac{P(x)}{Q(x)} > 0$ for all $x > M$,

- $\frac{R(x)}{S(x)} > 0$ for all $x > N$.

If we let $K = \max\{M, N\}$, then clearly both the sum $\frac{P(x)}{Q(x)} + \frac{R(x)}{S(x)}$ and the product $\frac{P(x)}{Q(x)} \cdot \frac{R(x)}{S(x)}$ are positive for all $x > K$, and hence they belong to $\mathbb{Q}(\alpha)^+$.

The property that $\mathbb{Q}(\alpha)$ is the disjoint union $\mathbb{Q}(\alpha)^- \cup \{0\} \cup \mathbb{Q}(\alpha)^+$ is a straight consequence of the fact that every non-zero fraction $\frac{P(x)}{Q(x)}$ is either eventually positive or eventually negative. □

Let us now verify that $\mathbb{Q}(\alpha)$ is in fact the smallest non-Archimedean field.

PROPOSITION 1.10. *Let* $\mathbb{F}$ *be a non-Archimedean field. Then* $\mathbb{Q}(\alpha)$ *is isomorphic to an ordered subfield of* $\mathbb{F}$.

PROOF. Pick any infinite positive element $\xi \in \mathbb{F}$, and define the function $\Psi : \mathbb{Q}(\alpha) \to \mathbb{F}$ by letting

$$\Psi\left(\frac{a_0 + a_1\alpha + \ldots + a_n\alpha^n}{b_0 + b_1\alpha + \ldots + b_m\alpha^m}\right) = \frac{a_0 + a_1\xi + \ldots + a_n\xi^n}{b_0 + b_1\xi + \ldots + b_m\xi^m}.$$

(Recall that we can always assume $\mathbb{Q} \subseteq \mathbb{F}$.) It is verified in a straightforward manner that Ψ is a 1-1 homomorphism of ordered fields. □

---

[2] Clearly, $\frac{P(\alpha)}{Q(\alpha)} \in \mathbb{Q}(\alpha)^+ \Leftrightarrow \lim_{x \to +\infty} \frac{P(x)}{Q(x)} > 0$, but we do not want to introduce the notion of limit at this stage.

4. Proper extensions of the real line

In analysis, the most used extension of the real field is given by the complex numbers $\mathbb{C}$. In the rest of this chapter, we focus on those extensions of $\mathbb{R}$ that are *ordered* fields.

THEOREM 1.11. *Any ordered field $\mathbb{F}$ that properly extends the real field $\mathbb{R}$ is non-Archimedean.*

PROOF. Fix an element $\xi \in \mathbb{F}$ that does not belong to $\mathbb{R}$. Without loss of generality we can assume $\xi > 0$. In case $\xi \geq n$ for all $n \in \mathbb{N}$, then ξ is already a counter-example to the Archimedean property of $\mathbb{F}$. So, let us suppose that $0 < \xi < n$ for some $n \in \mathbb{N}$. Then the set

$$X = \{\, x \in \mathbb{R} \mid \xi < x \,\}$$

of the *real* upper bounds of ξ is nonempty and bounded below. By the *completeness property* of $\mathbb{R}$, we can take the least upper bound

$$r = \inf X \in \mathbb{R}.$$

Notice that $r \neq \xi$, because $\xi \notin \mathbb{R}$. We claim that $\varepsilon = r - \xi \neq 0$ is infinitesimal.[3]

Fix $n \in \mathbb{N}$. By the properties of least upper bound, we have that:

- $\xi \geq r - \frac{1}{n}$, and hence $\varepsilon \leq \frac{1}{n}$;
- $r + \frac{1}{n} > \xi$, and hence $\varepsilon > -\frac{1}{n}$.

As the above inequalities hold for all $n \in \mathbb{N}$, we conclude that ε is a non-zero infinitesimal.   □

By the same construction as the one given in Section 3, it can be shown that there exists the smallest ordered field that properly extends the real line, namely:

$$\mathbb{R}(\alpha) = \left\{ \frac{a_0 + a_1\alpha + \ldots + a_n\alpha^n}{b_0 + b_1\alpha + \ldots + b_m\alpha^m} \,\middle|\, a_i, b_j \in \mathbb{R}, a_n, b_m \neq 0 \right\}$$

where the set of *positive elements* $\mathbb{R}(\alpha)^+$ is defined by setting

$$\frac{a_0 + a_1\alpha + \ldots + a_n\alpha^n}{b_0 + b_1\alpha + \ldots + b_m\alpha^m} \in \mathbb{R}(\alpha)^+ \iff \frac{a_n}{b_m} > 0.$$

---

[3] Notice that we cannot assume $\xi < r$. For instance, if $\xi = 1 + \delta$ for some positive infinitesimal δ, then $r = \inf\{s \in \mathbb{R} \mid \xi < s\} = 1 < \xi$.

5. Standard parts

Thanks to infinitesimal numbers, it is possible to formalize a notion of "closeness".

DEFINITION 1.12. *We say that the numbers ξ and ζ are* infinitely close *if they have infinitesimal distance, that is, $\xi - \zeta$ is infinitesimal. In this case we write $\xi \sim \zeta$.*

It is easily seen that $\sim$ is an equivalence relation. The next theorem gives a picture of the order-structure of the ordered fields that extend $\mathbb{R}$, and shows a canonical representation of the finite numbers.

THEOREM 1.13 (Standard Part Theorem).
Let $\mathbb{F}$ be an ordered field that properly extends $\mathbb{R}$. Then every finite number ξ is infinitely close to a unique real number $r \sim \xi$, called the standard part *of ξ. We denote $r = \mathrm{st}(\xi)$.*

PROOF. We proceed similarly as in the proof of Theorem 1.11. Since ξ is finite, the set of its *real* upper bounds

$$X = \{x \in \mathbb{R} \mid x > \xi\}$$

is nonempty and bounded below. Let $r = \inf X \in \mathbb{R}$ be its least upper bound. We claim that $r \sim \xi$.

Let $n \in \mathbb{N}$ be fixed. By the properties of least upper bound, we have that:

- $\xi \geq r - \frac{1}{n}$, hence $r - \xi \leq \frac{1}{n}$.
- $r + \frac{1}{n} > \xi$, hence $r - \xi > -\frac{1}{n}$.

We conclude that $r - \xi$ is infinitesimal, as desired. $\square$

Finally, the standard part is unique because two real numbers that are infinitely close are necessarily equal.

COROLLARY 1.14 (Canonical Form).
Let $\mathbb{F}$ be an ordered field that properly extends $\mathbb{R}$. Then every finite number $\xi \in \mathbb{F}$ has a unique representation in the canonical form:

$$\xi = \mathrm{st}(\xi) + \varepsilon$$

where $\mathrm{st}(\xi) \in \mathbb{R}$ is a real number, and $\varepsilon = \xi - \mathrm{st}(\xi)$ is infinitesimal.

It is useful to extend the notion of a standard part also to infinite numbers, by setting:

- $\mathrm{st}(\xi) = +\infty$ if ξ is infinite and positive;
- $\mathrm{st}(\xi) = -\infty$ if ξ is infinite and negative.

Accordingly, we will adopt the usual conventional algebra on the extended real line $\mathbb{R} \cup \{-\infty, +\infty\}$:

- $r + \infty = +\infty$ and $r - \infty = -\infty$ for all $r \in \mathbb{R}$;
- $(+\infty) + (+\infty) = +\infty$ and $(-\infty) + (-\infty) = -\infty$;
- $r \cdot (+\infty) = +\infty$ and $r \cdot (-\infty) = -\infty$ for all positive $r \in \mathbb{R}$;
- $r \cdot (+\infty) = -\infty$ and $r \cdot (-\infty) = +\infty$ for all negative $r \in \mathbb{R}$;
- $(+\infty) \cdot (+\infty) = (-\infty) \cdot (-\infty) = +\infty$ and $(+\infty) \cdot (-\infty) = (-\infty) \cdot (+\infty) = -\infty$.

Standard parts satisfy compatibility properties with respect to sums, products and quotients.

PROPOSITION 1.15. *Let $\mathbb{F}$ be an ordered field that properly extends $\mathbb{R}$. For all numbers $\xi, \zeta \in \mathbb{F}$, the following equalities hold:*

(1) $\operatorname{st}(\xi + \zeta) = \operatorname{st}(\xi) + \operatorname{st}(\zeta)$;

(2) $\operatorname{st}(\xi \cdot \zeta) = \operatorname{st}(\xi) \cdot \operatorname{st}(\zeta)$;

(3) $\operatorname{st}(\frac{\xi}{\zeta}) = \frac{\operatorname{st}(\xi)}{\operatorname{st}(\zeta)}$ *whenever ζ is not infinitesimal;*

with the only exceptions of the following indeterminate *forms:*

$$(+\infty) + (-\infty); \quad \infty \cdot 0; \quad \frac{\infty}{\infty}; \quad \frac{0}{0}.$$

PROOF. All proofs are straightforward applications of the elementary properties of Proposition 1.6. As an example, let us see (2). Write $\xi = r + \varepsilon$ and $\zeta = s + \delta$ in their canonical forms, where $r, s \in \mathbb{R}$ and ε, δ are infinitesimals. Then $\xi \cdot \zeta = r \cdot s + \vartheta$, where $\vartheta = r \cdot \delta + s \cdot \varepsilon + \varepsilon \cdot \delta \sim 0$ is infinitesimal, as the sum of three infinitesimal quantities. $\square$

Similarly as with "classic" *limits*, in the indeterminate forms above, the resulting standard parts could be any element $l \in \mathbb{R} \cup \{-\infty, +\infty\}$, with the only restriction of sign compatibility. Indeed, for every given l, one can easily find numbers $\xi_i, \zeta_i \in \mathbb{F}$ such that

- $\operatorname{st}(\xi_1) = +\infty$, $\operatorname{st}(\zeta_1) = -\infty$ and $\operatorname{st}(\xi_1 + \zeta_1) = l$;
- $\operatorname{st}(\xi_2) = \infty$, $\operatorname{st}(\zeta_2) = 0$ and $\operatorname{st}(\xi_2 \cdot \zeta_2) = l$;
- $\operatorname{st}(\xi_3) = \infty$, $\operatorname{st}(\zeta_3) = \infty$ and $\operatorname{st}(\frac{\xi_3}{\zeta_3}) = l$;
- $\operatorname{st}(\xi_4) = 0$, $\operatorname{st}(\zeta_4) = 0$ and $\operatorname{st}(\frac{\xi_4}{\zeta_4}) = l$.

Standard parts also satisfy a compatibility property with respect to order. (The order relation on the real numbers is extended to $\mathbb{R} \cup \{-\infty, +\infty\}$ in the obvious way by setting $-\infty \leq l \leq +\infty$ for all $l \in \mathbb{R}$.)

PROPOSITION 1.16. *Let* $\mathbb{F}$ *be an ordered field that properly extends* $\mathbb{R}$, *and let* $\xi, \zeta \in \mathbb{F}$. *Then* $\mathrm{st}(\xi) < \mathrm{st}(\zeta) \Rightarrow \xi < \zeta$ *(and hence,* $\xi \leq \zeta \Rightarrow$ $\mathrm{st}(\xi) \leq \mathrm{st}(\zeta))$.

We remark that the above implication cannot be reversed. *E.g.,* if ξ is finite and $\varepsilon > 0$ is infinitesimal, then clearly $\xi < \xi + \varepsilon$, while $\mathrm{st}(\xi) = \mathrm{st}(\xi + \varepsilon)$.

6. Monads and galaxies

Similarly to the notion of being "infinitely close" (see Definition 1.12). one can also consider the relation of *finite closeness*:

$$\xi \sim_f \zeta \quad \text{if and only if} \quad \xi - \zeta \quad \text{is finite.}$$

It is readily seen that also $\sim_f$ is an equivalence relation. In the literature, the equivalence classes relative to the two relations of closeness $\sim$ and $\sim_f$, are called monads and galaxies, respectively.

DEFINITION 1.17. *The* monad *of a number* ξ *is the set of all numbers that are infinitely close to it:*[4]

$$\mathfrak{mon}(\xi) = \{ \zeta \mid \zeta \sim \xi \}.$$

The galaxy *of a number* ξ *is the set of all numbers that are finitely close to it:*

$$\mathfrak{gal}(\xi) = \{ \zeta \mid \zeta \sim_f \xi \}.$$

So, $\mathfrak{mon}(0)$ is the set of infinitesimal numbers, and $\mathfrak{gal}(0)$ is the set of finite numbers.

In the following proposition, the main basic properties of monads and galaxies are itemized.

PROPOSITION 1.18.
   (1) *Any two monads (or galaxies) are either equal or disjoint;*
   (2) *The monad (or the galaxy) of a point* ξ *is the coset of* ξ *modulo* $\mathfrak{mon}(0)$ *(modulo* $\mathfrak{gal}(0)$, *respectively). That is:*
$$\mathfrak{mon}(\xi) = \{ \xi + \varepsilon \mid \varepsilon \in \mathfrak{mon}(0) \}; \quad \mathfrak{gal}(\xi) = \{ \xi + \zeta \mid \zeta \in \mathfrak{gal}(0) \}.$$
   (3) *Monads and galaxies are convex sets.*[5]

---

[4] The name *monad* was introduced by A. Robinson in honor of Leibniz' philosophy. In fact, Robinson indicated Leibniz' use of infinitesimal quantities as one of the inspiring sources for his "discovery" of nonstandard analysis.

[5] Recall that a subset A of an ordered set $(X, <)$ is *convex* if it includes all intervals whose endpoints are in A. That is, if $a, a' \in A$ and $a < x < a'$, then also $x \in A$.

PROOF. (1) simply says that monads and galaxies are equivalence classes. (2) is straightforward from the definitions. As for (3), notice first that both the infinitesimal numbers $\mathbf{mon}(0)$ and the finite numbers $\mathfrak{gal}(0)$ are convex sets. Then, by (2), all monads and galaxies are convex because they are translates of convex sets.   □

CHAPTER 2

Alpha-Calculus

The intuitive notions of "infinitely small" and "infinitely large" quantities are central in the development of differential calculus. The current "ϵ-δ approach", as elaborated by K. Weierstrass in the second half of the 19th century, incorporates those notions in an indirect manner, grounding on the notion of *limit*.

As seen in the previous chapter, ordered fields that properly extend $\mathbb{R}$ are suitable for a formalization of "small quantity" and of "large quantity" as an *infinitesimal* and as an *infinite* number, respectively. Thus one is naturally lead to the following question.

> *Can one develop a differential calculus on a suitable ordered field $\mathbb{F} \supset \mathbb{R}$ in a similar fashion as "classic" calculus is developed on the real field $\mathbb{R}$?*

The major obstacle is the fact that real analysis is grounded on the *completeness property* of $\mathbb{R}$, while every ordered field that properly extends the real line is necessarily not complete. In this chapter we show that this problem can be overcome by considering suitable extensions of $\mathbb{R}$ with additional properties, namely the *hyperreal fields* $\mathbb{R}^*$.

By means of the field operations, every rational function can be naturally extended from the real field to any field the extends $\mathbb{R}$. However, calculus essentially requires transcendental functions, such us *sine, cosine, logarithm, exponential, etc.* Thus a fundamental requirement for a "good" field $\mathbb{R}^* \supset \mathbb{R}$ should be the possibility of canonically extending *every* function $f : \mathbb{R} \to \mathbb{R}$ to a function $f^* : \mathbb{R}^* \to \mathbb{R}^*$ (and similarly for multivariable functions $f : \mathbb{R}^k \to \mathbb{R}$). Of course, those extensions should preserve all the basic properties of the original functions. For instance, the extensions $\sin^*, \cos^*, \exp^* : \mathbb{R}^* \to \mathbb{R}^*$ should satisfy the familiar equalities $\sin^*(\xi + \zeta) = \sin^*(\xi) \cdot \cos^*(\zeta) + \cos^*(\xi) \cdot \sin^*(\zeta)$ and $\exp^*(\xi + \zeta) = \exp^*(\xi) \cdot \exp^*(\zeta)$ for all numbers $\xi, \zeta \in \mathbb{R}^*$, and so forth.

In the same way, we would need a canonical way of extending every subset $A \subseteq \mathbb{R}^k$ to a subset $A^* \subseteq (\mathbb{R}^*)^k$ in such a way that the same basic properties are satisfied. For instance, if $(0, 1)$ is the set of all real numbers between 0 and 1, then $(0, 1)^*$ should be the set of all $\xi \in \mathbb{R}^*$ that lie between 0 and 1; and if $A = \mathrm{ran}(f)$ is the range of a function f, then $A^* = \mathrm{ran}(f^*)$ should be the range of the extension f^*; and so forth. In general, our goal is to have real functions f and real subsets A to be indistinguishable from the corresponding extensions f^* and A^*, as far as their "elementary properties" are concerned.[1]

We itemize below these requirements for our wanted ordered field $\mathbb{R}^* \supset \mathbb{R}$:

- Every real subset $A \subseteq \mathbb{R}^k$ is extended to a subset $A^* \subseteq (\mathbb{R}^*)^k$;
- Every real function $f : A \to B$ is extended to a function $f^* : A^* \to B^*$;
- The following *transfer principle* holds:
  If P is an "elementary property" of real functions $f_1, \ldots, f_n$ and real subsets $S_1, \ldots, S_k$, then

$$P(f_1, \ldots, f_n, S_1, \ldots, S_k) \iff P(f_1^*, \ldots, f_n^*, S_1^*, \ldots, S_n^*)$$

By postulating only three elementary axioms that rule a notion of "Alpha-limit" for real sequences, we will introduce a field which essentially realizes the three properties itemized above. Such a field $\mathbb{R}^*$ will allow for a full differential calculus, namely the *Alpha-Calculus*.

It is worth remarking that – as a consequence of the *transfer principle* – such an extended calculus with infinite and infinitesimal numbers is equivalent to the usual calculus, in the sense that it proves exactly the same "elementary properties" of the real line, including all the fundamental classic theorems.

1. The axioms of Alpha-Calculus

We introduce Alpha-Calculus axiomatically, by postulating elementary properties that rule the *Alpha-limit* of a sequence of real numbers. Informally, one may think of the Alpha-limit as the value taken at a given "ideal infinite" natural number. The name Alpha-Calculus is

---

[1] Here by real subset A we mean a set of real tuples $A \subseteq \mathbb{R}^k$, and by real function we mean a function $f : A \to B$ where A and B are real subsets.

justified by the crucial role played by the "new" number α (read "Alpha").

Alpha-Calculus Theory ACT

(ACT0) Existence Axiom. *Every real sequence $\varphi : \mathbb{N} \to \mathbb{R}$ has a unique "Alpha-limit", denoted by $\lim_{n\uparrow\alpha} \varphi(n)$ or, more simply, by $\lim_\alpha \varphi$.*

(ACT1) Real Number Axiom. *If $c_r(n) = r$ is the constant sequence with value a real number r, then $\lim_{n\uparrow\alpha} c_r(n) = r$.*

(ACT2) Alpha Number Axiom. *The Alpha-limit of the identity sequence $\imath(n) = n$ is a "new" number denoted by α, that is, $\lim_{n\uparrow\alpha} n = \alpha \notin \mathbb{N}$.*

(ACT3) Field Axiom. *The set of all Alpha-limits of real sequences*

$$\mathbb{R}^* = \left\{ \lim_{n\uparrow\alpha} \varphi(n) \,\middle|\, \varphi : \mathbb{N} \to \mathbb{R} \right\}$$

is a field, called the hyperreal field, *where:*

- $\lim_{n\uparrow\alpha} \varphi(n) + \lim_{n\uparrow\alpha} \psi(n) = \lim_{n\uparrow\alpha}(\varphi(n) + \psi(n))$
- $\lim_{n\uparrow\alpha} \varphi(n) \cdot \lim_{n\uparrow\alpha} \psi(n) = \lim_{n\uparrow\alpha}(\varphi(n) \cdot \psi(n))$

The intuitive content of the axioms is quite self-evident. The *Existence Axiom* is the starting assumption of our theory: one can take the Alpha-limit of *any* real sequence. The *Real Number Axiom* says that constant sequences have the expected Alpha-limit. The *Alpha Number Axiom* provides the existence of a new number α that will play the central role in the theory. Finally, the *Field Axiom* says that the family $\mathbb{R}^*$ of Alpha-limits have the structure of a field, where sums and products are coherent with pointwise sums and products of sequences, respectively.

By a well-known construction, if one starts from the set of rational numbers and considers suitable equivalence classes of Cauchy sequences, one obtains the real numbers. So, in a precise sense, the real numbers can be seen as a sort of "ideal limits" of sequences of rational numbers. Somewhat similarly, we are now considering the Alpha-limits of sequences of real numbers to obtain the *hyperreal numbers* $\mathbb{R}^*$ of Alpha-Calculus.

In order to avoid misunderstandings, it is worth stressing right away that Alpha-limits are a very peculiar notion of a limit.[2] Probably, the most striking difference with respect to classic limits is given by the fact that they do always exist. Moreover, we will show that sequences that are different at all points must have different Alpha-limits. In consequence, the Alpha-limit of any infinitesimal sequence that never vanishes is a non-zero infinitesimal number, and so one has plenty of infinitesimal (and infinite) numbers in hand.

A natural question immediately arises:

> *Can one safely assume that the axioms of Alpha-Calculus Theory ACT are not contradictory?*

The answer is *yes*, because we can construct a *model* (actually, many models) that satisfies all the given axioms.

In order to clarify the meaning of this latter statement, let us recall a familiar example. One possible way of introducing (classic) calculus is by axiomatic method: this is done by postulating the existence of the real numbers as a *complete ordered field*. If one wants to prove that such an approach is safe, then one has to show that it admits *models*. To this end, one usually considers either equivalence classes of *Cauchy sequences* of rational numbers, or *Dedekind cuts* of rational numbers. Then one defines the field operations and the order relation on the constructed sets, and finally one checks that the resulting structures satisfy all properties of a complete ordered field.

As we will see at the end of this chapter (see §11), also models of the Alpha-Calculus can be constructed in an algebraic manner. In a few words, one considers the *ring of real sequences* $\mathfrak{Fun}(\mathbb{N}, \mathbb{R})$, takes its quotient modulo a *non-principal maximal ideal* $\mathfrak{m}$, and defines the Alpha-limit of a sequence as its canonical projection. Then it is verified that all axioms of Alpha-Calculus Theory are fulfilled.[3]

It is worth remarking that in order to develop calculus, one does not need to know about the construction of the real numbers by means of Cauchy sequences or by means of Dedekind cuts. In the same way, in order to develop Alpha-Calculus, one does not need to know about the construction of its models. All the reader is required to do at this stage is just to take the axioms of ACT as "true".

---

[2] In §6.5, we will show that the Alpha-limit is an actual limit with respect to suitable topologies; however, this fact is just a curiosity that is not used in the practice of Alpha-Calculus. See also §3.7 for a discussion of similarities and differences between Alpha-limits and "classic" limits of real sequences.

[3] The *Real Number Axiom* needs the cosets of constant sequences to be identified with the corresponding real numbers.

In the sequel, we will use the Greek letters $\varphi, \psi, \vartheta, \ldots$ to denote real sequences.

2. First properties of Alpha-Calculus

All proofs in this section are entirely elementary and, besides the axioms, they only require some little familiarity with the properties of field.

We begin with the really peculiar property that pointwise different sequences must have different Alpha-limits. We remark that this can be seen as a strong non-triviality condition, allowing for plenty of "new" numbers in our theory.

PROPOSITION 2.1. *If $\varphi(n) \neq \psi(n)$ for all n, then the Alpha-limits* $\lim_{n\uparrow\alpha} \varphi(n) \neq \lim_{n\uparrow\alpha} \psi(n)$.

PROOF. For all n, let $\vartheta(n)$ be the reciprocal of $\varphi(n) - \psi(n) \neq 0$, so that $\vartheta(n) \cdot (\varphi(n) - \psi(n)) = 1$. By taking the Alpha-limits, we get $(\lim_\alpha \vartheta) \cdot (\lim_\alpha \varphi - \lim_\alpha \psi) = 1$. In particular, $\lim_\alpha \varphi \neq \lim_\alpha \psi$. □

Changing finitely many values to a given sequence leaves its Alpha-limit unaltered.

PROPOSITION 2.2. *If $\varphi(n) = \psi(n)$ for all but finitely many n, then* $\lim_{n\uparrow\alpha} \varphi(n) = \lim_{n\uparrow\alpha} \psi(n)$.

PROOF. Let $\{n_1, \ldots, n_k\} = \{n \mid \varphi(n) \neq \psi(n)\}$. Notice that

$$(\varphi(n) - \psi(n)) \cdot (n - n_1) \cdots (n - n_k) = 0.$$

So, by taking the Alpha-limits, we obtain:

$$(\lim_\alpha \varphi - \lim_\alpha \psi) \cdot (\boldsymbol{\alpha} - n_1) \cdots (\boldsymbol{\alpha} - n_k) = 0.$$

As $\boldsymbol{\alpha} \notin \mathbb{N}$, it is $\boldsymbol{\alpha} - n_i \neq 0$ for all i, and hence it must be $\lim_\alpha \varphi = \lim_\alpha \psi = 0$, as desired. □

As a straight consequence, we can strengthen Proposition 2.1 to the following:

PROPOSITION 2.3. *If $\varphi(n) \neq \psi(n)$ for all but finitely many n, then* $\lim_{n\uparrow\alpha} \varphi(n) \neq \lim_{n\uparrow\alpha} \psi(n)$.

Proof. Consider the sequence

$$\psi'(n) = \begin{cases} \psi(n) & \text{if } \varphi(n) \neq \psi(n) \\ \psi(n) + 1 & \text{otherwise.} \end{cases}$$

Then $\varphi(n) \neq \psi'(n)$ for all n, and so $\lim_\alpha \varphi \neq \lim_\alpha \psi'$, by Proposition 2.1. On the other hand, by the previous proposition one has $\lim_\alpha \psi' = \lim_\alpha \psi$ because $\psi(n) = \psi'(n)$ for all but finitely many n. We conclude that $\lim_\alpha \varphi \neq \lim_\alpha \psi$, as desired.   □

COROLLARY 2.4. *The number* $\alpha \notin \mathbb{R}$.

Proof. For every given $r \in \mathbb{R}$, the identity sequence $\imath(n) = n$ can equal r in at most one point (when $r \in \mathbb{N}$). So, by the previous proposition, $\alpha = \lim_{n \uparrow \alpha} n \neq \lim_{n \uparrow \alpha} r = r$.   □

Another important consequence of the axioms **ACT** is the following.

PROPOSITION 2.5. *Assume that for every* n, *the sequence* $\varphi(n)$ *equals one of the values* $\psi_1(n), \ldots, \psi_k(n)$. *Then there exists an index* i *such that* $\lim_{n \uparrow \alpha} \varphi(n) = \lim_{n \uparrow \alpha} \psi_i(n)$.

Proof. By the hypothesis, the product

$$(\varphi(n) - \psi_1(n)) \cdots (\varphi(n) - \psi_k(n)) = 0$$

is the sequence constantly equal to zero, and so

$$0 = \lim_\alpha \left[(\varphi - \psi_1) \cdot \ldots \cdot (\varphi - \psi_k) \right] = \lim_\alpha (\varphi - \psi_1) \cdot \ldots \cdot \lim_\alpha (\varphi - \psi_k).$$

Then there must be an index i such that $\lim_\alpha(\varphi - \psi_i) = 0$, and we obtain the thesis $\lim_\alpha \varphi = \lim_\alpha \psi_i$.   □

By taking as $\psi_i(n) = r_i$ the constant sequences with value r_i, we get the

COROLLARY 2.6. *If the sequence* $\varphi(n)$ *only takes the finitely many values* $r_1, \ldots, r_k$, *then* $\lim_{n \uparrow \alpha} \varphi(n) = r_i$ *for some* i.

Another consequence of the above proposition is particularly relevant.

COROLLARY 2.7 (Definition by cases).
If a sequence φ is defined by cases:

$$\varphi(n) = \begin{cases} \psi_1(n) & \text{if property } P_1 \text{ holds} \\ \cdots \\ \psi_k(n) & \text{if property } P_k \text{ holds} \end{cases}$$

then $\lim_{n\uparrow\alpha} \varphi(n)$ equals one of the Alpha-limits $\lim_{n\uparrow\alpha} \psi_i(n)$.

Next, we prove two more properties that will be useful in the sequel.

PROPOSITION 2.8. *Suppose that $\lim_{n\uparrow\alpha} \varphi(n) = 0$. If a sequence ψ vanishes where φ does, that is, if $\varphi(n) = 0 \Rightarrow \psi(n) = 0$, then also $\lim_{n\uparrow\alpha} \psi(n) = 0$.*

PROOF. Let

$$\vartheta(n) = \begin{cases} \varphi(n) & \text{if } \varphi(n) \neq 0 \\ 1 & \text{if } \varphi(n) = 0 \end{cases}$$

Since $\vartheta(n) \neq 0$ for all n, its Alpha-limit $\xi = \lim_\alpha \vartheta \neq 0$. Now, it is readily verified that the sequence $(\vartheta(n) - \varphi(n)) \cdot \psi = 0$ is constantly equal to zero and so, by taking the Alpha-limits, $(\xi - 0) \cdot \lim_\alpha \psi = 0$. Since $\xi \neq 0$, it must be $\lim_\alpha \psi = 0$.   □

Alpha-limits are coherent with compositions.

PROPOSITION 2.9. *Let $f : A \to B$ be a function where $A, B \subseteq \mathbb{R}$, and let $\varphi, \varphi' : \mathbb{N} \to A$ be two sequences taking values in A. Then*

$$\lim_{n\uparrow\alpha} \varphi(n) = \lim_{n\uparrow\alpha} \varphi'(n) \implies \lim_{n\uparrow\alpha} f(\varphi(n)) = \lim_{n\uparrow\alpha} f(\varphi'(n)).$$

PROOF. Let $\psi = \varphi - \varphi'$ and $\vartheta = (f \circ \varphi) - (f \circ \varphi')$. Clearly $\vartheta(n) = 0$ whenever $\psi(n) = 0$. By the hypothesis, $\lim_\alpha \psi = 0$ and so, by the previous proposition, we obtain that also $\lim_\alpha \vartheta = 0$, and the assertion follows.   □

We close this section by showing that the *hyperreal field* $\mathbb{R}^*$ comes with a natural order.

THEOREM 2.10. *The hyperreal field $\mathbb{R}^*$ is an* ordered field *whose positive part is $(\mathbb{R}^*)^+ = \{\lim_\alpha \varphi \mid \varphi(n) > 0 \text{ for all } n\}$.*

PROOF. It directly follow from the *Field Axiom* that $(\mathbb{R}^*)^+$ is closed under sums and products, and that

$$(\mathbb{R}^*)^- = \left\{ -\lim_\alpha \varphi \mid \varphi(n) > 0 \text{ for all } n \right\} = \left\{ \lim_\alpha \psi \mid \psi(n) < 0 \text{ for all } n \right\}.$$

We now have to show that the three sets $(\mathbb{R}^*)^-, \{0\}, (\mathbb{R}^*)^+$ form a partition of $\mathbb{R}^*$. Notice first that they are pairwise disjoint by Proposition 2.1. For every sequence $\varphi : \mathbb{N} \to \mathbb{R}$, define:

$$\varphi^+(n) = \begin{cases} \varphi(n) & \text{if } \varphi(n) > 0 \\ 1 & \text{otherwise.} \end{cases} \qquad \varphi^-(n) = \begin{cases} \varphi(n) & \text{if } \varphi(n) < 0 \\ -1 & \text{otherwise.} \end{cases}$$

Clearly

$$\varphi(n) \cdot (\varphi(n) - \varphi^+(n)) \cdot (\varphi(n) - \varphi^-(n)) = 0,$$

so $\lim_\alpha \varphi \neq 0$ implies that either $\lim_\alpha \varphi = \lim_\alpha \varphi^+ \in (\mathbb{R}^*)^+$ or $\lim_\alpha \varphi = \lim_\alpha \varphi^- \in (\mathbb{R}^*)^-$, as desired. $\qquad\square$

EXERCISE 2.11. *If the sequence $\varphi(n)$ is finite-to-one[4], then its Alpha-limit $\lim_{n\uparrow\alpha} \varphi(n) \notin \mathrm{ran}(\varphi)$ is a "new" element.*

SOLUTION. By the hypothesis, for every fixed k the sequence φ takes the value $\varphi(k)$ only finitely many times. In other words φ is different from the constant sequence $c_{\varphi(k)}$ for all but finitely many n. So, the Alpha-limits $\lim_\alpha \varphi \neq \lim_\alpha c_{\varphi(k)} = \varphi(k)$ are different. We conclude that $\lim_\alpha \varphi \notin \mathrm{ran}(\varphi)$. $\qquad\square$

3. Hyper-extensions of sets of reals

There is a canonical way of extending each set of real numbers to a set of hyperreal numbers.

DEFINITION 2.12. *The* hyper-extension $A^* \subseteq \mathbb{R}^*$ *of a set $A \subseteq \mathbb{R}$ is the set of all Alpha-limits of real sequences that take values in A:*

$$A^* = \left\{ \lim_{n\uparrow\alpha} \varphi(n) \;\middle|\; \varphi : \mathbb{N} \to A \right\}.$$

Notice that trivially $\emptyset^* = \emptyset$. Notice also that hyper-extensions are actually supersets. Indeed,

$$A = \left\{ \lim_{n\uparrow\alpha} c_a(n) \;\middle|\; a \in A \right\} \subseteq A^*.$$

---

[4]A function $f : A \to B$ is *finite-to-one* if each value is taken finitely many times, that is, if all preimages $f^{-1}(b) = \{a \in A \mid f(a) = b\}$ are finite.

PROPOSITION 2.13. *Let $A, B \subseteq \mathbb{R}$ be sets of real numbers. Then $A^* = B^*$ if and only if $A = B$.*

PROOF. One implication is trivial. If $A \not\subseteq B$, pick an element $a \in A$ with $a \notin B$. Then for every sequence $\varphi : \mathbb{N} \to B$, we have $\varphi(n) \neq a$ for all n and then $\lim_\alpha \varphi \neq a$. This proves that $a \notin B^*$, hence $A^* \not\subseteq B^*$. Similarly, if $B \not\subseteq A$ then $B^* \not\subseteq A^*$, We conclude that $A^* \neq B^*$ whenever $A \neq B$, as desired. □

PROPOSITION 2.14. *Let $A \subseteq \mathbb{R}$ be a set of real numbers. Then $A = A^*$ if and only if A is finite.*

PROOF. Assume first that $A = \{a_1, \ldots, a_k\}$ is finite. Then every sequence $\varphi : \mathbb{N} \to A$ can only take (some of) the finitely many values $a_1, \ldots, a_k$. By Proposition 2.6, it must be $\lim_\alpha \varphi = a_i$ for some i, and so $\lim_\alpha \varphi \in A$.

If A is infinite, pick a 1-1 sequence $\varphi : \mathbb{N} \to A$. Then, for every $a \in A$, we have that $\varphi(n) \neq a$ for all but at most one n. So, Proposition 2.2 applies and we have that $\lim_\alpha \varphi \neq a$. We conclude that $\lim_\alpha \varphi \in A^* \setminus A$. □

Hyper-extensions preserve the basic set operations.

PROPOSITION 2.15. *Let $A, B \subseteq \mathbb{R}$ be sets of real numbers. Then:*
(1) $(A \cap B)^* = A^* \cap B^*$;
(2) $A \subseteq B \Leftrightarrow A^* \subseteq B^*$;
(3) $(A \cup B)^* = A^* \cup B^*$;
(4) $(A^c)^* = (A^*)^c$ *where* $A^c = \mathbb{R} \setminus A$ *is the complement* ;
(5) $(A \setminus B)^* = A^* \setminus B^*$.

PROOF. (1) is straightforward from the definitions and from the following equality:

$$\{\varphi \mid \varphi : \mathbb{N} \to A \cap B\} = \{\psi \mid \psi : \mathbb{N} \to A\} \cap \{\vartheta \mid \vartheta : \mathbb{N} \to B\}.$$

(2). By (1) and by Proposition 2.13 we have:

$$A^* \subseteq B^* \Leftrightarrow A^* = A^* \cap B^* = (A \cap B)^* \Leftrightarrow A = A \cap B \Leftrightarrow A \subseteq B.$$

(3). If at least one of the two sets A, B is empty, then the conclusion is trivial. Otherwise, fix elements $a_0 \in A$ and $b_0 \in B$. For every sequence $\varphi : \mathbb{N} \to A \cup B$, set:

$$\varphi'(n) = \begin{cases} \varphi(n) & \text{if } \varphi(n) \in A \\ a_0 & \text{if } \varphi(n) \in B \setminus A \end{cases} \; ; \quad \varphi''(n) = \begin{cases} \varphi(n) & \text{if } \varphi(n) \in B \\ b_0 & \text{if } \varphi(n) \in A \setminus B \end{cases}$$

Clearly, $(\varphi(n) - \varphi'(n)) \cdot (\varphi(n) - \varphi''(n)) = 0$, and so either $\lim_\alpha \varphi = \lim_\alpha \varphi' \in A^*$ or $\lim_\alpha \varphi = \lim_\alpha \varphi'' \in B^*$. Thus $\lim_\alpha \varphi \in A^* \cup B^*$ and the inclusion $(A \cup B)^* \subseteq A^* \cup B^*$ is proved. The reverse inclusion is readily verified, because both $A^*, B^* \subseteq (A \cup B)^*$ by (2).

(4). Notice that a set X is the complement of A^* if and only if $X \cap A^* = \emptyset$ and $X \cup A^* = \mathbb{R}^*$. By using (1) and (3), it is easily checked that $X = (A^c)^*$ satisfies the required properties. Indeed,

$$(A^c)^* \cap A^* = (A^c \cap A)^* = \emptyset^* = \emptyset \quad \text{and} \quad (A^c)^* \cup A^* = (A^c \cup A)^* = \mathbb{R}^*.$$

(5). Notice that $A \setminus B = A \cap B^c$, and apply (1) and (4).   □

No new real numbers are found in hyper-extensions (see condition (3) below).

PROPOSITION 2.16. *For every $A, B \subseteq \mathbb{R}$:*

(1) $A \subseteq B \Leftrightarrow A \subseteq B^*$;

(2) $A \cap B = \emptyset \Leftrightarrow A \cap B^* = \emptyset$;

(3) $A^* \cap \mathbb{R} = A$.

PROOF. (1). If $A \subseteq B$, then $A \subseteq A^* \subseteq B^*$. Conversely, assume that $A \not\subseteq B$ and pick $a \in A \cap B^c$. Then $a = \lim_{n \uparrow \alpha} c_a(n) \in (B^c)^* = (B^*)^c$. We conclude that $a \in A \cap (B^*)^c$, and hence $A \not\subseteq B^*$.

(2). It directly follows from (1), by noticing that

$$A \cap B = \emptyset \Leftrightarrow A \subseteq B^c \Leftrightarrow A \subseteq (B^c)^* = (B^*)^c \Leftrightarrow A \cap B^* = \emptyset.$$

(3). By (2), one has the following equivalences:

$$\emptyset = A \cap (\mathbb{R} \cap A^c) \Leftrightarrow A^* \cap (\mathbb{R} \cap A^c) = (A^* \cap \mathbb{R}) \cap A^c = \emptyset \Leftrightarrow A^* \cap \mathbb{R} \subseteq A.$$

The other inclusion $A \subseteq A^* \cap \mathbb{R}$ is obvious.   □

Particularly important are the hyper-extensions of numerical sets, namely:

- The set of *hypernatural* numbers $\mathbb{N}^*$;
- The set of *hyperinteger* numbers $\mathbb{Z}^*$;
- The set of *hyperrational* numbers $\mathbb{Q}^*$;
- The set of *hyperreal* numbers $\mathbb{R}^*$.

Further on in §8, we will see that the above hyper-extensions share the same "elementary" properties as the corresponding sets of numbers.

4. The Alpha-measure and the qualified sets

In this section we introduce the "Alpha-measure", a useful and powerful tool that will be used repeatedly in the sequel. The Alpha-measure allows one to identify the "qualified" sets, namely those subsets of $\mathbb{N}$ that have measure 1. In a precise sense, the qualified sets are those sets of indexes that really count, because the Alpha-limit of a sequence is determined by the values taken on a qualified set.

DEFINITION 2.17. *The* Alpha-measure *is the function*

$$\mu_\alpha : \mathcal{P}(\mathbb{N}) \to \{0, 1\}$$

defined by setting

$$\mu_\alpha(A) = \begin{cases} 1 & \text{if } \alpha \in A^* \\ 0 & \text{if } \alpha \notin A^* \end{cases}$$

The intuition is that in any given sequence, a set of indexes is relevant if and only if its Alpha-measure is 1.

Notice that the Alpha-measure μ_α on $\mathbb{N}$ is strictly related to the *Dirac measure* δ_α on $\mathbb{N}^*$: in fact, we have $\mu_\alpha(A) = \delta_\alpha(A^*)$ for all $A \subseteq \mathbb{N}$.

PROPOSITION 2.18.

(1) *If $F \subset \mathbb{N}$ is a finite set then $\mu_\alpha(F) = 0$;*

(2) *Let $\mu_\alpha(A) = 1$. Then for every partition $A = A_1 \cup \ldots \cup A_k$ into finitely many pieces, there exists one and only one piece A_j such that $\mu_\alpha(A_j) = 1$.*

(3) *If $\mu_\alpha(A_j) = 1$ for $j = 1, \ldots, k$, then $\mu_\alpha(\bigcap_{j=1}^k A_j) = 1$.*

PROOF. (1). If $F = \{n_1 < \ldots < n_k\} \subset \mathbb{N}$ is finite, then $^*F = F$ by Proposition 2.14, and hence $\alpha \notin {}^*F$.

(2). From $A_i \cap A_j = \emptyset$ for $i \neq j$, it follows that $A_i^* \cap A_j^* = (A_i \cap A_j)^* = \emptyset^* = \emptyset$. Moreover, $A^* = (A_1 \cup \ldots \cup A_k)^* = A_1^* \cup \ldots \cup A_k^*$. This shows that $A^* = A_1^* \cup \ldots \cup A_k^*$ is a partition, and so α belongs to exactly one piece A_j^*.

(3). Just notice that $\alpha \in A_1^* \cap \ldots \cap A_k^* = (A_1 \cap \ldots \cap A_k)^*$. □

As a straight consequence of the above proposition, we obtain that μ_α is actually a (finitely additive) measure.

THEOREM 2.19. *μ_α is a finitely additive non-atomic probability measure defined on all subsets of $\mathbb{N}$, that is:*[5]

---

[5] A measure μ is called *non-atomic* if $\mu(\{x\}) = 0$ for all x.

(1) $\mu_\alpha(\mathbb{N}) = 1$;

(2) If $A \cap B = \emptyset$ then $\mu_\alpha(A \cup B) = \mu_\alpha(A) + \mu_\alpha(B)$;

(3) $\mu_\alpha(\{n\}) = 0$ for all $n \in \mathbb{N}$.

PROOF. (1) is trivial and (3) directly follows from (1) of the previous proposition.

(2). If $\alpha \notin (A \cup B)^* = A^* \cup B^*$, then trivially $\alpha \notin A^*$ and $\alpha \notin B^*$, hence $0 = \mu_\alpha(A \cup B) = \mu_\alpha(A) + \mu_\alpha(B)$. Now assume that $\alpha \in A^* \cup B^*$. Since $A^* \cap B^* = {}^*(A \cap B) = \emptyset^* = \emptyset$, we must have that either $\alpha \in A^*$ and $\alpha \notin B^*$, or $\alpha \notin A^*$ and $\alpha \in B^*$. In both cases, $1 = \mu_\alpha(A \cup B) = \mu_\alpha(A) + \mu_\alpha(B)$. □

We remark that the Alpha-measure is not σ-additive, as otherwise we would have $1 = \mu_\alpha(\mathbb{N}) = \mu_\alpha(\bigcup_{n \in \mathbb{N}}\{n\}) = \sum_{n \in \mathbb{N}} \mu_\alpha(\{n\}) = 0$, a contradiction.

DEFINITION 2.20. Let $P(n)$ be any property that depends on the natural number n. We say that P holds almost everywhere (a.e. for short) or that P holds for almost all n, if the set of natural numbers that satisfy it has Alpha-measure 1, i.e. if $\mu_\alpha(\{n \in \mathbb{N} \mid P(n)\}) = 1$.

E.g., given two sequences $\varphi(n), \psi(n)$, we say that $\varphi(n) = \psi(n)$ a.e. to mean that the property $P(n)$ stating that φ and ψ take the same value at n holds for almost all n, that is, $\mu_\alpha(\{n \mid \varphi(n) = \psi(n)\}) = 1$.

As we have seen above, finite sets of natural numbers have Alpha-measure 0, and hence all cofinite sets have Alpha-measure 1.[6] In consequence:

- If a property P holds "eventually" (that is, it holds for all but finitely many n) then P holds a.e..

DEFINITION 2.21. A set $A \subseteq \mathbb{N}$ is qualified if $\mu_\alpha(A) = 1$. We denote by $\mathcal{Q}$ the family of qualified sets.

The following properties are just reformulations of what proved above. (All proofs are straightforward and are omitted.)

THEOREM 2.22. The family $\mathcal{Q}$ of qualified sets satisfies the following properties:[7]

---

[6] A cofinite set is a set whose complement is finite.

[7] A family of subsets satisfying the six properties itemized in this theorem is called a non-principal ultrafilter on $\mathbb{N}$ (see §11.1).

(1) $\emptyset \notin \mathcal{Q}$ and $\mathbb{N} \in \mathcal{Q}$;

(2) $\mathcal{Q}$ is closed upward: $B \supseteq A \in \mathcal{Q} \Rightarrow B \in \mathcal{Q}$;

(3) $\mathcal{Q}$ is closed under intersections:
$A_1, \ldots, A_k \in \mathcal{Q} \Rightarrow A_1 \cap \ldots \cap A_k \in \mathcal{Q}$;

(4) In every finite partition $A = A_1 \cup \ldots \cup A_k$ of a set $A \in \mathcal{Q}$ there exists one and only one piece $A_j \in \mathcal{Q}$;

(5) $A \notin \mathcal{Q} \Leftrightarrow A^c \in \mathcal{Q}$;

(6) If F is finite, $F \notin \mathcal{Q}$.

Particularly relevant is property (5), stating that a set is *not* qualified if and only if its complement is qualified.

We now prove the main property that justifies the name "qualified". It states that the Alpha-limit of a sequence only depends on the values taken on a qualified set.

THEOREM 2.23. *Let φ, ψ be real sequences. Then*

$$\varphi(n) = \psi(n) \text{ a.e.} \iff \lim_{n\uparrow\alpha} \varphi(n) = \lim_{n\uparrow\alpha} \psi(n).$$

PROOF. Let $X = \{n \mid \varphi(n) = \psi(n)\}$. Assume first that $\varphi(n) = \psi(n)$ a.e., that is, $\alpha \in X^*$. Then there exists a sequence $\xi(n)$ such that $\lim_\alpha \xi = \alpha$ and $\xi(n) \in X$ for all n, and so $\varphi(\xi(n)) = \psi(\xi(n))$ for all n. Since $\lim_\alpha \xi = \alpha = \lim_\alpha \imath$ where $\imath(n) = n$ is the identity sequence, by Proposition 2.9 it follows that

$$\begin{aligned} \lim_{n\uparrow\alpha} \varphi(n) &= \lim_{n\uparrow\alpha} \varphi(\imath(n)) = \lim_{n\uparrow\alpha} \varphi(\xi(n)) \\ &= \lim_{n\uparrow\alpha} \psi(\xi(n)) = \lim_{n\uparrow\alpha} \psi(\imath(n)) = \lim_{n\uparrow\alpha} \psi(n). \end{aligned}$$

Conversely, if $X = \{n \mid \varphi(n) = \psi(n)\}$ is *not* qualified, that is, if $\alpha \notin X^*$, then $\alpha \in (X^*)^c = (X^c)^*$. So, there exists a sequence $\zeta(n)$ such that $\lim_{n\uparrow\alpha} \zeta(n) = \alpha$ and $\zeta(n) \notin X$ for all n; then $\varphi(\zeta(n)) \neq \psi(\zeta(n))$ for all n. Now recall that two sequences that are different at all points have different Alpha-limits (see Proposition 2.1). Again by Proposition 2.9, we have:

$$\begin{aligned} \lim_{n\uparrow\alpha} \varphi(n) &= \lim_{n\uparrow\alpha} \varphi(\imath(n)) = \lim_{n\uparrow\alpha} \varphi(\zeta(n)) \\ &\neq \lim_{n\uparrow\alpha} \psi(\zeta(n)) = \lim_{n\uparrow\alpha} \psi(\imath(n)) = \lim_{n\uparrow\alpha} \psi(n). \end{aligned}$$

$\square$

As a straightforward consequence, we obtain that the same property of Theorem 2.23 also holds for the membership relation.

THEOREM 2.24. *Let φ be a real sequence and $A \subseteq \mathbb{R}$. Then*

$$\varphi(n) \in A \text{ a.e.} \iff \lim_{n \uparrow \alpha} \varphi \in A^*.$$

PROOF. $\Rightarrow$ By the hypothesis, there exists a sequence $\psi : \mathbb{N} \to A$ such that $\varphi(n) = \psi(n)$ a.e.. Then $\lim_\alpha \varphi = \lim_\alpha \psi \in A^*$.

$\Leftarrow$ By the hypothesis and the definition of A^*, there exists a sequence $\psi : \mathbb{N} \to A$ such that $\lim_\alpha \varphi = \lim_\alpha \psi$. Then $\varphi(n) = \psi(n) \in A$ a.e.. $\qquad\square$

5. The transfer principle, informally

Thanks to the notions of qualified set and of "almost all" we can now formulate a *transfer principle* that plays the role of a general rule of the thumb in our theory.

> **Transfer Principle.** *An "elementary property" P is satisfied by real numbers $\varphi_1(n), \ldots, \varphi_k(n)$ for almost all n if and only if P is satisfied by the Alpha-limits $\lim_{n \uparrow \alpha} \varphi_1(n), \ldots, \lim_{n \uparrow \alpha} \varphi_k(n)$.*
>
> $$P(\varphi_1(n), \ldots, \varphi_k(n)) \text{ a.e.} \iff P(\lim_{n \uparrow \alpha} \varphi_1(n), \ldots, \lim_{n \uparrow \alpha} \varphi_k(n)).$$

Usually, when applying the right-handed implication $\Longrightarrow$ one simply talks about *transfer*, and when applying the converse left-handed implication $\Longleftarrow$ one refers to *backward transfer*.

For now, we remark that the above *transfer principle* can only be taken at an informal intuitive level. Its content will be made precise only in Chapter 5, where the notion of an "elementary property" will be given a rigorous definition by using the formalism of first-order logic.

A first observation that one can make right away is the fact that *not* every property P can be "elementary". For example, the property $P(x)$ saying that "x is a natural number" is trivially satisfied by all values $\imath(n) = n$ of the identity sequence, while it does not extend to its Alpha-limit $\lim_{n \uparrow \alpha} n = \alpha \notin \mathbb{N}$.

Notice that Theorem 2.23 is the first fundamental example of the *transfer principle* where one takes the equality relation "$x_1 = x_2$" as the "elementary property" $P(x_1, x_2)$.

We now prove other fundamental instances of the *transfer principle* where the considered "elementary properties" are inequalities.

THEOREM 2.25 (*Transfer of inequalities*).
Let φ, ψ be arbitrary real sequences. Then

(1) $\varphi(n) \neq \psi(n)$ a.e. $\iff \lim_{n\uparrow\alpha} \varphi(n) \neq \lim_{n\uparrow\alpha} \psi(n)$

(2) $\varphi(n) < \psi(n)$ a.e. $\iff \lim_{n\uparrow\alpha} \varphi(n) < \lim_{n\uparrow\alpha} \psi(n)$

(3) $\varphi(n) \leq \psi(n)$ a.e. $\iff \lim_{n\uparrow\alpha} \varphi(n) \leq \lim_{n\uparrow\alpha} \psi(n)$

PROOF. (1). The result directly follows from the equivalence of Theorem 2.23, by noticing that $\varphi(n) \neq \psi(n)$ a.e. if and only if it not the case that $\varphi(n) = \psi(n)$ a.e..

(2). By the definition of order on $\mathbb{R}$ and $\mathbb{R}^*$, and by Theorem 2.24, we have the following chain of equivalences: $\varphi(n) < \psi(n)$ a.e. $\Leftrightarrow$ $(\psi - \varphi)(n) \in \mathbb{R}^+$ a.e. $\Leftrightarrow \lim_\alpha(\psi - \varphi) = \lim_\alpha \psi - \lim_\alpha \varphi \in (\mathbb{R}^*)^+ \Leftrightarrow \lim_\alpha \varphi < \lim_\alpha \psi$.

(3). Similarly as above, $\varphi(n) \leq \psi(n)$ a.e. $\Leftrightarrow \psi(n) - \varphi(n) \in \mathbb{R}^+ \cup \{0\}$ a.e. $\Leftrightarrow \lim_\alpha \psi - \lim_\alpha \varphi \in (\mathbb{R}^+ \cup \{0\})^* = (\mathbb{R}^*)^+ \cup \{0\} \Leftrightarrow \lim_\alpha \varphi \leq \lim_\alpha \psi$. □

6. Hyper-extensions of functions

Let $f : A \to B$ be a function where A, B are subsets of $\mathbb{R}$. By composing with f, every sequence $\varphi : \mathbb{N} \to A$ that takes values in A is turned into a sequence $f \circ \varphi : \mathbb{N} \to B$ that takes values in B:

There is a canonical way to extend every real function to a "hyper-real" function.

DEFINITION 2.26. *Let $f : A \to B$ be a function where $A, B \subseteq \mathbb{R}$. Its hyper-extension $f^* : A^* \to B^*$ is defined by setting, for every sequence $\varphi : \mathbb{N} \to A$,*

$$f^*\left(\lim_{n\uparrow\alpha} \varphi(n)\right) = \lim_{n\uparrow\alpha} f(\varphi(n)).$$

Recall that, by Proposition 2.9, if $\varphi, \varphi' : \mathbb{N} \to A$ are two sequences that take the same Alpha-limit $\lim_{n\uparrow\alpha} \varphi(n) = \lim_{n\uparrow\alpha} \varphi'(n)$ then $\lim_{n\uparrow\alpha} f(\varphi(n)) = \lim_{n\uparrow\alpha} f(\varphi'(n))$. In consequence, the definition above is well-posed.

In some sense, f^* is a sort of "continuous" extension of f, in that it commutes with Alpha-limits. Notice that f^* is an actual extension of the function f because, for every $a \in A$,

$$f^*(a) = f^*\left(\lim_{n\uparrow\alpha} c_a(n)\right) = \lim_{n\uparrow\alpha}(f \circ c_a)(n) = \lim_{n\uparrow\alpha} c_{f(a)}(n) = f(a).$$

PROPOSITION 2.27.

(1) If $\imath : \mathbb{N} \to \mathbb{N}$ is the identity then $\imath^* : \mathbb{N}^* \to \mathbb{N}^*$ is the identity;

(2) If f, g are real functions then $(f \circ g)^* = f^* \circ g^*$ (provided the composition $f \circ g$ is defined);

(3) A real function $f : A \to B$ is 1-1 if and only if its hyper-extension $f^* : A^* \to B^*$ is 1-1;

(4) A real function $f : A \to B$ is onto if and only if its hyper-extension $f^* : A^* \to B^*$ is onto;

(5) If the real function f is defined on X, then

$$\{f(x) \mid x \in X\}^* = \{f^*(\xi) \mid \xi \in X^*\}.$$

In particular, $\mathrm{ran}(f)^* = \mathrm{ran}(f^*)$.

PROOF. (1) is straightforward from the definitions.

(2). Let $f : A \to B$ and $\varphi : \mathbb{N} \to A$. Then $(g \circ f)^*(\lim_\alpha \varphi) = \lim_\alpha \left((g \circ f) \circ \varphi\right) = \lim_\alpha \left(g \circ (f \circ \varphi)\right) = g^*\left(\lim_\alpha(f \circ \varphi)\right) = g^*\left(f^*(\lim_\alpha \varphi)\right)$.

(3). We recall that f^* is an extension of f. So, if f^* is 1-1, then also its restriction $f^*_{\restriction A} = f$ is 1-1. Conversely, let $\xi = \lim_\alpha \varphi$ and $\zeta = \lim_\alpha \psi$ where $\varphi, \psi : \mathbb{N} \to A$, and assume that $f^*(\xi) = f^*(\zeta)$, that is $\lim_\alpha f \circ \varphi = \lim_{n\uparrow\alpha} f \circ \psi$. Then $f(\varphi(n)) = f(\psi(n))$ a.e., and since f is 1-1, it follows that $\varphi(n) = \psi(n)$ a.e., and hence that $\xi = \zeta$.

(5). Let $Y = \{f(x) \mid x \in X\}$. If $\eta \in Y^*$, that is, if $\eta = \lim_{n\uparrow\alpha} \vartheta(n)$ for some sequence $\vartheta : \mathbb{N} \to Y$, then for every n there exists an element $\varphi(n) \in X$ such that $\vartheta(n) = f(\varphi(n))$. If $\xi = \lim_\alpha \varphi \in X^*$, we have that $f^*(\xi) = \lim_\alpha f \circ \varphi = \lim_\alpha \vartheta = \eta$. Conversely, if $\xi = \lim_\alpha \varphi$ for some $\varphi : \mathbb{N} \to X$ then $f(\varphi(n)) \in Y$ for all n, and so $f^*(\xi) = \lim_{n\uparrow\alpha} f \circ \varphi \in Y^*$.

(4). It directly follows from (5) because f is onto $\Leftrightarrow \mathrm{ran}(f) = A \Leftrightarrow [\mathrm{ran}(f)]^* = \mathrm{ran}(f^*) = A^* \Leftrightarrow f^*$ is onto. $\qquad\square$

COROLLARY 2.28. *If a real function f takes only finitely many values, then* $\operatorname{ran}(f^*) = \operatorname{ran}(f)$. *In particular, if f is constant then also f^* is constant.*

PROOF. By the previous proposition, $\operatorname{ran}(f^*) = \operatorname{ran}(f)^*$; and since $\operatorname{ran}(f)$ is finite, its hyper-image $\operatorname{ran}(f)^* = \operatorname{ran}(f)$, by Proposition 2.14. □

PROPOSITION 2.29. *Let f and g be real functions defined on X. Then*

(1) $\{x \in X \mid f(x) = g(x)\}^* = \{\xi \in X^* \mid f^*(\xi) = g^*(\xi)\}$

(2) $\{x \in X \mid f(x) \neq g(x)\}^* = \{\xi \in X^* \mid f^*(\xi) \neq g^*(\xi)\}$

(3) $\{x \in X \mid f(x) < g(x)\}^* = \{\xi \in X^* \mid f^*(\xi) < g^*(\xi)\}$

(4) $\{x \in X \mid f(x) \le g(x)\}^* = \{\xi \in X^* \mid f^*(\xi) \le g^*(\xi)\}$

PROOF. (1). Consider the set $E = \{x \in X \mid f(x) = g(x)\}$. We want to show that $\xi \in E^* \Leftrightarrow f^*(\xi) = g^*(\xi)$ for all $\xi \in X^*$. Let $\xi = \lim_\alpha \varphi$ where $\varphi : \mathbb{N} \to X$. Then $\xi \in E^* \Leftrightarrow \varphi(n) \in E$ *a.e.* $\Leftrightarrow$ $f(\varphi(n)) = g(\varphi(n))$ *a.e.* $\Leftrightarrow f^*(\xi) = \lim_\alpha f \circ \varphi = \lim_\alpha g \circ \varphi = g^*(\xi)$.

(2). It directly follows from (1), by noticing that

$$\{\xi \in X^* \mid f^*(\xi) \neq g^*(\xi)\} = (E^*)^c = (E^c)^*.$$

(3). Let $D = \{x \in X \mid f(x) < g(x)\}$. Similarly as above, given an arbitrary $\xi = \lim_\alpha \varphi \in A^*$ where $\varphi : \mathbb{N} \to X$, we have the following equivalences: $\xi \in D^* \Leftrightarrow f(\varphi(n)) < g(\varphi(n))$ *a.e.* $\Leftrightarrow$ (by *transfer* of inequalities) $f^*(\xi) = \lim_\alpha f \circ \varphi < \lim_\alpha g \circ \varphi = g^*(\xi)$.

(4). It follows from (1) and (3), because

$$\{x \in X \mid f(x) \le g(x)\}^* = (D \cup E)^* = D^* \cup E^*.$$

□

We conclude this section by formulating another example of the *transfer principle*. It is a particular case of the previous proposition where one considers constant functions.

PROPOSITION 2.30 (*Transfer for real functions*).
Let $f : A \to B$ be a real function, let $r \in \mathbb{R}$ be a real number, and let $\bowtie$ be any of the relations: $=, \neq, \le, <, \ge, >$. *Then*

$$f(a) \bowtie r \text{ for all } a \in A \iff f^*(\xi) \bowtie r \text{ for all } \xi \in A^*.$$

7. Some more relevant basic properties

As we have seen, the hyperreal field $\mathbb{R}^*$ is a special ordered field that properly extends the real line. In consequence, $\mathbb{R}^*$ is non-Archimedean, and one can distinguish between infinitesimal, finite and infinite hyperreal numbers. The following is a straight application of the *transfer principle* for inequalities.

PROPOSITION 2.31. *The number* α *is infinite, that is,* $\alpha > k$ *for all* $k \in \mathbb{N}$.

PROOF. Let $\imath : \mathbb{N} \to \mathbb{N}$ be the identity map. For any given $k \in \mathbb{N}$, trivially $\imath(n) > k$ for all $n > k$. So, $\imath(n) > k$ a.e. and by *transfer* $\alpha = \lim_{n\uparrow\alpha} n > \lim_{n\uparrow\alpha} c_k(n) = k$.   □

In the sequel, we will consider the following prototype of an infinitesimal number.

DEFINITION 2.32. *The special number "Eta" is defined as*

$$\eta = 1/\alpha$$

As the reciprocal of the infinite number α, clearly η is a non-zero infinitesimal obtained as the Alpha-limit of the sequence $n \mapsto 1/n$.

The *transfer principle* of inequalities is a fundamental tool in the practice of Alpha-Calculus. As an example, suppose we want to show that the following hyperreal numbers are displayed in increasing order.

$$0 < \; 1 - \cos(\eta) \; < \; \eta \; < \; \sin(2\eta) \; < \; \frac{2 + 5\alpha}{3 + 2\alpha} \; < \; \frac{3 + 5\eta}{1 + 7\eta}$$

By *transfer*, it is enough to notice that the inequalities below hold for all sufficiently large n (we omit the proof that this is actually the case).

$$0 \; < \; 1 - \cos(1/n) \; < \; 1/n \; < \; \sin(2/n) \; < \; \frac{2 + 5n}{3 + 2n} \; < \; \frac{3 + 5/n}{1 + 7/n}$$

Again by *transfer* of inequalities, it is verified that hyper-extensions of intervals are intervals.

PROPOSITION 2.33. *Let* $(a, b)_{\mathbb{R}} = \{x \in \mathbb{R} \mid a < x < b\}$ *be an open interval of real numbers. Then its hyper-extension*

$$[(a, b)_{\mathbb{R}}]^* = \{\xi \in \mathbb{R}^* \mid a < \xi < b\} = (a, b)_{\mathbb{R}^*}.$$

Similar characterizations also hold for the other kinds of real intervals: $[a, b]$, $(a, b]$, $[a, b)$, $(a, +\infty)$, $[a, +\infty)$, $(-\infty, b)$, $(-\infty, b]$.

PROOF. Let us start by checking that for every $\xi \in \mathbb{R}^*$, we have $\xi \in (a, +\infty)^* \Leftrightarrow \xi > a$. If $\xi = \lim_\alpha \varphi$, then $\xi \in (a, +\infty)^*$ if and only if $\varphi(n) \in (a, +\infty)$ a.e., that is, if $\varphi(n) > a$ a.e.. By *transfer*, this last condition holds if and only if $\xi = \lim_\alpha \varphi > \lim_\alpha c_a = a$, as desired.[8] In the same manner, one proves that $\xi \in (-\infty, b)^* \Leftrightarrow \xi < b$.

As hyper-extensions commute with intersections, we have that

$$(a, b)^* = [(a, +\infty) \cap (-\infty, b)]^* = (a, +\infty)^* \cap (-\infty, b)^*$$

and the equivalence $\xi \in (a, b)^* \Leftrightarrow a < \xi < b$ follows from what proved above.

The other proofs for intervals of the form $[a, b]$, $(a, b]$, $[a, b)$, $(a, +\infty)$, $[a, +\infty)$, $(-\infty, b)$,$(-\infty, b]$ are entirely similar.   □

As a consequence of the above proposition, it is readily seen that the *finite numbers* $\mathbb{R}^*_{\text{fin}}$ of the hyperreal field are given by:

$$\mathbb{R}^*_{\text{fin}} = \bigcup_{n \in \mathbb{N}} (-n, n)^*.$$

On the other hand, the set of all hyperreal numbers is given by:

$$\mathbb{R}^* = \left(\bigcup_{n \in \mathbb{N}} (-n, n) \right)^*.$$

Since $\mathbb{R}^*_{\text{fin}} \neq \mathbb{R}^*$, this example shows that while hyper-extensions commute with *finite* unions and intersections, they do *not* commute with *countable* unions and intersections.

Basic equalities and inequalities of real numbers extend to the corresponding equalities and inequalities of hyperreal numbers. A few examples are itemized below, and many more can be easily conceived by the reader.

From this point on, we will abuse notation and write f to denote both a real function and its hyper-extension f^*, when confusion is unlikely.

PROPOSITION 2.34. *For all $\xi, \zeta \in \mathbb{R}^*$, the following identities hold:*

(1) $\sin(\xi \pm \zeta) = \sin(\xi) \cos(\zeta) \pm \cos(\xi) \sin(\zeta)$;

(2) $\cos(\xi \pm \zeta) = \cos(\xi) \cos(\zeta) \mp \sin(\xi) \sin(\zeta)$;

(3) $e^\xi \cdot e^\zeta = e^{\xi + \zeta}$;

(4) $(e^\xi)^\zeta = e^{\xi \cdot \zeta}$;

(5) $e^\xi > 0$;

---

[8] Recall that c_a denotes the constant sequence with value a.

(6) *If $\xi, \zeta > 0$ then $\log(\xi \cdot \zeta) = \log(\xi) + \log(\zeta)$;*

(7) *If $\xi, \zeta > 0$, $\log(\xi^\zeta) = \zeta \cdot \log(\xi)$;*

(8) *If $\xi > 0$, then $\xi > 1 \Leftrightarrow \log(\xi) > 0$.*

All proofs are straightforward, and are omitted.

The monotonicity properties of functions are preserved under hyper-extensions. Again, the *transfer* principles plays a central role in the proofs.

PROPOSITION 2.35. *A real function $f : A \to B$ is increasing (or decreasing, or non-increasing, or non-decreasing) if and only if its hyper-extension $f^* : A^* \to B^*$ is increasing (or decreasing, or non-increasing, or non-decreasing, respectively).*

PROOF. Assume that f is increasing. (The proofs of the other cases are almost identical.) Let $\xi = \lim_{n \uparrow \alpha} \varphi(n)$ and $\zeta = \lim_{n \uparrow \alpha} \psi(n)$ where $\varphi, \psi : \mathbb{N} \to A$, and suppose that $\xi < \zeta$. By *backward transfer*, we have that $\varphi(n) < \psi(n)$ a.e., and hence $f(\varphi(n)) < f(\psi(n))$ a.e.. So, again by *transfer*, we conclude that $f^*(\xi) = \lim_{n \uparrow \alpha} f(\varphi(n)) < \lim_{n \uparrow \alpha} f(\psi(n)) = f^*(\zeta)$, as desired.   □

8. Hyper-extensions of sets of numbers

In this section we focus on the fundamental properties of hyper-extensions of sets of numbers. In particular, we show that the basic properties of natural, integer, rational and real numbers transfer to the corresponding hyper-extensions.

PROPOSITION 2.36.

(1) $\mathbb{N}$ *is an* initial segment *of $\mathbb{N}^*$, that is, if $\nu \in \mathbb{N}^* \setminus \mathbb{N}$, then $\nu > k$ for all $k \in \mathbb{N}$;*

(2) *The set $\mathbb{N}^*$ is* unbounded *in $\mathbb{R}^*$, that is, for every $\xi \in \mathbb{R}^*$ there exists $\nu \in \mathbb{N}^*$ with $\nu > \xi$;*

(3) *The hyperintegers $\mathbb{Z}^*$ are an ordered ring ;*

(4) *The order on $\mathbb{Z}^*$ is discrete: Every hyperinteger $\nu \in \mathbb{Z}^*$ has successor $\nu + 1$ and predecessor $\nu - 1$, that is, there are no $\mu \in \mathbb{Z}^*$ such that $\nu < \mu < \nu + 1$ or $\nu - 1 < \mu < \nu$;*

(5) *For every $\xi \in \mathbb{R}^*$ there exists a unique $\nu \in \mathbb{Z}^*$, called the* hyperinteger part *of ξ, such that $\nu \le \xi < \nu + 1$;*

(6) *The hyperrational numbers $\mathbb{Q}^*$ are an ordered subfield of $\mathbb{R}^*$;*

(7) *Both the hyperrational numbers $\mathbb{Q}^*$ and the hyperirrational numbers $\mathbb{R}^* \setminus \mathbb{Q}^*$ are dense in $\mathbb{R}^*$, that is, for all $\xi < \zeta$ in $\mathbb{R}^*$ there exist $\sigma \in \mathbb{Q}^*$ and $\tau \notin \mathbb{Q}^*$ with $\xi < \sigma < \zeta$ and $\xi < \tau < \zeta$.*

PROOF. (1). Let $\nu \in \mathbb{N}^*$, and assume that $\nu \leq k$ for some $k \in \mathbb{N}$. Pick a sequence $\varphi : \mathbb{N} \to \mathbb{N}$ such that $\lim_{n\uparrow\alpha} \varphi(n) = \nu$. Then, by *backward transfer*, we have $\varphi(n) \leq k$ a.e., and so we can pick a sequence $\varphi' : \mathbb{N} \to \{1, \ldots, k\}$ such that $\varphi'(n) = \varphi(n)$ a.e.. Since φ' takes only (some of) the finitely values $1, \ldots, k$, by Corollary 2.6 it must be $\lim_{n\uparrow\alpha} \varphi'(n) = i$ for some $1 \leq i \leq k$. In particular, $\nu = \lim_\alpha \varphi = \lim_\alpha \varphi' \in \mathbb{N}$.

(2). Let $\xi = \lim_\alpha \varphi$. For each n, pick $\psi(n) \in \mathbb{N}$ such that $\psi(n) > \varphi(n)$. Then, by *transfer*, the number $\nu = \lim_\alpha \psi \in \mathbb{N}^*$ has the desired property that $\nu > \lim_\alpha \varphi = \xi$.

(4). Given two hyperintegers $\nu < \mu$, pick sequences $\varphi, \psi : \mathbb{N} \to \mathbb{Z}$ such that $\lim_\alpha \varphi = \nu$ and $\lim_\alpha \psi = \mu$. By *backward transfer*, $\varphi(n) < \psi(n)$ a.e., and hence $\varphi(n) + 1 \leq \psi(n)$ a.e.. By *transfer*, we obtain $\nu + 1 = \lim_\alpha(\varphi + c_1) \leq \lim_\alpha \psi = \mu$, where c_1 is the constant sequence with value 1. The proof for predecessors is entirely similar.

(5). Let $\xi = \lim_\alpha \varphi$. For each $n \in \mathbb{N}$, let $\vartheta(n) \in \mathbb{Z}$ be the integer part of $\varphi(n)$, that is, $\vartheta(n)$ is the unique integer such that $\vartheta(n) \leq \varphi(n) < \varphi(n) + 1$. By *transfer*, the hyperinteger $\nu = \lim_{n\uparrow\alpha} \vartheta(n) \in \mathbb{Z}^*$ has the desired property $\nu \leq \xi < \nu + 1$. The uniqueness of ν follows from the discreteness property of $\mathbb{Z}^*$ (proved in (4)).

(7). Let $\xi = \lim_\alpha \varphi_1$ and $\zeta = \lim_\alpha \varphi_2$ be hyperreal numbers with $\xi < \zeta$. By *backward transfer*, the set $D = \{n \mid \varphi_1(n) < \varphi_2(n)\}$ is qualified. By the density of $\mathbb{Q}$ and $\mathbb{R} \setminus \mathbb{Q}$ in $\mathbb{R}$, for each $n \in X$ we can pick elements $x_n \in \mathbb{Q}$ and $y_n \notin \mathbb{Q}$ such that $\varphi_1(n) < x_n, y_n < \varphi_2(n)$. Now pick ψ and ϑ any two sequences such that $\psi(n) = x_n$ and $\vartheta(n) = y_n$ for all $n \in D$, respectively. By *transfer*, it is easily checked that $\sigma = \lim_\alpha \psi \in \mathbb{Q}^*$ and $\tau = \lim_\alpha \vartheta \notin \mathbb{Q}^*$ satisfy $\xi < \sigma, \tau < \zeta$.

(3) and (6) are straightforward, and are omitted.   □

PROPOSITION 2.37. *A set $B \subseteq \mathbb{R}$ is unbounded if and only if B^* contains an infinite number.*

PROOF. If B is unbounded, then pick a sequence $\varphi(n) \in B$ such that $|\varphi(n)| > n$. Then $\xi = \lim_\alpha \varphi \in B^*$ is infinite.

Conversely, let $\xi \in B^*$ be infinite, and pick a sequence $\varphi : \mathbb{N} \to B$ such that $\lim_\alpha \varphi = \xi$. Then $\{\varphi(n) \mid n \in \mathbb{N}\} \subseteq B$ is unbounded. Indeed, if by contradiction there exists a real number $M > 0$ such that

$|\varphi(n)| \leq M$ for all n, then also the Alpha-limit $|\xi| = |\lim_\alpha \varphi| \leq M$, against the hypothesis. □

A relevant example of a property that is *not* preserved under hyper-extension is *completeness*. Indeed, recall that the hyperreal field $\mathbb{R}^*$ is not complete because it is non-Archimedean (see Proposition 1.7).[9]

We close this section by proving a result on the cardinality of hyper-extensions. We recall that the *continuum* $\mathfrak{c} = 2^{\aleph_0}$ is the (uncountable) cardinality of the real numbers $\mathbb{R}$.

PROPOSITION 2.38. *Hypernatural, hyperinteger, hyperrational and hyperreal numbers, they all have the cardinality of the continuum:*

$$|\mathbb{N}^*| = |\mathbb{Z}^*| = |\mathbb{Q}^*| = |\mathbb{R}^*| = \mathfrak{c}$$

PROOF. It is well-known that the integers and the rational numbers are countable, so we can pick bijections $f : \mathbb{N} \to \mathbb{Z}$ and $g : \mathbb{N} \to \mathbb{Q}$. The corresponding hyper-extensions $f^* : \mathbb{N}^* \to \mathbb{Z}^*$ and $g^* : \mathbb{N}^* \to \mathbb{Q}^*$ are bijections too, and so $|\mathbb{N}^*| = |\mathbb{Z}^*| = |\mathbb{Q}^*|$. Now, trivially $|\mathbb{Q}^*| \leq |\mathbb{R}^*|$. Moreover, by definition, $\mathbb{R}^* = \{\lim_\alpha \varphi \mid \varphi : \mathbb{N} \to \mathbb{R}\}$, and so $|\mathbb{R}^*| \leq |\mathfrak{Fun}(\mathbb{N}, \mathbb{R})| = \mathfrak{c}^{\aleph_0} = \mathfrak{c}$. We are left to show that $\mathfrak{c} \leq |\mathbb{Q}^*|$.

To this end fix a positive infinitesimal, *e.g.* the number $\eta = 1/\alpha$. By the density property of $\mathbb{Q}^*$ in $\mathbb{R}^*$ (see Proposition 2.36), we know that for every $r \in \mathbb{R}$ we can pick an hyperrational number $\xi_r \in \mathbb{Q}^*$ such that $r - \eta < \xi_r < r + \eta$, and hence $\xi_r \sim r$. The correspondence $r \mapsto \xi_r$ yields a 1-1 map from $\mathbb{R}$ into $\mathbb{Q}^*$, and this proves desired inequality $\mathfrak{c} \leq |\mathbb{Q}^*|$. □

9. The Qualified Set Axiom

By identifying a property P of natural numbers with its *extension*, namely with the set of elements that satisfy it, one has a natural way of extending P to a property P^* of hypernatural numbers.

DEFINITION 2.39. *Let P be a property of natural numbers. We will say that P^* is satisfied by a hypernatural number $\nu \in \mathbb{N}^*$, and write $P^*(\nu)$, if $\nu \in \{n \in \mathbb{N} \mid P(n)\}^*$.*

---

[9] The non-elementarity of the completeness property is given by the fact that it does not talk about elements of the object under consideration, namely the real numbers, but rather about *subsets* of real numbers. This kind of properties are named *second-order* in mathematical logic.

Notice that the property P^* extends property P; indeed, if $n \in \mathbb{N}$ then $P(n) \Leftrightarrow P^*(n)$.

E.g., we will say that

- the hypernatural number ν is *even** if $\nu \in \{n \in \mathbb{N} \mid n \text{ is even}\}^*$;
- ν is a *cube** if $\nu \in \{n^3 \mid n \in \mathbb{N}\}^*$;
- ν is a *prime** if $\nu \in \{p \in \mathbb{N} \mid p \text{ is prime}\}^*$;

and so forth.[10]

EXERCISE 2.40. *Prove the following equivalences:*

(1) $\nu \in \mathbb{N}^*$ *is even* if and only if there exists $\mu \in \mathbb{N}^*$ such that $\nu = 2\mu$;

(2) $\nu \in \mathbb{N}^*$ *is a cube* if and only if there exists $\mu \in \mathbb{N}^*$ such that $\nu = \mu^3$;

(3) $\nu \in \mathbb{N}^*$ *is prime* if and only there exist no hypernatural numbers $\mu_1, \mu_2 > 1$ such that $\nu = \mu_1 \mu_2$;

(4) $\nu, \mu \in \mathbb{N}^*$ *are coprime* if their only common divisor in $\mathbb{N}^*$ is 1; that is, for every $\sigma, \nu_1, \mu_1 \in \mathbb{N}^*$, if $\nu = \sigma\nu_1$ and $\mu = \sigma\mu_1$ then $\sigma = 1$.

EXERCISE 2.41. *Prove that every positive hyperrational number $\xi \in \mathbb{Q}^*$ has a unique representation in reduced form; that is, there exist unique $\nu, \mu \in \mathbb{N}^*$ such that $\xi = \frac{\nu}{\mu}$ and ν, μ are coprime*.*

Remark that – according to the above definition – the number α satisfies a property P^* if and only if $P(n)$ holds *a.e.*. In other words:

$$P^*(\alpha) \iff \alpha \in \{n \in \mathbb{N} \mid P(n)\}^* \iff \{n \in \mathbb{N} \mid P(n)\} \text{ is qualified.}$$

So, in this sense, the properties of α are precisely the properties determined by the qualified sets. Now, a question arises naturally:

Which are the properties of α?

Clearly, the property "$\alpha > k$" is satisfied for any given $k \in \mathbb{N}$ because $n > k$ holds *a.e.*. However, we disclose that this is essentially all one can prove about α by the axioms of Alpha-Theory (see Theorem 2.57).

Although the sole property that α be infinite is sufficient to develop Alpha-Calculus, nevertheless there are applications in which it would

---

[10] If the property P is expressed in the first-order language that will be presented in the logic section §5.1, then P^* retains the same formal expression. Some examples are found in Exercise 2.40.

be useful to also assume additional properties. For example, in the theory of *numerosity* (see Part 5), it will be convenient to assume that α be both a multiple of k and a k-th power of every natural number k.

Divisibility Property. *For every* $k \in \mathbb{N}$, $\frac{\alpha}{k} \in \mathbb{N}^*$.

Root Property. *For every* $k \in \mathbb{N}$, $\sqrt[k]{\alpha} \in \mathbb{N}^*$.

Notice that the *Divisibility Property* is equivalent to the property that the set of k-multiples $\{km \mid m \in \mathbb{N}\}$ is qualified for all $k \in \mathbb{N}$; and notice that the *Root Property* is equivalent to the property that the set of k-th powers $\{m^k \mid m \in \mathbb{N}\}$ is qualified for all $k \in \mathbb{N}$.

EXERCISE 2.42. *Prove the following properties:*

(1) *If the set* $Q_1 = \{m! \mid m \in \mathbb{N}\}$ *is qualified then the* Divisibility Property *holds.*

(2) *If the set* $Q_2 = \{m^{m!} \mid m \in \mathbb{N}\}$ *is qualified then the* Root Property *holds.*

(3) *If the set* $Q_3 = \{m!^{m!} \mid m \in \mathbb{N}\}$ *is qualified then both the* Divisibility *and the* Root Properties *hold.*

In the practice, depending on the given context, one looks for a convenient choice of an infinite set $Q \subset \mathbb{N}$, and postulates the following additional axiom.

(QSA)$_Q$ Qualified Set Axiom. *The set* Q *is qualified, that is,* $\alpha \in Q^*$.

It is worth stressing right away that, for *any* choice of the infinite set Q, one can consistently add **(QSA)$_Q$** to the axioms of Alpha-Calculus axioms.[11] We remark that the smallest the set Q, the more information one gets about α. Clearly, the choice of an appropriate Q will depend on the applications that one has in mind. For example, by the properties in Exercise 2.42, one directly gets the following

COROLLARY 2.43. *Assume that the Qualified Set Axiom* **(QSA)$_Q$** *holds for an infinite set* $Q \subseteq \{m!^{m!} \mid m \in \mathbb{N}\}$. *Then both the* Divisibility *and the* Root Properties *hold.*

---

[11] This fact will be proved further on in this chapter, in Theorem 2.56.

Now, suppose we are given an infinite (countable) list of properties of natural numbers:

$$P_1(n), \ P_2(n), \ \ldots \ , \ P_k(n), \ \ldots$$

We ask the following: *"By postulating $(QSA)_Q$ for a suitable choice of the set Q, can one prove that $P_k^*(\alpha)$ holds for all k?"* Clearly, this cannot be obtained for arbitrary lists of properties. Indeed, if for some k the conjunction "$P_1(n)$ and $\ldots$ and $P_k(n)$" is satisfied only by finitely many n, then its negation "not $(P_1(n)$ and $\ldots$ and $P_k(n))$" holds *a.e.*. In consequence, one has "not $(P_1^*(\alpha)$ and $\ldots$ and $P_k^*(\alpha))$", and therefore at least one of the $P_i^*(\alpha)$ fails. However, the next result shows that this is essentially the only possible counter-example.

THEOREM 2.44. *Assume that the properties $\{P_k(n) \mid k \in \mathbb{N}\}$ are "mutually compatible", in the sense that for every k the following set is infinite:*

$$A_k \ = \ \{n \in \mathbb{N} \mid P_1(n) \ and \ldots \ and \, P_k(n)\}.$$

Then there exists an infinite $Q \subseteq \mathbb{N}$ such that $(QSA)_Q$ implies $P_k^*(\alpha)$ for all k.

PROOF. For any $k \in \mathbb{N}$, let

$$a_k \ = \ \min\{n > k \mid P_1(n) \ and \ \ldots \ and \, P_k(n)\}.$$

By the hypothesis of mutual compatibility the above set is nonempty, and so the definition of a_k is well-posed. Moreover, the following set is infinite, as it contains arbitrarily large numbers:

$$Q \ = \ \{a_k \mid k \in \mathbb{N}\}.$$

We now want to show that for every k, the axiom $(QSA)_Q$ implies $P_k^*(\alpha)$. To verify this, notice that $P_k(a_i)$ holds for all $i \geq k$. But then the set $\{n \in \mathbb{N} \mid P_k(n)\}$ is qualified, because it contains the intersection $Q \cap [a_k, +\infty)_{\mathbb{N}}$ of two qualified sets. □

By using the same argument as above, one easily checks the following general property.

EXERCISE 2.45. *Let $P_1(n), \ P_2(n), \ \ldots \ , \ P_k(n), \ \ldots$ be an infinite list of properties of the natural numbers. Suppose there exists a qualified set Q such that:*

$$\forall k \ \exists n_k \ \forall n \geq n_k \ (n \in Q \ \Rightarrow \ P_k(n)).$$

Then $P_k^*(\alpha)$ holds for all k.

10. Rings and ideals

Let us quickly recall a few general notions from algebra that will be used in the sequel.[12]

An *ideal* $\mathfrak{i}$ of a commutative ring R is a *proper* subset that is closed under opposites and sums, and such that $a \cdot x \in \mathfrak{i}$ for all $a \in R$ and $x \in \mathfrak{i}$.

The *quotient ring* $R/\mathfrak{i}$ is defined as the set of all *cosets*

$$[a]_{\mathfrak{i}} = \{b \in R \mid b - a \in \mathfrak{i}\}.$$

In other words, $R/\mathfrak{i} = R/\equiv$ is the quotient set modulo the equivalence relation $b \equiv a \Leftrightarrow b - a \in \mathfrak{i}$ induced by $\mathfrak{i}$. It is easily verified that the ring operations of R are inherited by the quotient $R/\mathfrak{i}$.

An ideal $\mathfrak{m}$ is called *maximal* if it is maximal with respect to inclusion, that is, if there are no ideals $\mathfrak{i}$ such that $\mathfrak{m} \subsetneq \mathfrak{i}$. A useful characterization is the fact that an ideal $\mathfrak{m}$ of a unital commutative ring R is maximal if and only if the corresponding quotient $R/\mathfrak{m}$ is a field, namely the *residue field*.

In the sequel, we will focus on the set of all *real sequences*:

$$\mathfrak{Fun}(\mathbb{N}, \mathbb{R}) = \{\varphi \mid \varphi : \mathbb{N} \to \mathbb{R}\}.$$

This set comes naturally with the structure of a *commutative ring*, where sums and products of sequences are defined pointwise. Precisely, for every $\varphi, \psi \in \mathfrak{Fun}(\mathbb{N}, \mathbb{R})$, one defines $\varphi + \psi$ and $\varphi \cdot \psi$ by putting for all $n \in \mathbb{N}$:

$$(\varphi + \psi)(n) = \varphi(n) + \psi(n)$$
$$(\varphi \cdot \psi)(n) = \varphi(n) \cdot \psi(n)$$

A relevant example of an ideal of $\mathfrak{Fun}(\mathbb{N}, \mathbb{R})$ is given by the set of all sequences that eventually vanish:

$$\mathfrak{i}_0 = \{\varphi \in \mathfrak{Fun}(\mathbb{N}, \mathbb{R}) \mid \exists k \, \forall n > k \; \varphi(n) = 0\}.$$

EXERCISE 2.46. *Verify that $\mathfrak{i}_0$ is an ideal.*

SOLUTION. For every sequence φ, denote by

$$Z(\varphi) = \{n \mid \varphi(n) = 0\}$$

its *zero-set*. Notice that $\varphi \in \mathfrak{i}_0$ if and only if $Z(\varphi)$ is *cofinite*, that is, its complement is finite or empty. The closure of $\mathfrak{i}_0$ under opposites is trivial, because $Z(\varphi) = Z(-\varphi)$. Now notice that $Z(\varphi_1 + \varphi_2) \supseteq Z(\varphi_1) \cap Z(\varphi_2)$. Since the intersection of two cofinite sets is cofinite,

---

[12] A classic reference is the popular S. Lang's textbook [**67**], where all the algebraic notions and properties presented here can be found.

and the superset of a cofinite set is trivially cofinite, we obtain the implication $\varphi_1, \varphi_2 \in i_0 \Rightarrow \varphi_1 + \varphi_2 \in i_0$. Finally, $Z(\varphi) \subseteq Z(\psi \cdot \varphi)$, and so $\varphi \in i_0 \Rightarrow \psi \cdot \varphi \in i_0$. $\qquad \square$

We remark that i_0 is *not* maximal. *E.g.*, let $\varphi(n) = 1 + (-1)^n$ and $\psi(n) = 1 + (-1)^{n+1}$. Then both $\varphi, \psi \notin i_0$ while their product $\varphi \cdot \psi = c_0$ is the null sequence; so, $[\varphi], [\psi] \neq [0]$ but $[\varphi] \cdot [\psi] = [0]$. In particular, R/i_0 is not a field because it admits zero divisors, and hence i_0 is not a maximal ideal.

Easy examples of maximal ideals of $\mathfrak{Fun}(\mathbb{N}, \mathbb{R})$ are given by the *principal ideals* $\mathfrak{m}_k$, where k is a fixed natural number:

$$\mathfrak{m}_k = \{\varphi \in \mathfrak{Fun}(\mathbb{N}, \mathbb{R}) \mid \varphi(k) = 0\}.$$

EXERCISE 2.47. *Verify that the principal ideals* $\mathfrak{m}_k$ *are maximal.*

SOLUTION. By contradiction, assume that there exists an ideal i that properly includes $\mathfrak{m}_k$. Pick $\vartheta \in i \setminus \mathfrak{m}_k$, then $\vartheta(k) \neq 0$. If τ is the sequence where $\tau(k) = 1/\vartheta(k)$ and $\tau(n) = 1$ for $n \neq k$, then the product $\vartheta \cdot \tau = c_1 \in i$. As a consequence, for all $\varphi : \mathbb{N} \to \mathbb{R}$, we have that $\varphi = \varphi \cdot c_1 \in i$, and we would conclude that $i = \mathfrak{Fun}(\mathbb{N}, \mathbb{R})$, a contradiction. $\qquad \square$

For every $k \in \mathbb{N}$, let us denote by χ_k^c the characteristic function of the complement of the singleton $\{k\}$, that is,

$$\chi_k^c(n) = \begin{cases} 1 & \text{if } n \neq k \\ 0 & \text{if } n = k. \end{cases}$$

A useful characterization of the principal ideals is the following.

PROPOSITION 2.48. *For every* $k \in \mathbb{N}$, *the principal ideal* $\mathfrak{m}_k$ *is the unique ideal of* $\mathfrak{Fun}(\mathbb{N}, \mathbb{R})$ *that contains* χ_k^c.

PROOF. By definition, $\chi_k^c \in \mathfrak{m}_k$ because $\chi_k^c(k) = 0$. Conversely, suppose that i is an ideal of $\mathfrak{Fun}(\mathbb{N}, \mathbb{R})$ with $\chi_k^c \in i$. Then for every sequence φ such that $\varphi(k) = 0$ we have $\varphi = \varphi \cdot \chi_k^c \in i$, and so $\mathfrak{m}_k \subseteq i$. By the maximality of $\mathfrak{m}_k$, it follows that $\mathfrak{m}_k = i$. $\qquad \square$

PROPOSITION 2.49. *The following properties are equivalent for any maximal ideal* $\mathfrak{m}$ *of* $\mathfrak{Fun}(\mathbb{N}, \mathbb{R})$:

(1) $\mathfrak{m}$ *is non-principal;*

(2) $\mathfrak{m} \supset i_0$;

(3) *The embedding* $\pi : \mathbb{R} \to \mathfrak{Fun}(\mathbb{N}, \mathbb{R})/\mathfrak{m}$ *given by* $\pi(r) = [c_r]_\mathfrak{m}$
is not onto. In other words, the residue field $\mathfrak{Fun}(\mathbb{N}, \mathbb{R})/\mathfrak{m}$
properly extends the copy of $\mathbb{R}$ *as given by* $\{[c_r]_\mathfrak{m} \mid r \in \mathbb{R}\}$.

PROOF. $(1) \Rightarrow (2)$. Given $\varphi \in \mathfrak{i}_0$, let

$$\{n \mid \varphi(n) \neq 0\} = \{n_1, \ldots, n_h\}.$$

Then the product $\chi_{n_1}^c \cdot \ldots \cdot \chi_{n_h}^c \cdot \varphi = c_0$ is the null sequence, and hence it trivially belongs to $\mathfrak{m}$. Now, for every i, the coset $[\chi_{n_i}^c]_\mathfrak{m} \neq [0]_\mathfrak{m}$, that is, $\chi_{n_1}^c \notin \mathfrak{m}$, otherwise we would have $\mathfrak{m} = \mathfrak{m}_{n_i}$, against the hypothesis. Since the quotient $\mathfrak{Fun}(\mathbb{N}, \mathbb{R})/\mathfrak{m}$ is a field, and hence it does not contain zero divisors, it follows that $[\varphi]_\mathfrak{m} = [0]_\mathfrak{m}$, that is, $\varphi \in \mathfrak{m}$.

$(1) \Rightarrow (3)$. We claim that $[\imath]_\mathfrak{m} \neq [c_r]_\mathfrak{m}$ for all $r \in \mathbb{R}$, where $\imath(n) = n$ is the identity sequence. Assume by contradiction that $\imath - c_r \in \mathfrak{m}$ for some $r \in \mathbb{R}$, and consider the sequence $\tau(n) = 1/(n - r)$ for $n \neq r$, and $\tau(n) = 1$ if $n = r$. If $r \notin \mathbb{N}$, then the product $(\imath - c_r) \cdot \tau = c_1 \in \mathfrak{m}$, a contradiction. If $r = n_0 \in \mathbb{N}$, then $(\imath - c_r) \cdot \tau = \chi_{n_0}^c \in \mathfrak{m}$, and so $\mathfrak{m} = \mathfrak{m}_{n_0}$ would be principal, against the hypothesis.

$(3) \Rightarrow (1)$. If $\mathfrak{m} = \mathfrak{m}_k$ is principal, then for every φ we have that $[\varphi]_\mathfrak{m} = [c_{\varphi(k)}]_\mathfrak{m}$.

$(2) \Rightarrow (1)$. Let $\mathfrak{m} = \mathfrak{m}_k$ be principal. Then $\mathfrak{i}_0 \not\subseteq \mathfrak{m}$ because, *e.g.*, $(c_1 - \chi_k^c) \in \mathfrak{i}_0$ but $(c_1 - \chi_k^c) \notin \mathfrak{m}_k$. $\square$

Recall that an ideal $\mathfrak{p}$ is called *prime* if $x \cdot y \in \mathfrak{p}$ implies that $x \in \mathfrak{p}$ or $y \in \mathfrak{p}$. Any maximal ideal is prime, but there are prime ideals that are not maximal; however, in all rings of sequences that take values in a field, the two notions coincide.

PROPOSITION 2.50. *Let K be a field. Then every prime ideal of the ring of sequences* $\mathfrak{Fun}(\mathbb{N}, K)$ *is maximal.*

PROOF. Let $\mathfrak{p}$ be a prime ideal of $\mathfrak{Fun}(\mathbb{N}, K)$. To prove its maximality, we have to show that the quotient ring $\mathfrak{Fun}(\mathbb{N}, K)/\mathfrak{p}$ is a field. In other words, we have to see that every non-zero $[\varphi]_\mathfrak{p} \in \mathfrak{Fun}(\mathbb{N}, K)/\mathfrak{p}$ has an inverse, that is, for every $\varphi \notin \mathfrak{p}$ there exists ψ such that $[\varphi \cdot \psi]_\mathfrak{p} = [c_1]_\mathfrak{p}$. To this end, let us define the following sequence:

$$\psi(n) = \begin{cases} \frac{1}{\varphi(n)} & \text{if } \varphi(n) \neq 0 \\ 0 & \text{otherwise}. \end{cases}$$

Clearly, the product $\varphi \cdot (\varphi \cdot \psi - c_1) = c_0$ is the null sequence, and so it trivially belongs to $\mathfrak{p}$. As $\varphi \notin \mathfrak{p}$, by the primality of $\mathfrak{p}$ it follows that $\varphi \cdot \psi - c_1 \in \mathfrak{p}$, as desired. $\square$

We conclude this section by showing that non-principal maximal ideals actually exist. To this end, let us recall an important general result in algebra, whose proof makes an essential use of *Zorn's Lemma*, an equivalent formulation of the *Axiom of Choice*.[13]

THEOREM 2.51. *Every ideal can be extended to a maximal ideal.*

PROOF. Given an ideal $\mathfrak{i}$ of the ring R, consider the family

$$\mathcal{F} = \{\mathfrak{j} \mid \mathfrak{j} \supseteq \mathfrak{i} \text{ is an ideal of } R\}.$$

Trivially, $\mathcal{F} \neq \emptyset$ because $\mathfrak{i} \in \mathcal{F}$. In order to apply *Zorn's Lemma* we need to show that every chain $\{\mathfrak{j}_x \mid x \in X\}$ of the partially ordered set $\langle \mathcal{F}, \subseteq \rangle$ has an upper bound. (A *chain* of $\mathcal{F}$ is a subset that is linearly ordered by inclusion.) This can be easily seen by considering the union $\mathfrak{j} = \bigcup_{x \in X} \mathfrak{j}_x$, and by noticing that the property of a chain imply that $\mathfrak{j}$ is an ideal. Indeed, if $\xi, \eta \in \mathfrak{j}$, we can pick indexes x and y such that $\xi \in \mathfrak{i}_x$ and $\eta \in \mathfrak{j}_y$, respectively. By the property of chain, it is $x \leq y$ or $y \leq x$, and so it makes sense to pick $z = \max\{x, y\}$. Then both $\xi, \eta \in \mathfrak{j}_z$ and $\xi + \eta \in \mathfrak{j}_z \subseteq \mathfrak{j}$. The other properties of an ideal are proved in a similar manner. $\square$

11. Models of Alpha-Calculus

The goal of this section is to put Alpha-Calculus on a firm footing. By only using a few basic facts from algebra – most notably the existence of non-principal maximal ideals in the ring of real sequences $\mathfrak{Fun}(\mathbb{N}, \mathbb{R})$ – we will be able to construct and characterize all models of Alpha-Calculus. To begin with, we have to make it explicit what we mean by a "model".

DEFINITION 2.52. *A model of Alpha-Calculus consists of a map*

$$\varphi \longmapsto \lim_{n \uparrow \alpha} \varphi(n)$$

that assigns an Alpha-limit $\lim_{n \uparrow \alpha} \varphi(n)$ *to each real sequence* $\varphi : \mathbb{N} \to \mathbb{R}$ *in such a way that axioms (ACT1), (ACT2) and (ACT3) are realized.*

---

[13] With the expression "essential use" we mean that Theorem 2.51 cannot be proved without the Axiom of Choice. Indeed, it is known that there are models of *Zermelo-Fraenkel set theory* ZF where one finds rings without maximal ideals. A classic reference on the axiom of choice and related principles is T. Jech's monograph [**57**].

The above notion of model can be equivalently given in purely algebraic terms by means of suitable ring homomorphisms.

DEFINITION 2.53. *An* Alpha-morphism

$$J : \mathfrak{Fun}(\mathbb{N}, \mathbb{R}) \twoheadrightarrow \mathbb{R}^*$$

is a ring homomorphism from the ring of real sequences $\mathfrak{Fun}(\mathbb{N}, \mathbb{R})$ *onto a field* $\mathbb{R}^*$ *such that:*

(AM1) $J(c_r) = r$ *for every* $r \in \mathbb{R}$ *;*

(AM2) $\mathbb{R}^* \supsetneq \mathbb{R}$ *is a field that properly extends the reals;*

(AM3)   *For every* $\varphi, \psi : \mathbb{N} \to \mathbb{R}$ *:*
   $- J(\varphi + \psi) = J(\varphi) + J(\psi)$ *;*
   $- J(\varphi \cdot \psi) = J(\varphi) \cdot J(\psi)$ *.*

THEOREM 2.54. *Given a model of Alpha-Calculus, let*

$$\mathbb{R}^* = \left\{ \lim_{n \uparrow \alpha} \varphi(n) \,\middle|\, \varphi : \mathbb{N} \to \mathbb{R} \right\}.$$

Then the "Alpha-limit map" $J : \mathfrak{Fun}(\mathbb{N}, \mathbb{R}) \twoheadrightarrow \mathbb{R}^*$ *defined by putting* $J : \varphi \mapsto \lim_{n \uparrow \alpha} \varphi(n)$ *is an Alpha-morphism.*

Conversely, if $J : \mathfrak{Fun}(\mathbb{N}, \mathbb{R}) \twoheadrightarrow \mathbb{R}^*$ *is an Alpha-morphism, then the notion of Alpha-limit obtained by putting* $\lim_{n \uparrow \alpha} \varphi(n) = J(\varphi)$ *for every* $\varphi : \mathbb{N} \to \mathbb{R}$ *is a model of Alpha-Calculus.*

PROOF. In one direction, notice that by the *Field Axiom*, the set $\mathbb{R}^*$ of all Alpha-limits is a field. Moreover, by the *Real Number Axiom* jointly with the *Field Axiom*, such a field $\mathbb{R}^*$ includes $\mathbb{R}$ as a subfield. Clearly, properties (AM1) and (AM3) are just reformulations of (ACT1) and (ACT3), respectively. Finally, since $\lim_{n \uparrow \alpha} n = \alpha \notin \mathbb{R}$, the inclusion $\mathbb{R}^* \supsetneq \mathbb{R}$ is proper, and also (AM2) follows.

Conversely, we already noticed that the *Field Axiom* and the *Real Number Axiom* are straight reformulations of the properties (AM1) and (AM3) of an Alpha-morphism. So, we are left to verify the *Alpha Number Axiom* (ACT2), that is, the property that the image $J(\imath) \notin \mathbb{N}$ of identity sequence $\imath(n) = n$ does *not* belong to $\mathbb{N}$. To this end, fix a real sequence ϑ such that $J(\vartheta) \notin \mathbb{R}$ (this is possible by (AM2)). Given $k \in \mathbb{N}$, pick any real sequence φ_k such that $\varphi_k(n) = 1/(n-k)$ for $n \neq k$. Then for all n,

$$[(\varphi_k(n) \cdot (n-k)) - 1] \cdot (\vartheta(n) - \vartheta(k)) = 0.$$

By the properties (AM1) and (AM3), it follows that

$$[J(\varphi_k) \cdot J(\imath - c_k) - 1] \cdot (J(\vartheta) - \vartheta(k)) = 0.$$

Now $J(\vartheta) \neq \vartheta(k)$ because $J(\vartheta) \notin \mathbb{R}$, and so $J(\varphi_k) \cdot J(\imath - c_k) = 1$. In particular $J(\imath - c_k) = J(\imath) - k \neq 0$, as desired. $\square$

We are now ready to prove a characterization theorem for Alpha-morphisms (and hence for the models of ACT) by showing a precise correspondence with the non-principal maximal ideals of $\mathfrak{Fun}(\mathbb{N}, \mathbb{R})$.

THEOREM 2.55 (Characterization Theorem).
A map $J : \mathfrak{Fun}(\mathbb{N}, \mathbb{R}) \twoheadrightarrow \mathbb{R}^*$ onto a field $\mathbb{R}^*$ is an Alpha-morphism if and only if there exist

- *a non-principal maximal ideal $\mathfrak{m}$ of $\mathfrak{Fun}(\mathbb{N}, \mathbb{R})$;*
- *a field isomorphism $\theta : \mathfrak{Fun}(\mathbb{N}, \mathbb{R})/\mathfrak{m} \to \mathbb{R}^*$ where $\theta(\pi(c_r)) = r$ for all $r \in \mathbb{R}$;*

such that the following diagram commutes:

$$\mathfrak{Fun}(\mathbb{N}, \mathbb{R}) \xrightarrow{\quad J \quad} \mathbb{R}^*$$
$$[\,\cdot\,]_\mathfrak{m} \searrow \qquad \nearrow \theta$$
$$\mathfrak{Fun}(\mathbb{N}, \mathbb{R})/\mathfrak{m}$$

Moreover, for every infinite $Q \subseteq \mathbb{N}$, the resulting model of ACT also satisfies $(QSA)_Q$ if and only if $\{\varphi \mid \varphi(n) = 0 \text{ for all } n \in Q\} \subseteq \mathfrak{m}$.

PROOF. If we are given a commutative diagram as above, it is readily seen from the hypotheses that J satisfies properties (AM1) and (AM3). Moreover, $\mathbb{R}^*$ includes $\mathbb{R} = \{\theta([c_r]_\mathfrak{m}) \mid r \in \mathbb{R}\}$ as a subfield. In order to prove (AM2), we are left to show that the inclusion $\mathbb{R} \subset \mathbb{R}^*$ is proper. To this end, we fix an arbitrary 1-1 sequence $\varphi(n)$ and show that $J(\varphi) \notin \mathbb{R}$. Assume by contradiction that $J(\varphi) = r \in \mathbb{R}$. Then

$$J(\varphi) = r \iff J(\varphi) = J(c_r) \iff [\varphi]_\mathfrak{m} = [c_r]_\mathfrak{m} \iff \varphi - c_r \in \mathfrak{m}.$$

Now, if $r \notin \operatorname{ran}(\varphi)$ then $\varphi - c_r$ never vanishes, and hence it is invertible and cannot belong to any ideal. If $r = \varphi(k)$ for a (necessarily unique) k, then pick any sequence ψ such that $\psi(n) = 1/(\varphi(n) - r)$ for all $n \neq k$. Since $\varphi - c_r \in \mathfrak{m}$, also $\sigma_k = \psi \cdot (\varphi - c_r) \in \mathfrak{m}$. Notice that $\sigma_k = \chi_k^c$ is the characteristic function of the complement of the singleton $\{k\}$. Now recall that such a sequence $\chi_k^c \in \mathfrak{m}$ if and only if $\mathfrak{m} = \mathfrak{m}_k$ is the principal maximal ideal on k (see Proposition 2.48). This contradicts the hypothesis on $\mathfrak{m}$.

Conversely, let $J : \mathfrak{Fun}(\mathbb{N}, \mathbb{R}) \twoheadrightarrow \mathbb{R}^*$ be an Alpha-morphism. If $\mathfrak{m} = \{\varphi \mid J(\varphi) = 0\}$ is the ideal given by the *kernel* of J, then by the *First Isomorphism Theorem* of algebra there exists an isomorphism $\theta : \mathfrak{Fun}(\mathbb{N}, \mathbb{R})/\mathfrak{m} \to \mathbb{R}^*$ such that $\theta \circ [\,\cdot\,]_\mathfrak{m} = J$. The ideal $\mathfrak{m}$ is maximal because the corresponding quotient is a field. We are left to show that $\mathfrak{m}$ is non-principal. If by contradiction $\mathfrak{m} = \mathfrak{m}_k$ for some k, then the mapping $[\varphi]_\mathfrak{m} \mapsto \varphi(k)$ would yield an isomorphism $\mathfrak{Fun}(\mathbb{N}, \mathbb{R})/\mathfrak{m} \cong \mathbb{R}$, as one can readily verify. But then $\mathbb{R}^* \cong \mathbb{R}$, and this is impossible because any ordered field that properly extends $\mathbb{R}$ is non-Archimedean (see Theorem 1.11), and hence it cannot be isomorphic to $\mathbb{R}$.

Now assume that $(\mathsf{QSA})_Q$ holds in the resulting model of Alpha-Calculus for an infinite set Q. If φ vanishes at all $n \in Q$ then $\varphi(n) = 0$ a.e. and hence, by *transfer*, $\lim_{n\uparrow\alpha} \varphi(n) = [\varphi]_\mathfrak{m} = 0$, that is, $\varphi \in \mathfrak{m}$. Conversely, fix an element $q_0 \in Q$, and consider the following sequence:

$$\psi(n) = \begin{cases} n & \text{if } n \in Q \\ q_0 & \text{otherwise.} \end{cases}$$

Clearly $\psi : \mathbb{N} \to Q$, and $\psi - \imath \in \mathfrak{m}$ because it vanishes at all $n \in Q$. Then

$$\boldsymbol{\alpha} = \lim_{n\uparrow\alpha} \imath(n) = [\imath]_\mathfrak{m} = [\psi]_\mathfrak{m} = \lim_{n\uparrow\alpha} \psi(n) \in Q^*,$$

that is, Q is qualified.   □

We finally obtain the following existence result.

THEOREM 2.56. *For every infinite $Q \subseteq \mathbb{N}$, there exist models of Alpha-Calculus Theory* ACT *plus the Qualified Set Axiom* $(\mathsf{QSA})_Q$.

PROOF. By the *Characterization Theorem*, we just need to show that there are non-principal maximal ideals $\mathfrak{m}$ that contain all sequences that vanish on Q. To this end, notice first that the following is an ideal

$$\mathfrak{i}_Q = \{\varphi \in \mathfrak{i}_0 \mid \varphi(n) = 0 \text{ for all } n \in Q\}.$$

By Theorem 2.51, $\mathfrak{i}_Q$ can be extended to a maximal ideal $\mathfrak{m}$.   □

We are now ready to prove a meta-theorem about the properties of $\boldsymbol{\alpha}$ that are demonstrated by our theory.

THEOREM 2.57. *Let $P(n)$ be a property of the natural numbers. Then Alpha-Calculus Theory* ACT *proves $P^*(\boldsymbol{\alpha})$ if and only if $P(n)$ is satisfied by all but finitely many natural n.*

PROOF. One direction is the property that all cofinite sets are qualified. To show the converse implication, we will use the general fact that all theorems of a theory necessarily hold in every model of its. Let P be a property that does not satisfy the required hypothesis, that is, assume that the set $\Lambda = \{n \mid \text{not } P(n)\}$ is infinite. By the previous theorem, there exists a model of ACT plus the Qualified Set Axiom $(QSA)_Q$ where $Q = \Lambda$. But then the property "not $P(n)$" holds almost everywhere, and so "not $P^*(\alpha)$" is true in that model. In consequence, $P^*(\alpha)$ is false in that model, and hence $P^*(\alpha)$ is not a theorem of ACT. □

To summarize, getting models of Alpha-Calculus is quite easy and only requires basic notions from algebra. Indeed, the construction consists in the following three steps.

(1) Pick a non-principal maximal ideal $\mathfrak{m}$ of the ring of real sequences $\mathfrak{Fun}(\mathbb{N}, \mathbb{R})$;

(2) Let the hyperreal field $\mathbb{R}^* = \mathfrak{Fun}(\mathbb{N}, \mathbb{R})/\mathfrak{m}$ be the residue field modulo $\mathfrak{m}$, where cosets of constant sequences are identified with the corresponding real numbers;

(3) Define the Alpha-limit of a real sequence $\varphi : \mathbb{N} \to \mathbb{R}$ by putting

$$\lim_{n \uparrow \alpha} \varphi(n) = [\varphi]_{\mathfrak{m}}.$$

CHAPTER 3

Infinitesimal Analysis by Alpha-Calculus

In this chapter we present the basic notions of calculus, prove a little selection of classic results, and work out a few examples, so as to give the flavor of the Alpha-Calculus in action. In particular, we will see how a direct use of infinitesimal and infinite numbers can completely replace the role of limits.

As a typical example of a notion that naturally arises in the framework of Alpha-Calculus, we introduce the grid integral, a generalization of the familiar Riemann integral that has the pedagogical advantage of making sense for *all* real functions.

1. The normal forms

In this first section, we show how one can use the "new" ideal number α to represent hyperreal numbers in a way that is useful and convenient in the practice. Notice that for every sequence $\varphi : \mathbb{N} \to \mathbb{R}$ we have:

$$\lim_{n \uparrow \alpha} \varphi(n) \;=\; \lim_{n \uparrow \alpha} \varphi(\imath(n)) \;=\; \varphi^*(\lim_{n \uparrow \alpha} \imath(n)) \;=\; \varphi^*(\alpha).$$

This suggests the following

DEFINITION 3.1. *We say that a hyperreal number $\xi \in \mathbb{R}^*$ is represented in a* normal form *(or has a* normal representation*) if it is written as $\varphi^*(\alpha)$, where $\lim_{n \uparrow \alpha} \varphi(n) = \xi$.*

Thus, $\sin^*(\alpha)$ is a normal form of the hyperreal number $\lim_{n \uparrow \alpha} \sin(n)$; $\sqrt[3]{\alpha}$ is a normal form of the number $\lim_{n \uparrow \alpha} \sqrt[3]{n}$; and so forth. Clearly, every hyperreal number has a normal form, but normal forms are not unique. In fact, let us recall that $\varphi^*(\alpha) = \lim_{n \uparrow \alpha} \varphi(n) = \lim_{n \uparrow \alpha} \psi(n) = \psi^*(\alpha)$ if and only if the two sequences φ and ψ agree *almost everywhere*.

For clarity, we itemize below a few basic properties of normal forms, which are just reformulations of axioms of ACT.

PROPOSITION 3.2.
(1) $\imath^*(\alpha) = \alpha$;
(2) $c_r^*(\alpha) = r$ *for all* $r \in \mathbb{R}$;
(3) $(\varphi + \psi)^*(\alpha) = \varphi^*(\alpha) + \psi^*(\alpha)$;
(4) $(\varphi \cdot \psi)^*(\alpha) = \varphi^*(\alpha) \cdot \psi^*(\alpha)$.

For the sake of simplicity, in the sequel we will abuse notation and sometimes omit asterisks on hyper-extensions of functions. Thus, often we will write f to also mean the hyper-extension f^* (when confusion is unlikely).

As a consequence of the equalities in the above proposition, no ambiguity arises when writing normal forms. For instance, notation $\cos\left(\frac{3+\sin(\alpha)}{e^\alpha}\right)$ could be in principle ambiguous, as it could be interpreted in two different ways, namely:

(1) $\vartheta(\alpha)$, where $\vartheta(n) = \cos\left(\frac{3+\sin(n)}{e^n}\right)$;

(2) $\cos(\xi)$, where $\xi = \frac{3+\varphi(\alpha)}{\psi(\alpha)}$, $\varphi(n) = \sin(n)$ and $\psi(n) = e^n$.

However, if $\phi = \frac{3+\varphi}{\psi}$, then it is readily verified that

$$\vartheta(\alpha) = (\cos \circ \phi)(\alpha) = \cos(\phi(\alpha)) = \cos\left(\frac{3 + \varphi(\alpha)}{\psi(\alpha)}\right) = \cos(\xi).$$

2. Infimum and supremum

As a special example of an ordered field that properly extends $\mathbb{R}$, the hyperreal field $\mathbb{R}^*$ of Alpha-Calculus satisfies the Standard Part Theorem 1.13. So, any finite $\xi \in \mathbb{R}^*$ is uniquely represented as the sum $r + \varepsilon$ of a real number $r = \text{st}(\xi) \in \mathbb{R}$, called the *standard part* of ξ, and of an infinitesimal number $\varepsilon = \xi - x \sim 0$. Recall that such a representation $\xi = r + \varepsilon$ is called *canonical form*. Two examples are the following:

- The canonical form of $\frac{5+\alpha}{7+2\alpha}$ is $\frac{1}{2} + \frac{3}{14+4\alpha}$;
- The canonical form of $\log(6 - \eta^3)$ is $\log 6 + \log\left(1 - \frac{\eta^3}{6}\right)$.[1]

---

[1] Recall that $\eta = \frac{1}{\alpha} = \lim_{n \uparrow \alpha} \frac{1}{n}$.

The notion of standard part was extended also to the infinite numbers by setting $\mathrm{st}(\xi) = +\infty$ if ξ is positive infinite, and $\mathrm{st}(\xi) = -\infty$ if ξ is negative infinite. The following useful characterizations of *least upper bound* (sup) and *greatest lower bound* (inf) hold.

PROPOSITION 3.3. *Let $A \subseteq \mathbb{R}$ be nonempty and let $l \in \mathbb{R} \cup \{-\infty, +\infty\}$. Then*

(1) $\sup A = l$ *if and only if $l \geq a$ for all $a \in A$ and $\mathrm{st}(\xi) = l$ for some $\xi \in A^*$.*

(2) $\inf A = l$ *if and only if $l \leq a$ for all $a \in A$ and $\mathrm{st}(\xi) = l$ for some $\xi \in A^*$.*

PROOF. We will only consider least upper bounds, because greatest lower bounds are treated similarly. Assume first that $l = \sup A \in \mathbb{R}$ is finite. Then for each positive $n \in \mathbb{N}$ there exists an element $\xi(n) \in A$ such that $l - 1/n \leq \xi(n) \leq l$. By *transfer* those inequalities pass to the Alpha-limit and we obtain $l - \eta \leq \xi \leq l$, where $\xi = \xi(\alpha) \in A^*$. Thus $\xi \sim l$ is the element we were looking for. If $\sup A = +\infty$, then for each $n \in \mathbb{N}$ we can pick an element $\xi(n) > n$ in A. Thus $\xi = \xi(\alpha) \in A^*$ and $\xi > \alpha$, and we have $\mathrm{st}(\xi) = +\infty = \sup A$ as desired.

Conversely, assume $l \geq a$ for all $a \in A$. If $l \in \mathbb{R}$ is finite and is not the least upper bound, then there exists a positive real number $r > 0$ such that $l - r \geq a$ for all $a \in A$. In particular, for each $\xi = \xi(\alpha) \in A^*$, we have $l - r \geq \xi(n)$ for all n and so, by *transfer*, $l - r \geq \xi$. We conclude that $\xi \nsim l$. Now assume that $\mathrm{st}(\xi) = +\infty$ for some $\xi \in A^*$. Then $\xi > r$ for every real $r \in \mathbb{R}$. By *backward transfer*, $\xi = \xi(\alpha)$ for some sequence $\xi : \mathbb{N} \to A$ such that $\xi(n) > r$ for all n. In particular, this shows that A is unbounded above, that is, $\sup A = +\infty$. $\square$

As it will be clear from the sequel, in our framework the classic notion of limit is completely replaced by that of standard part. In particular, differently from traditional calculus, a study of the limits of real sequences is not needed as a preliminary step towards a study of continuity and differentiability.

3. Continuity

The current ϵ-δ formalization of calculus – made rigorous by Weierstrass at the end of the 19th century – is grounded on the notion of *limit*, and excludes actual infinitesimal quantities as contradictory. On the opposite, in the Alpha-Calculus one defines all the basic notions of calculus by directly invoking infinitesimal and infinite numbers. As a result, the classic ϵ-δ definitions are reformulated in simpler terms.[2] We remark that – from a historical point of view – many of the definitions given below closely resemble (and in fact they are sometimes identical to) the original formulations, as adopted by fathers of calculus such as Leibniz, Euler, Cauchy and others.

The first relevant example is given by the notion of *continuity*. Informally, it is commonly said that a function f is continuous at a point x if $f(\xi)$ is "close" to $f(x)$ whenever ξ is "close" to x. If we replace the vague concept of "closeness" by the relation of being "infinitely close", we obtain the following rigorous definition.

DEFINITION 3.4. *Let $f : A \to \mathbb{R}$ be a real function defined on a neighborhood A of the point x. We say that f is continuous at x if for every $\xi \in A^*$,*

$$\xi \sim x \implies f(\xi) \sim f(x).^3$$

Thus, in order to test the continuity of a function f at a given point x, one has to estimate the size of $f(x + \varepsilon) - f(x)$ for a generic infinitesimal ε.

EXAMPLE 3.5. *The function $f(x) = x^2$ is continuous at all $x \in \mathbb{R}$.*

PROOF. $f(x + \varepsilon) - f(x) = (x + \varepsilon)^2 - x^2 = 2x\varepsilon + \varepsilon^2 \sim 0$ for all $\varepsilon \sim 0$. □

A "not so trivial" example is the following:

EXAMPLE 3.6. *The following function f is continuous at all irrational points $x \in \mathbb{R} \setminus \mathbb{Q}$ and discontinuous at all rational points $x \in \mathbb{Q}$:*

$$f(x) = \begin{cases} \frac{1}{q} & \text{if } x = \frac{p}{q} \in \mathbb{Q} \text{ is a reduced fraction with } p \geq 0 \text{ ;} \\ 0 & \text{if } x \in \mathbb{R} \setminus \mathbb{Q}. \end{cases}$$

PROOF. For simplicity, let us assume that $x \in \mathbb{Q}$ is positive (if $x \leq 0$ the proof is similar). Let $\xi \sim x \notin \mathbb{Q}$. If $\xi \notin \mathbb{Q}^*$ is hyperirrational, then $f(\xi) = 0 = f(x)$. If $\xi \in \mathbb{Q}^*$, let us write $\xi = \frac{\nu}{\mu}$ in reduced form (see Exercise 2.41), and assume by contradiction that $\mu \in \mathbb{N}$ is finite. If also $\nu \in \mathbb{N}_0$ were finite, then $\xi = \frac{\nu}{\mu} \in \mathbb{Q}$ and $\xi \sim x \Rightarrow x = \xi$ is rational. If ν were infinite, then also $\xi = \frac{\nu}{\mu} \in \mathbb{Q}^*$ would be infinite. Both cases contradict the hypothesis $\xi \sim x \notin \mathbb{Q}$. Thus μ must be infinite and $f(\xi) = \frac{1}{\mu} \sim 0 = f(x)$, as desired.

---

[2] Here "in simpler terms" has the precise meaning of "by using a smaller number of quantifiers".

[3] The equivalence of this definition of continuity with the familiar "standard" one will be proved in §11 of this chapter.

The other case $x \in \mathbb{Q}$ is easier. It is enough to show that there exists an element $\xi \notin \mathbb{Q}^*$ with $\xi \sim x$, so that $f(\xi) = 0 \nsim f(x)$. But this directly follows from the density property of the hyperirrational numbers in $\mathbb{R}^*$ (see (7) in Proposition 2.36). $\square$

We close this section by proving two fundamental results of calculus.

THEOREM 3.7 (Weierstrass – Extreme Value).
Let $f : [a, b] \to \mathbb{R}$ be a continuous function on a bounded and closed interval. Then f attains a least and a greatest value.

PROOF. Let $l = \inf\{f(x) \mid x \in [a, b]_\mathbb{R}\}$ (possibly $l = -\infty$). By the characterization given in Proposition 3.3, there exists $\xi \in [a, b]^*$ with $\mathrm{st}(f(\xi)) = l$. Since $[a, b]$ is closed and bounded, $\mathrm{st}(\xi) = x \in [a, b]$. Then, by continuity, $f(x) = \mathrm{st}(f(\xi)) = l$. In particular, $l \neq -\infty$ and f attains its least value at x. The existence of a greatest value is proved in the same way by considering $l' = \sup\{f(x) \mid x \in [a, b]\}$ in place of l. $\square$

A crucial step in the above proof is the implication "$\xi \in [a, b]^* \Rightarrow \mathrm{st}(\xi) \in [a, b]$". We remark that both the hypotheses that the interval $I = [a, b]$ be closed and bounded are needed. In fact, a bounded interval I is *not* closed if and only if there exist points $\xi \in I^*$ such that $\mathrm{st}(\xi) \notin I$ (*e.g.*, $(0, 1]$ is not closed because every positive infinitesimal $\varepsilon \in (0, 1]^*$, while $\mathrm{st}(\varepsilon) = 0 \notin (0, 1]$). Moreover, if I is unbounded, then there are infinite points $\xi \in I^*$; clearly, such points ξ are not infinitely close to any real point.

THEOREM 3.8 (Bolzano – Intermediate Value).
Let $f : [a, b] \to \mathbb{R}$ be a continuous function where $f(a) < f(b)$. Then for every intermediate value $f(a) < y < f(b)$ there exists $x \in (a, b)$ such that $f(x) = y$.

PROOF. For every $n \in \mathbb{N}$, let

$$A_n = \left\{ a + i \cdot \tfrac{b-a}{n} \mid i = 0, \ldots, n-1 \right\}$$

be the finite set of points yielding a partition of $[a, b]$ into n-many intervals of equal length $1/n$. Put

$$\varphi(n) = \max\{x \in A_n \mid f(x) < y\} \quad \text{and} \quad \psi(n) = \varphi(n) + \frac{b-a}{n}.$$

Notice that $\psi(n) \in [a, b]$ and $f(\psi(n)) \geq y$. Since $\psi(n) - \varphi(n) = \frac{1}{n}(b - a)$ for all positive n, by passing to Alpha-limits we obtain that $\zeta - \xi = \boldsymbol{\eta} \cdot (b - a) \sim 0$, where $\zeta = \psi(\boldsymbol{\alpha})$ and $\xi = \varphi(\boldsymbol{\alpha})$. In particular,

ξ and ζ have the same standard part $x = \mathrm{st}(\xi) = \mathrm{st}(\zeta)$. Moreover, since the interval $[a, b]$ is closed and bounded, from $\xi, \zeta \in [a, b]^*$ it follows that $x \in [a, b]$. Then, by continuity, $f(\xi) \sim f(x) \sim f(\zeta)$. Now, $f(\varphi(n)) < y$ and $f(\psi(n)) \geq y$ for all n imply that $f(\xi) < y \leq f(\zeta)$. We conclude that $f(x) = y$, as desired. $\qquad\square$

4. Uniform continuity

In our definition of continuity, it seems natural to extend the validity of the implication "$\xi \sim x \Rightarrow f^*(\xi) \sim f(x)$" also to the cases when $x \in \mathbb{R}^* \setminus \mathbb{R}$. What one obtains in this way is the familiar notion of *uniform continuity* (see below). This supports the intuition that – at least in our "non-standard" setting – uniform continuity is a more natural property than pointwise continuity.

DEFINITION 3.9. *A real function $f : A \to \mathbb{R}$ is uniformly continuous if for all $\xi, \zeta \in A^*$,*

$$\xi \sim \zeta \implies f^*(\xi) \sim f^*(\zeta).^4$$

As teachers know by experience, distinguishing between continuity and uniform continuity is sometimes a slippery matter. To this end, the above definition seems easier to grasp.

EXAMPLE 3.10. *The function $f : \mathbb{R} \to \mathbb{R}$ where $f(x) = x^2$ is not uniformly continuous.*

PROOF. Take ε any non-zero infinitesimal. Then $1/\varepsilon + \varepsilon \sim 1/\varepsilon$ while $f(1/\varepsilon + \varepsilon) - f(1/\varepsilon) = (1/\varepsilon + \varepsilon)^2 - (1/\varepsilon)^2 = 2 + \varepsilon^2$ is *not* infinitesimal. $\qquad\square$

In our framework, the proof of the fundamental Cantor's theorem is almost a triviality.

THEOREM 3.11 (Cantor).
Every continuous function $f : [a, b] \to \mathbb{R}$ on a closed and bounded interval is uniformly continuous.

PROOF. If $\xi, \zeta \in [a, b]^*$ are infinitely close, then $\mathrm{st}(\xi) = \mathrm{st}(\zeta) = x \in [a, b]$. By continuity at x, $f(\xi) \sim f(x) \sim f(\eta)$. $\qquad\square$

---

[4] The actual equivalence of this property with the "standard" definition of uniform continuity will be proved in §11 of this chapter.

5. Derivatives

Let f be a real function defined on a neighborhood of a point x.

DEFINITION 3.12. *We say that f has* derivative *at x if there exists a real number $f'(x) \in \mathbb{R}$ such that for all non-zero infinitesimals ε,*

$$\frac{f(x + \varepsilon) - f(x)}{\varepsilon} \sim f'(x).$$

Equivalently, f has derivative $f'(x)$ at x if for every infinitesimal ε there exists an infinitesimal τ such that

$$f(x + \varepsilon) = f(x) + f'(x) \cdot \varepsilon + \tau \cdot \varepsilon.$$

EXAMPLE 3.13. *The derivative of $f(x) = x^2$ is $f'(x) = 2x$.*

PROOF. For every given $x \in \mathbb{R}$ and for every infinitesimal $\varepsilon \neq 0$,

$$\frac{(x + \varepsilon)^2 - x^2}{\varepsilon} = \frac{2x\varepsilon + \varepsilon^2}{\varepsilon} = 2x + \varepsilon \sim 2x.$$

$\square$

The usual properties of derivatives, such as $(f + g)' = f' + g'$, $(f \cdot g)' = f'g + fg'$, $(1/f)' = -f'/f^2$, are all easily proved by directly applying the definition. As an example, let us see the following:

THEOREM 3.14 (Leibniz product rule).
Let f and g be real functions. If f has derivative at x, and g has derivative at $f(x)$, then the composition $g \circ f$ has derivative at x and $(g \circ f)'(x) = g'(f(x)) \cdot f'(x)$.

PROOF. Fix $\varepsilon \neq 0$ infinitesimal. By the continuity of f at x, we have that $f(x + \varepsilon) - f(x) = \eta \sim 0$. We can assume $\eta \neq 0$, otherwise trivially $f'(x) = g'(f(x)) = (g \circ f)'(x) = 0$. Then we have the following:

$$\frac{g(f(x + \varepsilon)) - g(f(x))}{\varepsilon} = \frac{g(f(x) + \eta) - g(f(x))}{\eta} \cdot \frac{f(x + \varepsilon) - f(x)}{\varepsilon}$$

$$\sim g'(f(x)) \cdot f'(x).$$

$\square$

Let us now see another application of the methods of Alpha-Calculus, and prove a central theorem about derivatives.

THEOREM 3.15 (Fermat).
Assume that the function $f : (a, b) \to \mathbb{R}$ has a derivative at all points. If f attains a greatest or a least value at $x \in (a, b)$, then $f'(x) = 0$.

PROOF. Suppose that f attains its least value at x (if $f(x)$ is the greatest value the proof is similar). Notice that $f(x)$ is also the least value taken by the hyper-extension $f : (a,b)^* \to \mathbb{R}^*$. In fact, given any sequence $\xi : \mathbb{N} \to (a,b)$, we have $f(\xi(n)) \leq f(x)$ for all n and so, by passing to Alpha-limits, we obtain that $(f \circ \xi)(\boldsymbol{\alpha}) = f(\xi(\boldsymbol{\alpha})) \leq f(x)$. Now fix an infinitesimal $\varepsilon > 0$. Since the considered interval (a,b) is open, both $x - \varepsilon, x + \varepsilon \in (a,b)^*$. Then $f(x) \leq f(x + \varepsilon)$ and $f(x) \leq f(x - \varepsilon)$, and hence:

$$f'(x_0) \sim \frac{f(x_0 + \varepsilon) - f(x_0)}{\varepsilon} \geq 0 \ \text{ and } \ f'(x_0) \sim \frac{f(x_0 - \varepsilon) - f(x_0)}{-\varepsilon} \leq 0.$$

We conclude that $f'(x_0) = 0$. $\qquad\square$

6. Limits

As shown in the previous sections, in our approach the notion of *limit* is not needed to develop the basics of calculus. In fact, the view of many mathematicians working in nonstandard analysis is that infinitesimal numbers do actually exist, the notion of a limit being just an awkward way not to explicitly mention them.

However, limits can be introduced in Alpha-Calculus as useful shorthands for properties of hyper-extensions of functions with respect to standard parts. Let us start with sequences.

Let $\langle a_n \mid n \in \mathbb{N} \rangle$ denote the real sequence $a : n \mapsto a_n$. With obvious notation, its hyper-extension $a^* : \mathbb{N}^* \to \mathbb{R}^*$ is written $\langle a_\nu \mid \nu \in \mathbb{N}^* \rangle$.

DEFINITION 3.16. *Let $l \in \mathbb{R} \cup \{-\infty, +\infty\}$. We say that $\lim_{n \to \infty} a_n = l$ if $\mathrm{st}(a_\nu) = l$ for all infinite $\nu \in \mathbb{N}^*$.*[5]

Notice that the above definition matches the intuition that l is the limit of a sequence $\langle a_n \mid n \in \mathbb{N} \rangle$ if the values a_n get "close" to l, as the indexes n get "large". To make this intuition rigorous, we used the formalized notions of "infinite closeness" and "infinitely large" number.

EXAMPLE 3.17. $\lim_{n \to +\infty} \sqrt{n+1} - \sqrt{n} = 0$.

PROOF. Take any infinite $\nu \in \mathbb{N}^*$. Then we have:

---

[5] The equivalence with the "standard" definition of limit will be proved in §11 in this chapter.

$$\sqrt{\nu+1} - \sqrt{\nu} = \frac{(\sqrt{\nu+1} - \sqrt{\nu})(\sqrt{\nu+1} + \sqrt{\nu})}{\sqrt{\nu+1} + \sqrt{\nu}}$$

$$= \frac{1}{\sqrt{\nu+1} + \sqrt{\nu}} \sim 0.$$

□

PROPOSITION 3.18. *If the sequence $\langle a_n \mid n \in \mathbb{N} \rangle$ is non-decreasing, then $\lim_{n\to\infty} a_n = \sup\{a_n \mid n \in \mathbb{N}\}$. Similarly, if the sequence is non-increasing, then $\lim_{n\to\infty} a_n = \inf\{a_n \mid n \in \mathbb{N}\}$.*

PROOF. Assume first that $\sup\{a_n \mid n \in \mathbb{N}\} = r \in \mathbb{R}$ is finite. As $a_n \leq r$ for all n, we have $a_\nu \leq r$ for all $\nu \in \mathbb{N}^*$, by *transfer*. To reach the desired conclusion, we have to exclude the possibility that $a_\mu \leq r'$ for some infinite $\mu \in \mathbb{N}^*$ and some real number $r' < r$. This cannot happen because also the hyper-extension $\langle a_\nu \mid \nu \in \mathbb{N} \rangle$ is non-increasing by Proposition 2.35. In fact, in this case we would have $a_n \leq a_\nu \leq r'$ for all n, and hence $\sup\{a_n \mid n \in \mathbb{N}\} \leq r' < r$, contradicting the hypothesis. The case of a non-decreasing sequence is entirely similar. □

The properties in the proposition below are particular cases of the *overspill* phenomenon, which will be introduced further on (see §6.1).

PROPOSITION 3.19 (*Overspill for real sequences*).
Let $\langle a_n \mid n \in \mathbb{N} \rangle$ and $\langle b_n \mid n \in \mathbb{N} \rangle$ be two real sequences, and let $\bowtie$ be any of the following relations: $=, \neq, \leq, <$. Then

(1) $a_n \bowtie b_n$ *infinitely often (that is, for infinitely many $n \in \mathbb{N}$)*
$\iff a_\nu \bowtie b_\nu$ *for some infinite $\nu \in \mathbb{N}^* \setminus \mathbb{N}$;*

(2) $a_n \bowtie b_n$ *eventually (that is, for all sufficiently large $n \in \mathbb{N}$)*
$\iff a_\nu \bowtie b_\nu$ *for all infinite $\nu \in \mathbb{N}^* \setminus \mathbb{N}$.*

PROOF. (1). Let $A = \{n \in \mathbb{N} \mid \varphi(n) = \psi(n)\}$. Then the hyper-extension $A^* = \{\nu \in \mathbb{N}^* \mid \varphi^*(\nu) = \psi^*(\nu)\}$ contains an infinite number if and only if A is unbounded (see Proposition 2.37). The other statements about $\neq, \leq$ and $<$ are proved similarly by considering the sets $B = \{n \in \mathbb{N} \mid \varphi(n) \neq \psi(n)\}$, $C = \{n \in \mathbb{N} \mid \varphi(n) \leq \psi(n)\}$, and $D = \{n \in \mathbb{N} \mid \varphi < \psi(n)\}$, respectively.

(2). It is the contrapositive of (1). Notice in fact that $\varphi(n) \bowtie \psi(n)$ *does not* hold eventually if and only if $\varphi(n) \not\bowtie \psi(n)$ holds infinitely often. By the previous point, the last statement is equivalent to the existence of an infinite $\nu \in \mathbb{N}^*$ such that $\varphi^*(\nu) \not\bowtie \psi^*(\nu)$, which in turn is equivalent to the negation of the property that $\varphi^*(\nu) \bowtie \psi^*(\nu)$ holds for all infinite $\nu \in \mathbb{N}^*$.   □

Let us now introduce the notion of limit for real functions. Simply, in this case one has to follow the intuition that $\lim_{x \to x_0} f(x) = l$ if the value $f(\xi)$ is "close" to l whenever ξ is "close" to x_0.

DEFINITION 3.20. *Let* $x_0, l \in \mathbb{R} \cup \{-\infty, +\infty\}$, *and let* f *be a function defined on a neighborhood of* x_0. *We say that* $\lim_{x \to x_0} f(x) = l$ *if* $f(\xi) \sim l$ *for all hyperreal numbers* $\xi \neq x_0$ *with* $\xi \sim x_0$.[6]

EXAMPLE 3.21. $\lim_{x \to 1+} \frac{x}{1-x} = -\infty$ *and* $\lim_{x \to 1-} \frac{x}{x-1} = +\infty$.[7]

PROOF. Take an infinitesimal $\varepsilon \neq 0$. Then

$$\frac{1+\varepsilon}{1-(1+\varepsilon)} = -\frac{1+\varepsilon}{\varepsilon} = -1 - \frac{1}{\varepsilon}.$$

If $\varepsilon > 0$ then the standard part of the above quantity is $-\infty$; and if $\varepsilon < 0$ then the standard part is $+\infty$.   □

EXAMPLE 3.22. $\lim_{x \to 0} \frac{\sin x}{x} = 0$.

PROOF. For every $0 < x < \pi/2$, it is well-known that the inequalities $\sin x \leq x \leq \tan x = \frac{\sin x}{\cos x}$ hold, and hence

$$\sqrt{1 - x^2} \leq \sqrt{1 - \sin^2 x} = \cos x \leq \frac{\sin x}{x} \leq 1.$$

Now let $\varepsilon \sim 0$ be a positive infinitesimal, and let $\varphi : \mathbb{N} \to \mathbb{R}$ be such that $\varphi(\alpha) = \varepsilon$. Since $0 < \varepsilon < \pi/2$, the following inequalities hold for almost all n:

$$\sqrt{1 - \varphi(n)^2} \leq \frac{\sin \varphi(n)}{\varphi(n)} \leq 1.$$

By passing to Alpha-limits, we get that $1 \sim \sqrt{1 - \varepsilon^2} \leq \frac{\sin \varepsilon}{\varepsilon} \leq 1$, and we conclude that $\frac{\sin \varepsilon}{\varepsilon} \sim 1$. For the case of negative infinitesimals ε, just recall that $\sin(-\varepsilon) = -\sin \varepsilon$, and repeat the argument above.   □

---

[6] As usual in the literature, we agree that A is a neighborhood of $-\infty$ if $(-\infty, r) \subseteq A$ for some $r \in \mathbb{R}$; and similarly for neighborhoods of $+\infty$. The equivalence of this definition of limit with the usual one will be given in §11 in this chapter.

[7] By writing $x \to x_0^+$ (or $x \to x_0^-$) we mean that only numbers $x > x_0$ (or $x < x_0$, respectively) must be considered.

7. Alpha-limit *versus* limit

Similarly to Weierstrass' ϵ-δ formalization of calculus, one could argue that also our Alpha-Calculus is grounded on a notion of limit for sequences, namely the Alpha-limit, but this idea could be misleading. While there are close relationships between the limit $\lim_{n\to\infty} \varphi(n)$ of a real sequence $\varphi(n)$ (when it does exist) and its Alpha-limit $\lim_{n\uparrow\alpha} \varphi(n)$, there are also substantial differences between the two notions.

Firstly, not every sequence has a limit, while Alpha-limits do always exist: all one can prove is that the standard part $\text{st}(\lim_{n\uparrow\alpha} \varphi(n))$ is a *limit point* of the sequence $\varphi(n)$. Probably, the most crucial diversity is that while pointwise different sequences may have the same "classic" limit, they necessarily have different Alpha-limits.

PROPOSITION 3.23. *Let* $\varphi : \mathbb{N} \to \mathbb{R}$ *be a real sequence, and let* $l \in \mathbb{R} \cup \{-\infty, +\infty\}$.

(1) *If* $\lim_{n\to\infty} \varphi(n) = l$ *then* $\text{st}(\lim_{n\uparrow\alpha} \varphi(n)) = l$;

(2) $\text{st}(\lim_{n\uparrow\alpha} \varphi(n))$ *is a limit point of* $\varphi(n)$, *that is, there exists a subsequence* $\varphi(n_k)$ *with* $\lim_{k\to\infty} \varphi(n_k) = \text{st}(\lim_{n\uparrow\alpha} \varphi(n))$.

PROOF. (1). Consider first the case when $l \in \mathbb{R}$ is finite. By the definition of limit, for every $\epsilon \in \mathbb{R}^+$ the following inequalities hold for all but finitely many n:

$$l - \epsilon \leq \varphi(n) \leq l + \epsilon.$$

By *transfer*, we obtain $l - \epsilon \leq \lim_{n\uparrow\alpha} \varphi(n) \leq l + \epsilon$, and since $\epsilon > 0$ is arbitrary, we conclude that $\text{st}(\lim_{n\uparrow\alpha} \varphi(n)) = l$. If $l = +\infty$, for every $M > 0$ the inequality $\varphi(n) \geq M$ holds for all but finitely many n, and hence $\lim_{n\uparrow\alpha} \varphi(n) \geq M$. Since $M > 0$ is arbitrary, the number $\lim_{n\uparrow\alpha} \varphi(n)$ is necessarily positive infinite. If $l = -\infty$ we proceed similarly.

(2). Assume first that the standard part of $\varphi(\alpha)$ is finite, that is, $\varphi(\alpha) \sim r$ for some real $r \in \mathbb{R}$. Then $|\varphi(\alpha) - r| < 1/k$ for every $k \in \mathbb{N}$ and so, by *backward transfer*, it must be $|\varphi(n) - r| < 1/k$ for almost all n; in particular $\Lambda_k = \{n \mid |\varphi(n) - r| < 1/k\}$ is infinite. By induction, we can define an increasing sequence of natural numbers $\langle n_k \mid k \in \mathbb{N} \rangle$ such that $n_k \in \Lambda_k$. It is readily seen that $\lim_{k\to\infty} \varphi(n_k) = r$. In case $l = +\infty$ (or $l = -\infty$), similarly as above one defines an increasing sequence $\langle n_k \mid k \in \mathbb{N} \rangle$ where $\varphi(n_k) > k$ (or $\varphi(n_k) < -k$) for all k, so that $\lim_{k\to\infty} \varphi(n_k) = +\infty$ (or $\lim_{k\to\infty} \varphi(n_k) = -\infty$, respectively). $\square$

8. The order of magnitude

A notion of order of magnitude, which is usually considered for functions or sequences, can also be defined for hyperreal numbers. (Actually, the definitions below also make sense in all ordered fields that properly extend $\mathbb{R}$.)

DEFINITION 3.24. *Let ξ, η be positive hyperreal numbers. We say that*

- ξ *and ζ have the same* order of magnitude *if both $\frac{\xi}{\zeta}$ and $\frac{\zeta}{\xi}$ are finite; in this case we write $\xi \simeq \zeta$;*
- ξ *has a* smaller order *than ζ, or ζ has a* greater order *than ξ, if $\frac{\xi}{\zeta} \sim 0$ is infinitesimal (equivalently, if $\frac{\zeta}{\xi}$ is infinite); in this case we write $\xi \ll \zeta$ or $\zeta \gg \xi$.*

All properties below directly follow from the definitions; proofs are left as exercises.

THEOREM 3.25. *Let ξ, η, ϑ be positive hyperreal numbers.*

(1) *If $\xi \simeq \zeta$ and $\zeta \simeq \vartheta$ then $\xi \simeq \vartheta$, that is, $\simeq$ is an equivalence relation;*

(2) *If $\xi \ll \zeta$ and $\zeta \ll \vartheta$ then $\xi \ll \vartheta$, that is, $\ll$ is a partial order;*

(3) *If $\xi \ll \zeta$ then $\xi + \zeta \simeq \zeta$;*

(4) *If ζ is infinite then $\xi \cdot \zeta \gg \xi$;*

(5) *If ζ is finite but not infinitesimal then $\xi \cdot \zeta \simeq \xi$;*

(6) *If ζ is infinitesimal then $\xi \cdot \zeta \ll \xi$;*

(7) *If $\zeta \ll \xi$ then $\frac{1}{\zeta} \gg \frac{1}{\xi}$;*

(8) *If $\xi \simeq \zeta$ and $\xi' \simeq \zeta'$ then $\xi \cdot \xi' \simeq \zeta \cdot \zeta'$.*

The next proposition illustrates a possible use of the notion of order of magnitude when computing the standard part of a number.

PROPOSITION 3.26. *Let ξ, ζ, ξ', ζ' be non-zero hyperreal numbers, and suppose that $|\xi'| \ll |\xi|$ and $|\zeta'| \ll |\zeta|$. Then*

$$\mathrm{st}\left(\frac{\xi + \xi'}{\zeta + \zeta'}\right) = \mathrm{st}\left(\frac{\xi}{\zeta}\right).$$

PROOF. It is enough to observe that

$$\frac{\xi + \xi'}{\zeta + \zeta'} = \frac{\xi}{\zeta} \cdot \frac{1 + \frac{\xi'}{\xi}}{1 + \frac{\zeta'}{\zeta}} \quad \text{where} \quad \mathrm{st}\left(\frac{1 + \frac{\xi'}{\xi}}{1 + \frac{\zeta'}{\zeta}}\right) = 1.$$

$\square$

Next, we will see examples where order of magnitudes are used to determine limits of real sequences. We need first a few basic properties.

PROPOSITION 3.27. *Let ω be an infinite positive number. Then*

(1) ω^ϵ *is infinite for every $\epsilon \in \mathbb{R}^+$;*

(2) $\omega^a \ll \omega^b$ *for every $a, b \in \mathbb{R}^+$ with $a < b$;*

(3) $\omega^a \ll b^\omega$ *for every $a, b \in \mathbb{R}^+$.*

Moreover, if ω is an infinite hypernatural number, then

(4) $b^\omega \ll \omega!$ *for every $b \in \mathbb{R}^+$;*

(5) $\omega! \ll \omega^\omega$.

PROOF. (1). Pick $n \in \mathbb{N}$ such that $1/n < \epsilon$. Notice that $\zeta = \omega^{1/n}$ is infinite, as otherwise $\omega = \zeta^n$ would be finite as well. Then also $\omega^\epsilon > \zeta$ is infinite.

(2) directly follows from (1) because $\omega^b/\omega^a = \omega^\epsilon$ where $\epsilon = b - a \in \mathbb{R}^+$.

(3) Given $a, b \in \mathbb{R}^+$, it is easily shown that for sufficiently large n, say for $n > k$, one has $n^a/b^n < 1/n$. Now let $\omega = \lim_{n \uparrow \alpha} \omega(n)$. Since $\omega > 0$ is infinite, it is $\omega(n) > k$ for almost all n, and so, for almost all n, the following inequality holds:

$$\frac{\omega(n)^a}{b^{\omega(n)}} < \frac{1}{\omega(n)}.$$

By *transfer*, we conclude that $\omega^a/b^\omega < 1/\omega \sim 0$, which proves the assertion.

(4) and (5) are proved similarly as above, by starting from the inequalities $b^n < n! < n^n$ that hold for all sufficiently large n. □

EXAMPLE 3.28. $\lim_{n\to\infty} \frac{n!+7n^3}{5n!+2n^7} = \frac{1}{5}$.

PROOF. Let ω be an arbitrary infinite hypernatural number ω. As shown above, $\omega! \gg 7\omega^3$ and $\omega! \gg 2\omega^7$. So, by Proposition 3.26, one has

$$\mathrm{st}\left(\frac{\omega! + 7\omega^3}{5\omega! + 2\omega^7}\right) = \mathrm{st}\left(\frac{\omega!}{5\omega!}\right) = \frac{1}{5}.$$

□

We remark that many limits of sequences can be computed by proceeding similarly as in the example above.

Given two hyperreal numbers, it might be useful to compare their orders by using real numbers.

DEFINITION 3.29. *Let ξ, ζ be positive hyperreal numbers and let $a \in \mathbb{R}$. We say that ξ has* order a *relative to ζ if $\xi \simeq \zeta^a$; and we say that ξ has* order $+\infty$ *relative to ζ if $\xi \gg \zeta^a$ for all $a \in \mathbb{R}^+$.*

Notice that finite numbers have order 0 relative to every infinite number; and that numbers that are not infinitesimal have order 0 relative to every infinitesimal.

EXAMPLE 3.30.

(1) *The number $5\alpha^2 + 9\alpha + 3\eta$ has order 2 relative to α, has order 1 relative to α^2, and has order -2 relative to $\eta = 1/\alpha$;*

(2) *The number $5\eta^3 + 7\eta^4$ has order 3 relative to η, and has order -3 relative to α;*

(3) *The number $\sqrt{\eta^3} + 7\eta^2$ has order $3/2$ relative to η, and has order $-3/2$ relative to α.*

We remark that there exist plenty of hyperreal numbers ξ, ζ such that ξ does not have order a relative to ζ for any $a \in \mathbb{R}$. Thus the following definition reveals useful:

DEFINITION 3.31. *Let ξ, ζ be positive hyperreal numbers. We say that ξ has* order a^+ *relative to ζ if $\zeta^a \ll \xi \ll \zeta^x$ for all real numbers $x > a$.*

Similarly, we say that ξ has order a^- *relative to ζ if $\zeta^x \ll \xi \ll \zeta^a$ for all real numbers $x < a$.*

EXAMPLE 3.32.

(1) *The number $\log \alpha$ has order 0^+ relative to α.*

(2) *The number $\frac{\alpha^2}{\log \alpha}$ has order 2^- relative to α.*

Notice that two numbers with the same order relative to α may have different orders of magnitude.

EXAMPLE 3.33. *The numbers*

$$\frac{\alpha^2}{\log \alpha} \ll \frac{\alpha^2}{\log \log \alpha}$$

have different orders of magnitude, but they both have order 2^- relative to α.

9. Hyperfinitely long sums and series

In this section, we consider a useful generalization of finite sums to sums indexed over initial segments of hypernatural numbers.

DEFINITION 3.34. *Let* $\langle a_n \mid n \in \mathbb{N} \rangle$ *be a sequence of real numbers, and let* $\nu \in \mathbb{N}^*$ *be a hypernatural number. The* hyperfinitely long sum $\sum_{i=1}^{\nu} a_i$ *is defined by setting:*

$$\sum_{i=1}^{\nu} a_i = \lim_{n \uparrow \alpha} \sum_{i=1}^{\varphi(n)} a_i$$

where $\varphi(n)$ *is a sequence of natural numbers such that* $\nu = \lim_{n \uparrow \alpha} \varphi(n)$.[8]

We remark that the above definition is well-posed because it does not depend on the choice of $\varphi(n)$. Indeed, assume that $\nu = \lim_{n \uparrow \alpha} \varphi(n) = \lim_{n \uparrow \alpha} \psi(n)$, and let $S_1(n) = \sum_{i=1}^{\varphi(n)} a_i$ and $S_2(n) = \sum_{i=1}^{\psi(n)} a_i$. Since $\varphi(n) = \psi(n)$ *a.e.*, also $S_1(n) = S_2(n)$ *a.e.*, and hence $\lim_{\alpha} S_1(n) = \lim_{\alpha} S_2(n)$.

We now extend the definition above by setting for all $\nu < \mu$ in $\mathbb{N}^*$:

$$\sum_{i=\mu}^{\nu} a_i = \sum_{i=1}^{\nu} a_i - \sum_{i=1}^{\mu-1} a_i.$$

A first example of a property of finite sums that is inherited by the hyperfinitely long ones, is the following.

EXAMPLE 3.35. *For* $\xi, \zeta \in \mathbb{R}^*$ *and* $\nu \in \mathbb{N}^*$, *the* Newton binomial formula *holds:*

$$(\xi + \zeta)^{\nu} = \sum_{i=0}^{\nu} \binom{\nu}{i} \cdot \xi^i \cdot \zeta^{\nu-i}.$$

PROOF. Let $\xi(n), \zeta(n)$ be real sequences with $\xi = \lim_{n \uparrow \alpha} \xi(n)$ and $\zeta = \lim_{n \uparrow \alpha} \zeta(n)$, and let $\nu(n)$ be a sequence of natural numbers with $\nu = \lim_{n \uparrow \alpha} \nu(n)$. By the "standard" *Newton formula*, for all n we have

$$(\xi(n) + \zeta(n))^{\nu(n)} = \sum_{i=0}^{\nu(n)} \binom{\nu(n)}{i} \cdot \xi(n)^i \cdot \zeta(n)^{\nu(n)-i}.$$

By passing to Alpha-limits, the assertion is proved. $\square$

---

[8] Hyperfinitely long sums are a particular case of *hyperfinite sums*, that will be defined in §4.8.

Here is a nice application of hyperfinitely long sums in the calculus of limits.

EXAMPLE 3.36. $\lim_{n\to\infty} \sqrt[n]{n} = 1$.

SOLUTION. Let ν be any infinite hypernatural number. As $\sqrt[\nu]{\nu} > 1$, we can write $\sqrt[\nu]{\nu} = 1 + \varepsilon$ for some positive $\varepsilon \in \mathbb{R}^*$. To reach the conclusion, we have to show that $\varepsilon \sim 0$. But this directly follows from Newton's binomial formula, because:

$$\nu = (1+\varepsilon)^\nu = \sum_{i=0}^\nu \binom{\nu}{i} \epsilon^i > \binom{\nu}{2} \epsilon^2 = \frac{\nu(\nu-1)}{2} \cdot \varepsilon^2 \Rightarrow$$

$$\Rightarrow \varepsilon < \sqrt{\frac{2}{\nu-1}} \sim 0.$$

$\square$

Having hyperfinitely long sums at hand, the following definition comes naturally.

DEFINITION 3.37. *Let $l \in \mathbb{R} \cup \{-\infty, +\infty\}$. We say that the* sum of *the series $\sum_{n=1}^\infty a_n = l$ if $\mathrm{st}\left(\sum_{i=1}^\nu a_i\right) = l$ for all infinite $\nu \in \mathbb{N}^*$. When $l \in \mathbb{R}$, we say that the series is* convergent.

Notice that $\sum_{n=1}^\infty a_n = l$ if and only if $\lim_{n\to\infty} \sum_{i=1}^n a_i = l$ (compare Definitions 3.16 and 3.34). In particular, if $a_i \geq 0$ for all n, then the sequence of partial sums $\langle \sum_{i=1}^n a_i \mid n \in \mathbb{N} \rangle$ is non-decreasing and so, by Proposition 3.18,

$$\sum_{n=1}^\infty a_n = \sup\left\{ \sum_{i=1}^n a_i \;\Big|\; n \in \mathbb{N} \right\}.$$

Below is the counterpart in Alpha-Calculus of a well-known criterion for the convergence of series.

THEOREM 3.38 (Cauchy's Criterion).
Let $\langle a_n \mid n \in \mathbb{N} \rangle$ be a real sequence where all $a_n \geq 0$. Then the following conditions are equivalent:

(1) *$\sum_{i=1}^\infty a_n$ converges;*
(2) *$\sum_{i=\mu}^\nu a_i$ is finite for all pairs $\mu < \nu$ of infinite hypernaturals;*
(3) *$\sum_{i=\mu}^\nu a_i$ is infinitesimal for all pairs $\mu < \nu$ of infinite hypernaturals.*

PROOF. (1) $\Rightarrow$ (3). Let $\sum_{i=1}^{\infty} a_i = r \in \mathbb{R}$. By definition, $\sum_{i=1}^{\nu} a_i = \xi \sim r$ and $\sum_{i=1}^{\mu-1} a_i = \zeta \sim r$. Then $\sum_{i=\mu}^{\nu} a_i = \xi - \zeta \sim r - r = 0$.

(3) $\Rightarrow$ (2) is trivial. We are left to show that (2) $\Rightarrow$ (1). Assume by contradiction that $\sum_{i=1}^{\infty} a_i = \sup\{\sum_{i=1}^{n} a_i \mid n \in \mathbb{N}\} = +\infty$. Then for each n we can pick $\nu(n) \in \mathbb{N}$ such that $\sum_{i=1}^{\nu(n)} a_i > (\sum_{i=1}^{n} a_i) + n$. If $\nu = \nu(\alpha)$, by *transfer* we obtain $\sum_{i=1}^{\nu} a_i > (\sum_{i=1}^{\alpha} a_i) + \alpha$. In particular, $\sum_{i=\alpha+1}^{\nu} a_i$ is infinite because larger than α. $\qquad\square$

10. The grid integral

Historically, the problem that led to the notion of an integral was the problem of evaluating areas. Let us consider the typical example.

Let a function $f : A \to \mathbb{R}$ be given, and assume for simplicity that f only takes non-negative values. Denote by $\mathcal{S}_A^f$ the surface in the Cartesian plane as determined by the graph of f:

$$\mathcal{S}_A(f) = \{(x, y) \in A \times \mathbb{R} \mid 0 \le y \le f(x)\}.$$

Now argue heuristically. Take a "very large" natural number n, and consider the corresponding *grid* of *ratio* $1/n$ on the interval $[-n, n]$.

DEFINITION 3.39. *Let $n \in \mathbb{N}$. The n-grid is the finite set*

$$\mathbb{H}(n) = \left\{ \pm\frac{i}{n} \,\middle|\, i = 0, 1, \ldots, n^2 \right\} \subset [-n, n].$$

Intuition suggests that the surface $\mathcal{S}_A(f)$ can be well approximated by a union of rectangles with bases of "small" length $1/n$, and with heights given by the values taken by f at the points in the grid $\mathbb{H}(n)$.

In Alpha-Calculus, we do have "very large" numbers at hand, namely the infinite numbers; actually, we have a canonical infinite natural number, namely α. So, it seems natural to (informally) propose the following definition:

$$\int_A f \, dx = \mathcal{S}_A(f) \sim \frac{1}{\alpha} \sum_{\xi \in \mathbb{H}(\alpha) \cap A^*} f^*(\xi).$$

We are going to show that, within the framework of Alpha-Calculus, we can in fact give a meaning to that formula and turn the heuristic argument above into a rigorous one.

DEFINITION 3.40. *Let $A \subseteq \mathbb{R}$ and let $f : A \to \mathbb{R}$. The corresponding* approximating sequence $\langle\, Sum^f_A(n) \mid n \in \mathbb{N}\,\rangle$ *is defined by setting:*

$$Sum^f_A(n) \;=\; \sum_{x \,\in\, \mathbb{H}(n) \cap A} f(x).$$

(In case $\mathbb{H}(n) \cap A = \emptyset$, we agree that $Sum^f_A(n) = 0$.) Notice that all sets $\mathbb{H}(n) \cap A$ are finite, and so the definition is well-posed.

DEFINITION 3.41. *Let $f : A \to \mathbb{R}$ be a real function defined on a subset $A \subseteq \mathbb{R}$. The* grid integral *of f over A is the following standard part:*

$$\int_A f(x)\, d_{\mathbb{H}}x \;=\; \mathrm{st}\left(\frac{1}{\alpha} \cdot Sum^f_A(\alpha)\right).$$

We remark that the above definition complies with the intuitive idea of an integral as an infinitely long sum of infinitesimal quantities. A relevant property of the grid integral is the fact that it is defined for *all* real functions f and for *all* (possibly unbounded) domains A.

Several basic properties of the grid integral are itemized below. All proofs follow from the definition in a straightforward manner, and are omitted.[9]

PROPOSITION 3.42.

(1) *If $f(x) \le g(x)$ for all $x \in A$, then $\int_A f(x)\, d_{\mathbb{H}}x \le \int_A g(x)\, d_{\mathbb{H}}x$;*

(2) $\left| \int_A f(x)\, d_{\mathbb{H}}x \right| \le \int_A |f(x)|\, d_{\mathbb{H}}x$;

(3) $\int_A (f(x) + g(x))\, d_{\mathbb{H}}x = \int_A f(x)\, d_{\mathbb{H}}x + \int_A g(x)\, d_{\mathbb{H}}x$;

(4) $\int_A \lambda \cdot f(x)\, d_{\mathbb{H}}x = \lambda \cdot \int_A f(x)\, d_{\mathbb{H}}x$ *for all $\lambda \in \mathbb{R}$;*

(5) $\int_A f(x)\, d_{\mathbb{H}}x = 0$ *whenever A is finite;*

(6) *If $A \cap B = \emptyset$, then $\int_{A \cup B} f(x)\, d_{\mathbb{H}}x = \int_A f(x)\, d_{\mathbb{H}}x + \int_B f(x)\, d_{\mathbb{H}}x$;*

(7) *If the symmetric difference $A \triangle B = (A \setminus B) \cup (B \setminus A)$ is finite, then $\int_A f(x)\, d_{\mathbb{H}}x = \int_B f(x)\, d_{\mathbb{H}}x$.*

As a consequence of the above properties, for every $a < b$ one has that

$$\int_{[a,b]} f(x)\, d_{\mathbb{H}}x \;=\; \int_{[a,b)} f(x)\, d_{\mathbb{H}}x \;=\; \int_{(a,b]} f(x)\, d_{\mathbb{H}}x \;=\; \int_{(a,b)} f(x)\, d_{\mathbb{H}}x.$$

---

[9] We follow the conventional algebra on $\mathbb{R} \cup \{-\infty, +\infty\}$ (see §1.5).

So, for intervals with end-points a and b, there is no ambiguity in adopting the familiar notation

$$\int_a^b f(x)\, d_{\mathbb{H}}x.$$

Let us now check that the grid integral gives the "right measure" to intervals.

PROPOSITION 3.43. *The grid integral $\int_a^b 1 \cdot d_{\mathbb{H}}x = b - a$.*

PROOF. Notice that $n(b-a) - 1 \le |\mathbb{H}(n) \cap (a,b)| \le n(b-a)$ for each $n \in \mathbb{N}$. Then, if $f(x) = 1$ is the constant function with value 1,

$$b - a - \frac{1}{n} \le \frac{1}{n} \cdot \left(\sum_{x \in \mathbb{H}(n) \cap (a,b)} f(x) \right) \le b - a.$$

By passing to the Alpha-limits we obtain that

$$b - a - \frac{1}{\alpha} \le \frac{1}{\alpha} \cdot \mathrm{Sum}_A^f(\alpha) \le b - a,$$

and hence, by taking the standard parts, we obtain that $\int_a^b 1 \cdot d_{\mathbb{H}}x = b - a$, as desired. $\qquad\square$

EXAMPLE 3.44. *The grid integral $\int_0^1 x^2\, d_{\mathbb{H}}x = 1/3$.*

PROOF. For every $n \in \mathbb{N}$,

$$\frac{1}{n} \sum_{x \in \mathbb{H}(n) \cap (0,1]} x^2 = \frac{1}{n} \cdot \sum_{i=1}^n \left(\frac{i}{n} \right)^2 = \frac{1}{n^3} \cdot \sum_{i=1}^n i^2 = \frac{n(n-1)(2n-1)}{6n^3}.$$

(We used the formula $1^2 + 2^2 + \ldots + n^2 = \frac{n(n-1)(2n-1)}{6}$.) Then:

$$\int_0^1 x^2\, d_{\mathbb{H}}x = \mathrm{st}\left(\frac{\alpha(\alpha-1)(2\alpha-1)}{6\alpha^3} \right) = \frac{1}{3}.$$

$\qquad\square$

Let us now see consider *improper* grid integrals over unbounded intervals, and use the familiar notation

- $\int_a^{+\infty} f(x)\, d_{\mathbb{H}}x = \int_{(a,+\infty)} f(x)\, d_{\mathbb{H}}x$;
- $\int_{-\infty}^b f(x)\, d_{\mathbb{H}}x = \int_{(-\infty,b)} f(x)\, d_{\mathbb{H}}x$;
- $\int_{-\infty}^{+\infty} f(x)\, d_{\mathbb{H}}x = \int_{\mathbb{R}} f(x)\, d_{\mathbb{H}}x$.

EXAMPLE 3.45. *The grid integral $\int_1^{+\infty} \frac{1}{x^2}\, d_{\mathbb{H}}x = 1$.*

PROOF. For every $n \in \mathbb{N}$,

$$\frac{1}{n} \sum_{x \in \mathbb{H}(n) \cap (1,+\infty)} \frac{1}{x^2} = \frac{1}{n} \sum_{i=n+1}^{n^2} \frac{1}{(i/n)^2} = n \cdot \sum_{i=n+1}^{n^2} \frac{1}{i^2}.$$

Now,

$$n \cdot \sum_{i=n+1}^{n^2} \frac{1}{i^2} \geq n \cdot \sum_{i=n+1}^{n^2-1} \frac{1}{i(i+1)} = n \cdot \sum_{i=n+1}^{n^2-1} \left(\frac{1}{i} - \frac{1}{i+1} \right) = \frac{n}{n+1} - \frac{1}{n}.$$

In the other direction,

$$n \cdot \sum_{i=n+1}^{n^2} \frac{1}{i^2} \leq n \cdot \sum_{i=n+1}^{n^2} \frac{1}{i(i-1)} = n \cdot \sum_{i=n+1}^{n^2} \left(\frac{1}{i-1} - \frac{1}{i} \right) = 1 - \frac{1}{n}.$$

By taking the Alpha-limits, we have that

$$1 \sim \frac{\alpha}{\alpha+1} - \frac{1}{\alpha} \leq \sum_{\xi \in \mathbb{H}(\alpha) \cap (1,+\infty)^*} \frac{1}{\xi^2} \leq 1 - \frac{1}{\alpha} \sim 1,$$

and hence $\int_1^{+\infty} \frac{1}{x^2} \, d_\mathbb{H} x = 1$. □

EXAMPLE 3.46. *The grid integral $\int_0^{+\infty} e^{-x} \, d_\mathbb{H} x = 1$.*

PROOF. For every $n \in \mathbb{N}$,

$$\frac{1}{n} \sum_{x \in \mathbb{H}(n) \cap [0,+\infty)} e^{-x} = \frac{1}{n} \sum_{i=0}^{n^2} e^{-\frac{i}{n}}.$$

So, by taking Alpha-limits,

$$\frac{1}{\alpha} \sum_{\xi \in \mathbb{H}(\alpha) \cap [0,+\infty)^*} e^{-\xi} = \eta \cdot \sum_{\nu=0}^{\alpha^2} (e^{-\eta})^\nu = \eta \cdot \sum_{\nu=0}^{\alpha^2-1} (e^{-\eta})^\nu + \eta \cdot (e^{-\eta})^{\alpha^2}$$

$$= \eta \cdot \frac{1 - (e^{-\eta})^{\alpha^2}}{1 - e^{-\eta}} + \frac{e^{-\alpha}}{\alpha} = \frac{\eta}{1 - e^{-\eta}} \cdot (1 - e^{-\alpha}) + \frac{e^{-\alpha}}{\alpha} \sim 1,$$

and hence $\int_0^{+\infty} e^{-x} \, d_\mathbb{H} x = 1$. □

Let us now turn to the fundamental theorems of integral calculus.

THEOREM 3.47 (Mean Value for Integrals).
If $f : [a,b] \to \mathbb{R}$ is continuous, then there exists $c \in (a,b)$ such that

$$f(c) = \frac{1}{b-a} \int_a^b f(x) \, d_\mathbb{H} x.$$

PROOF. By the *Extreme Value Theorem* 3.7, the continuous function f takes a least value m and a greatest value M on $[a, b]$. Then:

$$m \cdot (b - a) = \int_a^b m \cdot d_{\mathbb{H}} x \leq \int_a^b f(x) \, d_{\mathbb{H}} x \leq \int_a^b M \cdot d_{\mathbb{H}} x = M \cdot (b - a),$$

and so the number $\frac{1}{b-a} \int_a^b f \, d_{\mathbb{H}} x$ belongs to the interval $[m, M]$. The conclusion is obtained by applying the Intermediate Value Theorem 3.8. □

THEOREM 3.48 (Fundamental Theorem of Calculus).
Let $f : [a, b] \to \mathbb{R}$ be a continuous function. Then: $u(x) = \int_a^x f(t) \, d_{\mathbb{H}} t$ is an anti-derivative of f, that is, the derivative $u'(x) = f(x)$ for all $x \in (a, b)$.

PROOF. Fix $x \in (a, b)$ and an arbitrary infinitesimal $\varepsilon \neq 0$. Then $\varepsilon = \varepsilon(\boldsymbol{\alpha})$ for some sequence $\varepsilon(n) \neq 0$ for all n. Since (a, b) is open, without loss of generality we can also assume that all points $x + \varepsilon(n) \in (a, b)$. By the *Mean Value Theorem* seen above, for every $n \in \mathbb{N}$ there exists a suitable $\xi(n) \in (x, x + \varepsilon(n))$ such that

$$\frac{u(x + \varepsilon(n)) - u(x)}{\varepsilon(n)} = \frac{1}{\varepsilon(n)} \int_x^{x + \varepsilon(n)} f(t) \, d_{\mathbb{H}} t = f(\xi(n)).$$

Then, by taking $\xi = \lim_{n \uparrow \alpha} \xi(n)$, we obtain that

$$\frac{u(x + \varepsilon) - u(x)}{\varepsilon} = f^*(\xi).$$

It is left to show that $f^*(\xi) \sim f(x)$, but this directly follows from the continuity of f, because $\xi \in (x, x + \varepsilon) \Rightarrow \xi \sim x$. □

11. Equivalences with the "standard" definitions

In this section we will show that all our basic definitions (as given in this chapter) are actually equivalent to the usual "standard" ones. Moreover, we will prove that our grid integral coincides with the *Riemann integral* for all Riemann-integrable functions.

It is worth remarking that none of the results included in this section is needed when introducing Alpha-Calculus to freshmen. For them, in fact, there are no "standard" definitions to compare with!

Let us start with the definition of *limit of a real function* (see Definition 3.20). We consider first the finite case when both $l, x_0 \in \mathbb{R}$.

THEOREM 3.49. *Let $l, x_0 \in \mathbb{R}$, and let f be a real function defined on a neighborhood of $x_0 \in \mathbb{R}$. Then the following "standard" and "nonstandard" definitions of $\lim_{x \to x_0} f(x) = l$ are equivalent:*

(i) *For every positive $\epsilon \in \mathbb{R}$, there exists a positive $\delta \in \mathbb{R}$ such that for all $x \in \mathbb{R}$: $x_0 - \delta < x < x_0 + \delta \ \Rightarrow \ l - \epsilon < f(x) < l + \epsilon$;*

(ii) *$f(x_0 + \varepsilon) \sim l$ for every infinitesimal $\varepsilon \neq 0$.*

PROOF. For every positive $a \in \mathbb{R}$, let

$$X_a = \{x \in \mathbb{R} \mid x_0 - a < x < x_0 + a\} \, ; \, Y_a = \{x \in \mathbb{R} \mid l - a < f(x) < l + a\}.$$

Now fix a positive $r \in \mathbb{R}$. By the hypothesis, there exists a positive $\delta \in \mathbb{R}$ such that $X_\delta \subseteq Y_r$, and hence $X_\delta^* \subseteq Y_r^*$. Notice that if $\varepsilon \sim 0$, then trivially:

$$x_0 + \varepsilon \ \in \ (X_\delta)^* \ = \ \{\zeta \in \mathbb{R}^* \mid x_0 - \delta < \zeta < x_0 + \delta\}$$
$$\subseteq (Y_r)^* \ = \ \{\zeta \in \mathbb{R}^* \mid l - r < f(\zeta) < l + r\}.$$

Thus $l - r < f(x_0 + \varepsilon) < l + r$ for all positive $r \in \mathbb{R}$, and we conclude that $f(x_0 + \varepsilon) \sim l$.

Conversely, let us assume by contradiction that (i) fails. Then there exists a real number $r > 0$ such that for all $n \in \mathbb{N}$ one can pick an element $\xi(n) \in \mathbb{R}$ with $x_0 - 1/n < \xi(n) < x_0 + 1/n$ and $|f(\xi(n)) - l| \geq r$. By passing to the Alpha-limits, we obtain $x_0 - 1/\boldsymbol{\alpha} < \xi < x_0 + 1/\boldsymbol{\alpha}$ and $|f(\xi) - l| \geq r$, where $\xi = \xi(\boldsymbol{\alpha})$. In particular $\xi \sim x_0$ and $f(\xi) \not\sim l$, contradicting (ii). $\square$

Other cases of limits, are collected in the following

THEOREM 3.50. *Let the real function f be defined on the interval $(a, +\infty)$. For every $l \in \mathbb{R}$, the following "standard" and "nonstandard" definitions of $\lim_{x \to +\infty} f(x) = l$ are equivalent:*

(i) *For every positive $\epsilon \in \mathbb{R}$, there exists $M \in \mathbb{R}$ such that for all $x \in \mathbb{R}$, $x > M \ \Rightarrow \ l - \epsilon < f(x) < l + \epsilon$;*

(ii) *$f^*(\xi) \sim l$ for every positive infinite ξ.*

Moreover, also the following two definitions of $\lim_{x \to +\infty} f(x) = +\infty$ are equivalent:

(iii) *For every $L \in \mathbb{R}$, there exists $M \in \mathbb{R}$ such that for all $x \in \mathbb{R}$, $x > M \ \Rightarrow \ f(x) > L$;*

(iv) *$f(\xi)$ is positive infinite for every positive infinite ξ.*

PROOF. The proof follows the patterns of the previous theorem. Let us start with the implication $(i) \Rightarrow (ii)$. For every positive $r \in \mathbb{R}$, there exists $M \in \mathbb{R}$ such that

$$(M, +\infty) \subseteq Y_r = \{x \in \mathbb{R} \mid l - a < f(x) < l + a\}.$$

If ξ is positive infinite, then clearly

$$\xi \in (M, \infty)^* \subseteq Y_r^* = \{\zeta \in \mathbb{R}^* \mid l - r < f(\zeta) < l + r\}.$$

As this is true for all positive $r \in \mathbb{R}$, it follows that $f^*(\xi) \sim l$.

Conversely, suppose that (i) fails. Then there exists a positive $r \in \mathbb{R}$ such that for all $n \in \mathbb{N}$ one can pick an element $\xi(n) \in \mathbb{R}$ with $\xi(n) > n$ and $|f(\xi(n)) - l| \geq r$. If we take $\xi = \lim_{n \uparrow \alpha} \xi(n)$, by *transfer* we have $\xi > \alpha$ and $|f^*(\xi) - l| \geq r$. In particular ξ is a positive infinite number such that $f^*(\xi) \not\sim l$, contradicting (ii).

Now let us turn to $(iii) \Leftrightarrow (iv)$. Denote by $Z_L = \{x \in \mathbb{R} \mid f(x) > L\}$. By (iii), for every $L \in \mathbb{R}$ there exists $M \in \mathbb{R}$ such that $(M, +\infty) \subseteq Z_L$. If ξ is positive infinite, then clearly

$$\xi \in (M, +\infty)^* \subseteq (Z_L)^* = \{\zeta \in \mathbb{R}^* \mid f^*(\zeta) > L\}.$$

Thus $f^*(\xi) > L$ for every $L \in \mathbb{R}$, and $f^*(\xi)$ is positive infinite.

Conversely, assume that (iii) fails. Then there exists a real number L such that for all $n \in \mathbb{N}$ one can pick an element $\xi(n) \in \mathbb{R}$ such that $\xi(n) > n$ and $f(\xi(n)) \leq L$. By taking the Alpha-limits, we obtain $\xi > \alpha$ and $f^*(\xi) \leq L$, where $\xi = \xi(\alpha)$. In particular ξ is positive infinite while $f^*(\xi)$ is not infinite, contradicting (ii). $\quad\square$

Similar equivalences also hold between the "standard" and "non-standard" definitions of the remaining cases of limits, namely:

$$\lim_{x \to -\infty} f(x) = l \in \mathbb{R}; \quad \lim_{x \to +\infty} f(x) = -\infty$$

$$\lim_{x \to -\infty} f(x) = +\infty; \quad \lim_{x \to -\infty} f(x) = -\infty.$$

By checking the definitions of continuity, derivative, and limit as given in Alpha-Calculus (see Definitions 3.4, 3.12, 3.20, respectively), one readily verifies the following:

- f is *continuous* at x_0 if and only if $\lim_{x \to x_0} f(x) = f(x_0)$;
- f has *derivative* $f'(x_0)$ at x_0 if and only if $\lim_{x \to x_0} \frac{f(x) - f(x_0)}{x - x_0} = f'(x_0)$.

In consequence, Theorem 3.49 directly implies that also our definitions of continuity and derivative are actually equivalent to the "standard" ones.

The equivalences of the definitions of *uniform continuity* (see Definition 3.9) and of *limit of a sequence* (see Definition 3.16), are shown in a similar fashion, and we omit the proofs.

THEOREM 3.51. *Let $f : A \to \mathbb{R}$. Then the following "standard" and "non-standard" definitions of* uniform continuity *on A are equivalent:*

(i) *For every positive $\epsilon \in \mathbb{R}$, there exists a positive $\delta \in \mathbb{R}$ s.t. for all $x, y \in A$: $-\delta < x - y < \delta \Rightarrow -\epsilon < f(x) - f(y) < l$;*

(ii) *$f^*(\xi) \sim f^*(\zeta)$ for all $\xi, \zeta \in A^*$ such that $\xi \sim \zeta$.*

THEOREM 3.52. *Let $\langle a_n \mid n \in \mathbb{N} \rangle$ be a sequence. For every $l \in \mathbb{R}$ the following "standard" and "non-standard" definitions of $\lim_{n \to \infty} a_n = l$ are equivalent:*

(i) *For every positive $\epsilon \in \mathbb{R}$, there exists $N \in \mathbb{N}$ such that for all $n \in \mathbb{N}$, $n > N \Rightarrow l - \epsilon < a_n < l + \epsilon$;*

(ii) *$a_\nu \sim l$ for every infinite $\nu \in \mathbb{N}^*$.*

Moreover, also the following "standard" and "non-standard" definitions of $\lim_{n \to \infty} a_n = +\infty$ (or $\lim_{n \to \infty} a_n = -\infty$) are equivalent:

(iii) *For every $L \in \mathbb{R}$, there exists $N \in \mathbb{N}$ such that for all $n \in \mathbb{N}$: $n > N \Rightarrow a_n > L$ (or $n > N \Rightarrow a_n < L$);*

(iv) *a_ν is positive (negative) infinite for all infinite $\nu \in \mathbb{N}^*$.*

Finally, also our definition of *sum of a series* (see Definition 3.37) is equivalent to the standard one, because $\sum_{i=1}^{\infty} a_i = \lim_{n \to \infty} \sum_{i=1}^{n} a_i$.

EXERCISE 3.53. *By inverting the role of ϵ and δ in the definition of continuity of a function f at x_0, one can consider the following property:*

($\star$) *For every positive $\epsilon \in \mathbb{R}$, there exists a positive $\delta \in \mathbb{R}$ such that for all $x \in \mathbb{R}$, $|x - x_0| < \epsilon \Rightarrow |f(x) - f(x_0)| < \delta$.*

Show that ($\star$) does not depend on x_0 and it is equivalent to the following property: For every finite $\xi \in \mathbb{R}^*$, $f^*(\xi)$ is finite.[10]

---

[10] ($\star$) is the property of *local boundedness* of f. This example is taken from E. Nelson's paper [**77**].

12. Grid integral *versus* Riemann integral

As apparent from its definition, the grid integral $\int_A f(x)\, d_{\mathbb{H}}x$ has the pedagogical advantage of making sense for *all* domains $A \subseteq \mathbb{R}$ and for *all* functions $f : A \to \mathbb{R}$. Besides, it is consistent with the integrals as used in traditional courses of calculus. Indeed, the next theorem shows that the grid integral is a generalization of the familiar *Riemann integral*.

THEOREM 3.54. *Let f be a Riemann-integrable function over an interval $[a, b]$. Then its Riemann integral $\int_a^b f(x)\, dx$ coincides with the grid integral $\int_a^b f(x)\, d_{\mathbb{H}}x$.*

PROOF. Let us first fix the terminology. By *partition* of an interval $[a, b]$ we mean a finite sequence $a = x_0 < \ldots < x_s = b$. The *mesh* of a partition $\wp = \{x_1 < \ldots < x_s\}$ is the greatest length of a subinterval $[x_i, x_{i+1}]$ of the partition, that is:

$$\text{mesh}(\wp) \;=\; \max\{x_{i+1} - x_i \mid i = 1, \ldots, s - 1\}.$$

We recall that a function f is *Riemann-integrable* if there exists a real number $\ell = \int_a^b f(x)\, dx$ with the following property: For every positive $\epsilon \in \mathbb{R}$ there exists a positive $\delta \in \mathbb{R}$ such that for every partition $\wp = \{a = x_1 < \ldots < x_s = b\}$ with $\text{mesh}(\wp) < \delta$, and for every $y_1, \ldots, y_{s-1}$ with $y_i \in [x_i, x_{i+1}]$, the corresponding *Riemann sum* $\text{RS}_\wp = \sum_{i=1}^{s-1} f(y_i) \cdot (x_{i+1} - x_i)$ is such that $|\text{RS}_\wp - \ell| < \epsilon$.

Let $\ell = \int_a^b f(x)\, dx$ be the Riemann integral of f over $[a, b]$. Given $\epsilon > 0$, pick $\delta > 0$ as given by the definition. For every $n \in \mathbb{N}$ such that $n > \max\{1/\delta, 1/(b - a)\}$, let us consider the following partition:

$$\wp_n \;=\; (\mathbb{H}(n) \cap (a, b)) \cup \{a, b\} \;=\; \{a = x_1 < x_2 < \ldots < x_{s-1} < x_s = b\}.$$

Then take $y_i = x_i$ for $i = 1, \ldots, s - 2$ and $y_{s-1} = x_s$, and consider corresponding Riemann sum $\text{RS}_n = \text{RS}_{\wp_n}$:

$$\text{RS}_n \;=\; \sum_{i=2}^{s-2} f(x_i) \cdot (x_{i+1} - x_i) + f(a)(x_2 - a) + f(b)(b - x_{s-1}).$$

Notice that $1/n < b - a$ implies that the intersection $\mathbb{H}(n) \cap (a, b)$ is nonempty. Moreover:

$$\frac{1}{n} \cdot \left(\sum_{x \in \mathbb{H}(n) \cap (a,b)} f(x) \right) = \sum_{i=2}^{s-2} f(x_i) \cdot (x_{i+1} - x_i).$$

Now notice that $\mathrm{mesh}(\wp_n) = 1/n < \delta$, and so:

$$\left| \frac{1}{n} \cdot \left(\sum_{x \in \mathbb{H}(n) \cap (a,b)} f(x) \right) - \ell \right| = \left| \mathrm{RS}_n - \ell - f(a)(x_2 - a) - f(b)(b - x_{s-1}) \right|$$

$$\leq \left| \mathrm{RS}_n - \ell \right| + |f(a)|(x_1 - a) + |f(b)|(b - x_{s-1}) < \epsilon + \frac{|f(a)| + |f(b)|}{n}.$$

By taking the Alpha-limits, we finally obtain

$$\left| \int_a^b f(x)\, d_\mathbb{H}x - \ell \right| \sim \left| \frac{1}{\alpha} \cdot \left(\sum_{\xi \in \mathbb{H}(\alpha) \cap A^*} f^*(\xi) \right) - \ell \right| < \epsilon + \frac{|f(a)| + |f(b)|}{\alpha}.$$

As the above inequality holds for every positive $\epsilon \in \mathbb{R}$, we conclude that $\int_a^b f(x)\, d_\mathbb{H}x = \ell$, as desired. $\qquad\square$

13. Remarks and comments

1. Interesting historical remarks on the notion of continuity *versus* uniform continuity can be found in A. Robinson's comments about A.L. Cauchy's approach to analysis in Chapter X of [**83**].

A famous erroneous statement made by Cauchy in his celebrated textbook "*Cours d'Analyse de l'École Royale Polytechnique Sur le Calcul Infinitésimal*" (1821), is that the sum of a (pointwise) convergent series of continuous functions on an interval $[a, b]$ is continuous. Robinson suggests a possible "non-standard" interpretation of Cauchy's ideas that would make his proof correct. Namely, one could interpret Cauchy's vague notion of "*quantitées infiniment petites*" as infinitesimal hyperreal numbers, and analyze his subsequent definitions accordingly.

> For the notion of continuity, Cauchy's definition may thus be interpreted as stating that for $f(x)$ defined in the interval $a < x < b$, $f(x)$ is continuous in that interval if, for all infinitesimal α, the difference $f(x+\alpha) - f(x)$ is always (toujours) infinitesimal. If now we interpret 'always' as meaning 'for all standard x' then we obtain ordinary continuity in the interval but if by 'always' we mean 'for all x' then we obtain uniform continuity.

With this interpretation in mind, Cauchy's proof about convergent series becomes correct, because then one is in fact assuming the *uniform* continuity of the involved functions.

2. To the authors' knowledge, the modern idea of an infinitesimal calculus based on the existence of an "infinitely large" natural number, goes back to the work by C. Schmieden and D. Laugwitz [**85**], that appeared in the late 50's of the last century, a few years before nonstandard analysis was elaborated by A. Robinson. In that paper, they considered a new number Ω with the property that any "elementary property" P is true for Ω whenever P is true for all sufficiently large natural numbers.

This Ω-approach gives sound foundations to the use of infinitesimal numbers and permits the development of large parts of calculus (see [**68**]). Unfortunately, that theory turned out to be inadequate for stronger applications. One of the crucial drawbacks is that the resulting "nonstandard real numbers" $\mathfrak{R}$ contain zero divisors and they are only partially ordered. Indeed, the structure $\mathfrak{R}$ is obtained as the quotient $\mathfrak{Fun}(\mathbb{N}, \mathbb{R})/i_0$ of the real sequences modulo the (non-maximal) ideal i_0 of sequences that eventually vanish (see §2.10).

It is worth mentioning that the Ω-approach recently raised a new attention among researchers. Indeed, its more constructive flavour with respect to Robinsonian nonstandard analysis has been a sort of inspiration to formalizing nonstandard methods within the framework of intuitionistic and constructive mathematics (see, *e.g.*, the work by E. Palmgren [**79, 80**]; see also the second-order extension of the Ω-approach proposed by J.M. Henle in [**53**]).

Part 2

Alpha-Theory

The scope of Alpha-Calculus is limited to real functions, subsets of real numbers, and their hyper-extensions, and so it is too weak a theory for many applications.

As a first relevant example that brings beyond the scope of Alpha-Calculus, we can mention the *Dirac δ-function*. In some intuitive sense, this function could be identified with a function of the following form:

$$\delta(x) = \frac{\alpha}{\sqrt{\pi}} \cdot e^{-(\alpha x)^2} \quad \text{or} \quad \delta(x) = \frac{\alpha}{\pi} \cdot \frac{1}{1 + (\alpha x)^2}.$$

Although definable within Alpha-Calculus, we remark that these functions are *not* hyper-extensions f^* of any real function $f : \mathbb{R} \to \mathbb{R}$. In consequence, the methodology that we have been elaborating so far does not apply. As it will be shown further on (see §9.3), a possible definition of the *Dirac δ-function* can be given by working in the hyper-extension $\mathfrak{Fun}(\mathbb{R}, \mathbb{R})^*$ of the space of real functions, endowed with a suitable equivalence relation that identify functions such as the two displayed above. The Alpha-Theory will provide a convenient general framework where this kind of constructions – as well as many others – can be correctly formalized.

CHAPTER 4

Introducing the Alpha-Theory

Alpha-Theory is a natural generalization of Alpha-Calculus obtained by postulating the existence of an Alpha-limit for *arbitrary* (not necessarily real-valued) sequences. With respect to ACT, two more axioms are added that rule the Alpha-limits of sequences of sets.

1. The axioms of Alpha-Theory

Sticking to the everyday practice of many mathematicians, in the following we will distinguish between *sets* and *atoms*, the latter objects being primitive elements that are not sets. In particular, real numbers will be taken as atoms.

In the sequel, by *sequence* we mean any function defined on the set of natural numbers, with no restrictions on the set of its values.

Alpha-Theory AT

(AT0) Existence Axiom. *Every sequence $\varphi(n)$ has a unique "Alpha-limit" denoted by $\lim_{n\uparrow\alpha}\varphi(n)$ or, more simply, by $\lim_\alpha \varphi$. If $\varphi(n)$ is a sequence of atoms, then also $\lim_{n\uparrow\alpha}\varphi(n)$ is an atom.*

(AT1) Real Number Axiom. *If $c_r(n) = r$ is the constant sequence with value a real number r, then $\lim_{n\uparrow\alpha} c_r(n) = r$.*

(AT2) Alpha Number Axiom. *The Alpha-limit of the identity sequence $\imath(n) = n$ is a "new" number denoted by $\boldsymbol{\alpha}$, that is $\lim_{n\uparrow\alpha} n = \boldsymbol{\alpha} \notin \mathbb{N}$.*

(AT3) Field Axiom. *The set of all Alpha-limits of real sequences*

$$\mathbb{R}^* = \left\{ \lim_{n\uparrow\alpha} \varphi(n) \,\Big|\, \varphi : \mathbb{N} \to \mathbb{R} \right\}$$

is a field, called the hyperreal field, *where:*

- $\lim_{n\uparrow\alpha} \varphi(n) + \lim_{n\uparrow\alpha} \psi(n) = \lim_{n\uparrow\alpha}(\varphi(n) + \psi(n))$
- $\lim_{n\uparrow\alpha} \varphi(n) \cdot \lim_{n\uparrow\alpha} \psi(n) = \lim_{n\uparrow\alpha}(\varphi(n) \cdot \psi(n))$

(AT4) Internal Set Axiom. *If $c_\emptyset(n) = \emptyset$ is the constant sequence with value the empty set, then $\lim_{n\uparrow\alpha} c_\emptyset(n) = \emptyset$. If $\psi(n)$ is a sequence of nonempty sets, then*

$$\lim_{n\uparrow\alpha} \psi(n) = \left\{ \lim_{n\uparrow\alpha} \varphi(n) \,\Big|\, \varphi(n) \in \psi(n) \text{ for all } n \right\}.$$

(AT5) Composition Axiom. *If two sequences take the same Alpha-limit $\lim_{n\uparrow\alpha} \varphi(n) = \lim_{n\uparrow\alpha} \psi(n)$, then $\lim_{n\uparrow\alpha} f(\varphi(n)) = \lim_{n\uparrow\alpha} f(\psi(n))$ for all functions f such that the compositions $f \circ \varphi$ and $f \circ \psi$ make sense.*

A few comments are in order. To begin with, notice that axiom (AT0) is just a straight generalization of (ACT0) that extends the scope of the theory from real sequences to arbitrary sequences. The following three axioms (AT1), (AT2), (AT3) coincide with (ACT1), (ACT2) and (ACT3), respectively, and so the Alpha-Theory fully includes the Alpha-Calculus of ACT.

The new *Internal Set Axiom* (AT4) rules the Alpha-limits of sequences of sets.[1] Firstly, it states the obvious requirement that the Alpha-limit of the constant sequence with value the emptyset equals the emptyset. Then, it postulates that the *membership relation* is preserved under Alpha-limits, that is, $\lim_{n\uparrow\alpha} \varphi(n) \in \lim_{n\uparrow\alpha} \psi(n)$ whenever $\varphi(n) \in \psi(n)$ for all n. Finally, it prescribes the *transitivity property* that elements of Alpha-limits are themselves Alpha-limits.

The last *Composition Axiom* (AT5) states a natural coherence property of Alpha-limits with respect to compositions. We have seen that such a property is proved by the axioms of ACT for real sequences

---

[1] The adjective *internal* has been used here because objects that are Alpha-limits precisely correspond to the *internal elements* of *nonstandard analysis* (see Theorem 13.4).

$\varphi, \psi : \mathbb{N} \to \mathbb{R}$ and real functions $f : \mathbb{R} \to \mathbb{R}$ (see Proposition 2.9); however, it takes a new axiom to extend its validity to arbitrary sequences and functions.

REMARK 4.1. *The axioms of Alpha-Theory have a really broad range, in that they rule properties of all sequences of mathematical objects. As a consequence, for the sake of a full rigor, one should also specify the underlying axiomatic framework of mathematics. We will do this further on (see §14.1).*

Here, by *mathematical object* we mean any entity that is used in the ordinary practice of mathematics, namely numbers, sets, ordered tuples, functions, relations, and so forth. As suggested by the common use, in our approach we will distinguish between *sets* as legitimate collections of objects, and *atoms* as mathematical objects that are not sets. As a general rule, we will make the assumption that numbers be atoms, and that ordered tuples, relations and functions be sets. In fact, it is well-known in set theory that each of those notions admits a representation as a set. Precisely, we will agree on the following.

- *Real* and *hyperreal numbers* are atoms.
- *Ordered pairs* are the *Kuratowski pairs*:

$$(a, b) = \{\{a\}, \{a, b\}\}.$$

Inductively, ordered $(k + 1)$-tuples are defined by putting

$$(a_1, \ldots, a_k, a_{k+1}) = ((a_1, \ldots, a_k), a_{k+1}).$$

Kuratowski pairs $(a, b) = \{\{a\}, \{a, b\}\}$ are commonly used in set theory because they are sets that satisfy the defining property of ordered pairs, namely the equivalence:

$$(a, b) = (a', b') \iff a = a' \text{ and } b = b'.$$

Let us remark that, in the practice, one simply forgets about the formal definition of an ordered pair; all that matters is to recall the above characterizing property.[2]

- A *binary relation* $\mathcal{R}$ is identified with the set of ordered pairs that satisfy it:

$$\mathcal{R} = \{(a, a') \mid a\mathcal{R}a'\}.$$

---

[2] Similarly as done with the notions of set and atom, one could take also ordered tuples as primitive objects of our theory; however, in this case, one would need to postulate extra axioms in Alpha-Theory to rule their properties with respect to Alpha-limits.

One writes $\mathcal{R}(a, b)$ to mean $(a, b) \in \mathcal{R}$. (Similarly, for k-place relations.)

- A *function* f is identified with its *graph*

$$\{(a, b) \mid b = f(a)\}.$$

In other words, a function f is a binary relation with the "'functional'" property: "$(a, b), (a, b') \in f \Rightarrow b = b'$", and $f(a)$ denotes the unique element b such that $(a, b) \in f$.

When writing $f : A \to B$ we mean that f is a function whose domain is the set A and whose range is included in the set B.

REMARK 4.2. *In the practice, the "status" of atom or set may depend on the context. For instance, in the well-known construction of the real numbers as Dedekind cuts, real numbers are defined as suitable subsets of rational numbers. On the other hand, in the practice of real analysis one hardly considers a real number as a set. Similarly, when studying a topological space X, one takes its elements as atoms, even in the cases when they are in fact functions (function spaces) or sets (e.g., the Vietoris topology).*

2. First properties of Alpha-Theory

Since Alpha-Theory extends ACT, we already have at hand the notion of *qualified set* of natural numbers (see §2.4). In consequence, the expressions "for almost all n" and "almost everywhere" have a meaning also in this more general setting.

The "informal" *transfer principle* that was formulated in §2.5 for real sequences, can now be extended to arbitrary sequences of mathematical objects.[3]

Transfer Principle. *An "elementary property" P is satisfied by mathematical objects $\varphi_1(n), \ldots, \varphi_k(n)$ for almost all n if and only if P is satisfied by the Alpha-limits $\lim_{n\uparrow\alpha} \varphi_1(n), \ldots, \lim_{n\uparrow\alpha} \varphi_k(n)$:*

$$P(\varphi_1(n), \ldots, \varphi_k(n)) \text{ a.e.} \iff P(\lim_{n\uparrow\alpha} \varphi_1(n), \ldots, \lim_{n\uparrow\alpha} \varphi_k(n)).$$

---

[3] We recall that, for now, the above *transfer principle* can only be taken at an informal level; a rigorous definition of "elementary property" will be given in Chapter 5 by using the formalism of first-order logic.

In this section we present the first fundamental instances of the *transfer principle* that follow from the axioms of Alpha-Theory. The (rather tedious) proofs will be postponed until the next section.

To begin with, we observe that – without loss of generality – one can consider only "pure" sequences, that is, sequences that either always take atoms as values, or always take sets.

PROPOSITION 4.3. *Let* $\varphi(n)$ *be an arbitrary sequence. Then*

(1) $\varphi(n)$ *is an atom* a.e. $\Longleftrightarrow$ $\lim_{n\uparrow\alpha}\varphi(n)$ *is an atom;*

(2) $\varphi(n) = \emptyset$ a.e. $\Longleftrightarrow$ $\lim_{n\uparrow\alpha}\varphi(n) = \emptyset$;

(3) $\varphi(n)$ *is a nonempty set* a.e. $\Longleftrightarrow$ $\lim_{n\uparrow\alpha}\varphi(n)$ *is a nonempty set.*

Primary instances of "elementary" properties that are preserved by *transfer* are given by the *equality* and the *membership* relations.

THEOREM 4.4 (Transfer of equality and membership).
Let φ *and* ψ *be arbitrary sequences. Then:*

(1) $\varphi(n) = \psi(n)$ a.e. $\Longleftrightarrow$ $\lim_{n\uparrow\alpha}\varphi(n) = \lim_{n\uparrow\alpha}\psi(n)$;

(2) $\varphi(n) \in \psi(n)$ a.e. $\Longleftrightarrow$ $\lim_{n\uparrow\alpha}\varphi(n) \in \lim_{n\uparrow\alpha}\psi(n)$.

A straight consequence is the following.

PROPOSITION 4.5. $\lim_{n\uparrow\alpha}\{1,\ldots,n\} = \{1,\ldots,\alpha\}$.

PROOF. Let $\varphi : \mathbb{N} \to \mathbb{N}$ be an arbitrary sequence of natural numbers. By property (1) of the previous theorem, we have that $\lim_{n\uparrow\alpha}\varphi(n) \in \lim_{n\uparrow\alpha}\{1,\ldots,n\}$ if and only if $\varphi(n) \in \{1,\ldots,n\}$ a.e., that is, $\varphi(n) \le n$ a.e.. By Theorem 2.25 (3), the last property is equivalent to $\lim_{n\uparrow\alpha}\varphi(n) \le \alpha$, that is, $\lim_{n\uparrow\alpha}\varphi(n) \in \{1,\ldots,\alpha\}$. $\square$

Alpha-limits are coherent with k-sets.

THEOREM 4.6 (Transfer of k-sets).
Let $\varphi_1(n),\ldots,\varphi_k(n)$ *be arbitrary sequences. Then*

$$\lim_{n\uparrow\alpha}\{\varphi_1(n),\ldots,\varphi_k(n)\} = \left\{\lim_{n\uparrow\alpha}\varphi_1(n),\ldots,\lim_{n\uparrow\alpha}\varphi_k(n)\right\}.$$

As a corollary, one can extend the "definition by cases" property from real sequences to arbitrary sequences (see Corollary 2.7).

PROPOSITION 4.7. *If an arbitrary sequence $\psi(n)$ is defined by cases:*

$$\psi(n) = \begin{cases} \varphi_1(n) & \text{if property } P_1 \text{ holds} \\ \dots \\ \varphi_k(n) & \text{if property } P_k \text{ holds} \end{cases}$$

then $\lim_{n\uparrow\alpha} \psi(n)$ equals one of the Alpha-limits $\lim_{n\uparrow\alpha} \varphi_i(n)$.

PROOF. Since $\psi(n) \in \{\varphi_1(n), \dots, \varphi_k(n)\}$ for all n, $\lim_{n\uparrow\alpha} \psi(n) \in \lim_{n\uparrow\alpha}\{\varphi_1(n), \dots, \varphi_k(n)\} = \{\lim_{n\uparrow\alpha} \varphi_1(n), \dots, \lim_{n\uparrow\alpha} \varphi_k(n)\}$. $\qquad\square$

THEOREM 4.8.

(1) $\varphi(n) \subseteq \psi(n)$ a.e. $\iff \lim_\alpha \varphi \subseteq \lim_\alpha \psi$.

(2) $\vartheta(n) = \varphi(n) \cup \psi(n)$ a.e. $\iff \lim_\alpha \vartheta = \lim_\alpha \varphi \cup \lim_\alpha \psi$.

(3) $\vartheta(n) = \varphi(n) \cap \psi(n)$ a.e. $\iff \lim_\alpha \vartheta = \lim_\alpha \varphi \cap \lim_\alpha \psi$.

(4) $\vartheta(n) = \varphi(n) \setminus \psi(n)$ a.e. $\iff \lim_\alpha \vartheta = \lim_\alpha \varphi \setminus \lim_\alpha \psi$.

The axioms of Alpha-Theory guarantee that Alpha-limits are coherent with (Kuratowski) ordered pairs, as well as with Cartesian products.

THEOREM 4.9.

(1) $\vartheta(n) = (\varphi(n), \psi(n))$ a.e. $\iff \lim_\alpha \vartheta = (\lim_\alpha \varphi, \lim_\alpha \psi)$;

(2) $\vartheta(n)$ *is an ordered pair* a.e. $\iff \lim_\alpha \vartheta$ *is an ordered pair*.

(3) $\vartheta(n) = \varphi(n) \times \psi(n)$ a.e. $\iff \lim_\alpha \vartheta = \lim_\alpha \varphi \times \lim_\alpha \psi$.

More generally, the same results also hold for k-tuples and for Cartesian products with k-many factors.

Recall that the *domain* $\mathrm{dom}(\mathcal{R})$ and the *range* $\mathrm{ran}(\mathcal{R})$ of a binary relation $\mathcal{R}$ are defined by setting:

$$\mathrm{dom}(\mathcal{R}) = \{a \mid \exists b \ \mathcal{R}(a,b)\} ; \quad \mathrm{ran}(\mathcal{R}) = \{b \mid \exists a \ \mathcal{R}(a,b)\}.$$

THEOREM 4.10. *$\mathcal{R}_n$ is a binary relation with $\mathrm{dom}(\mathcal{R}_n) = A_n$ and $\mathrm{ran}(\mathcal{R}_n) = B_n$ for almost all n $\iff \mathcal{R} = \lim_{n\uparrow\alpha} \mathcal{R}_n$ is a binary relation with $\mathrm{dom}(\mathcal{R}) = \lim_{n\uparrow\alpha} A_n$ and $\mathrm{ran}(\mathcal{R}) = \lim_{n\uparrow\alpha} B_n$.*

Also the notion of function (that is, of binary relation with the "right-uniqueness" property) is preserved under Alpha-limits.

THEOREM 4.11. *f_n is a function with* $\mathrm{dom}(f_n) = A_n$ *and* $\mathrm{ran}(f_n) = B_n$ *for almost all* n $\Longleftrightarrow$ $F = \lim_{n\uparrow\alpha} f_n$ *is a function with* $\mathrm{dom}(F) = \lim_{n\uparrow\alpha} A_n$ *and* $\mathrm{ran}(F) = \lim_{n\uparrow\alpha} B_n$.

Moreover, if $\varphi(n) \in A_n$ *for all* n, *then*

$$\lim_{n\uparrow\alpha} f_n(\varphi(n)) = F\left(\lim_{n\uparrow\alpha} \varphi(n)\right).$$

PROPOSITION 4.12. *The functions f_n are 1-1 (or onto B_n) for almost all n if and only if the function $\lim_{n\uparrow\alpha} f_n$ is 1-1 (onto $\lim_{n\uparrow\alpha} B_n$, respectively).*

Contrary to the usual set-theoretic framework as given by Zermelo-Fraenkel set theory, in the Alpha-Theory one has $\in$-descending chains.

REMARK 4.13. *Consider the* von Neumann natural numbers:[4]
- $\overline{0} = \emptyset$,
- $\overline{n+1} = \overline{n} \cup \{\overline{n}\} = \{\overline{0}, \dots, \overline{n}\}$.

For every $k \in \mathbb{N}$, let

$$\overline{\alpha - k} = \lim_{n\uparrow\alpha} \overline{n - k},$$

where we agree that $\overline{n-k} = \emptyset$ for $k \geq n$. Then one has the descending chain:

$$\overline{\alpha} \ni \overline{\alpha - 1} \ni \dots \ni \overline{\alpha - k} \ni \overline{\alpha - (k+1)} \ni \dots$$

The proof of the above fact is straightforward from the definitions, and it is left to the reader as an exercise.[5]

3. Some detailed proofs

In this section we give detailed proofs of the results presented in the previous section. Although they are required for a rigorous development of the Alpha-Theory, the arguments are neither particularly interesting, nor necessary to understand the follow-up sections, and so they can be omitted at a first reading. In fact, the reader who is primarily interested in seeing Alpha-Theory in action, could safely skip this section, and directly take the results of the previous section as axioms of the theory.

---

[4] Here we are implicitly assuming the set of natural numbers $\mathbb{N}$ as given, along with the validity of the *induction principle*. A formalization of Alpha-Theory as an axiomatic set theory will be given in Chapter 14.

[5] See Proposition 14.4.

PROOF OF PROPOSITION 4.3. Let φ_1, φ_2 be the sequences such that

$$\varphi_1(n) = \begin{cases} \varphi(n) & \text{if } \varphi(n) \text{ is an atom} \\ 0 & \text{otherwise.} \end{cases}$$

$$\varphi_2(n) = \begin{cases} \varphi(n) & \text{if } \varphi(n) \text{ is a nonempty set} \\ \{0\} & \text{otherwise.} \end{cases}$$

Since $\varphi_1(n)$ is a sequence of atoms, by the *Existence Axiom* also $\lim_\alpha \varphi_1$ is an atom; and since $\varphi_2(n)$ is a sequence of nonempty sets, then by the *Internal Set Axiom* also $\lim_\alpha \varphi_2$ is a nonempty set. Now, if $\varphi(n)$ is an atom *a.e.* then $\varphi(n) = \varphi_1(n)$ *a.e.*, and so $\lim_\alpha \varphi = \lim_\alpha \varphi_1$ is atom. If $\varphi(n)$ is a nonempty set *a.e.* then $\varphi(n) = \varphi_2(n)$ *a.e.*, and so $\lim_\alpha \varphi = \lim_\alpha \varphi_2$ is a nonempty set. Finally, if $\varphi(n) = \emptyset$ *a.e.* then $\lim_\alpha \varphi = \lim_\alpha c_\emptyset = \emptyset$ by the *Internal Set Axiom*.

The converse implications directly follow from what proved above. Indeed, the three sets $A = \{n \mid \varphi(n) \text{ is an atom}\}$, $B = \{n \mid \varphi(n) = \emptyset\}$ and $C = \{n \mid \varphi(n) \text{ is a nonempty set}\}$ form a partition of $\mathbb{N}$, and so exactly one of those is qualified (see Theorem 2.22). If $\lim_\alpha \varphi$ is an atom, then neither B nor C can be qualified, and hence A must be qualified; and similarly for the other cases. □

Let us now prove the left-to-right implications of Theorem 4.4.

PROPOSITION 4.14. *Let $\varphi(n), \psi(n)$ be arbitrary sequences. Then*

(1) $\varphi(n) = \psi(n)$ *a.e.* $\implies \lim_{n\uparrow\alpha} \varphi(n) = \lim_{n\uparrow\alpha} \psi(n)$;

(2) $\varphi(n) \in \psi(n)$ *a.e.* $\implies \lim_{n\uparrow\alpha} \varphi(n) \in \lim_{n\uparrow\alpha} \psi(n)$.

PROOF. (1). The same argument used in the first part of the proof of Theorem 2.23 also applies to arbitrary sequences. Precisely, let $X = \{n \mid \varphi(n) = \psi(n)\}$. Since $\varphi(n) = \psi(n)$ *a.e.*, $\alpha \in X^*$ and so, by the *Internal Set Axiom*, there exists a sequence $\xi(n)$ such that $\lim_{n\uparrow\alpha} \xi(n) = \alpha$ and $\xi(n) \in X$ for all n, that is, $\varphi(\xi(n)) = \psi(\xi(n))$ for all n. Now, $\lim_\alpha \xi = \alpha = \lim_\alpha \imath$ where $\imath(n) = n$ is the identity sequence and so, by the *Composition Axiom*,

$$\lim_{n\uparrow\alpha} \varphi(n) = \lim_{n\uparrow\alpha} \varphi(\imath(n)) = \lim_{n\uparrow\alpha} \varphi(\xi(n))$$
$$= \lim_{n\uparrow\alpha} \psi(\xi(n)) = \lim_{n\uparrow\alpha} \psi(\imath(n)) = \lim_{n\uparrow\alpha} \psi(n).$$

(2). Let $\psi'(n)$ be the sequence defined by putting:

$$\psi'(n) = \begin{cases} \psi(n) & \text{if } \varphi(n) \in \psi(n) \\ \{\varphi(n)\} & \text{otherwise.} \end{cases}$$

Trivially $\varphi(n) \in \psi'(n)$ for all n and so, by the *Internal Set Axiom*, $\lim_{n\uparrow\alpha} \varphi(n) \in \lim_{n\uparrow\alpha} \psi'(n)$. By the hypothesis, we know that $\psi'(n) = \psi(n)$ a.e., so $\lim_{n\uparrow\alpha} \psi'(n) = \lim_{n\uparrow\alpha} \psi(n)$ by (1), and the conclusion follows. □

PROOF OF THEOREM 4.6. Let $\vartheta(n) = \{\varphi_1(n), \ldots, \varphi_k(n)\}$. By the *Internal Set Axiom*, the Alpha-limits $\lim_\alpha \varphi_1, \ldots, \lim_\alpha \varphi_k \in \lim_\alpha \vartheta$. Conversely, let $z \in \lim_\alpha \vartheta$; then $z = \lim_\alpha \zeta$ for some sequence ζ such that $\zeta(n) \in \vartheta(n)$ for all n. Now notice that $\mathbb{N} = \bigcup_{i=1}^k A_i$ where $A_i = \{n \mid \zeta(n) = \varphi_i(n)\}$. So, at least one A_i is qualified, and hence $z = \lim_\alpha \zeta = \lim_\alpha \varphi_i$, by (1) of the previous proposition. □

PROOF OF THEOREM 4.8. Let the sequence $\zeta(n)$ be given, and denote by

$$A = \{n \mid \zeta(n) \in \varphi(n)\}; \ B = \{n \mid \zeta(n) \in \psi(n)\}.$$

Recall the following properties of qualified sets (see Theorem 2.22):

- $A \cup B$ is qualified $\Leftrightarrow$ A is qualified or B is qualified;
- $A \cap B$ is qualified $\Leftrightarrow$ A is qualified and B is qualified;
- $A \setminus B$ is qualified $\Leftrightarrow$ A is qualified and B is *not* qualified.

As straight consequences, by Theorem 4.4 one obtains the following:

- If $\vartheta(n) = \varphi(n) \cup \psi(n)$:
  $\lim_\alpha \zeta \in \lim_\alpha \vartheta \Leftrightarrow \zeta(n) \in \vartheta(n)$ a.e. $\Leftrightarrow \zeta(n) \in \varphi(n)$ a.e. or $\zeta(n) \in \psi(n)$ a.e. $\Leftrightarrow \lim_\alpha \zeta \in \lim_\alpha \varphi$ or $\lim_\alpha \zeta \in \lim_\alpha \psi$.
- If $\vartheta(n) = \varphi(n) \cap \psi(n)$:
  $\lim_\alpha \zeta \in \lim_\alpha \vartheta \Leftrightarrow \zeta(n) \in \vartheta(n)$ a.e. $\Leftrightarrow \zeta(n) \in \varphi(n)$ a.e. and $\zeta(n) \in \psi(n)$ a.e. $\Leftrightarrow \lim_\alpha \zeta \in \lim_\alpha \varphi$ and $\lim_\alpha \zeta \in \lim_\alpha \psi$.
- If $\vartheta(n) = \varphi(n) \setminus \psi(n)$:
  $\lim_\alpha \zeta \in \lim_\alpha \vartheta \Leftrightarrow \zeta(n) \in \vartheta(n)$ a.e. $\Leftrightarrow \zeta(n) \in \varphi(n)$ a.e. and $\zeta(n) \notin \psi(n)$ a.e. $\Leftrightarrow \lim_\alpha \zeta \in \lim_\alpha \varphi$ and $\lim_\alpha \zeta \notin \lim_\alpha \psi$.

Thus properties (2), (3) and (4) are proved. As for (1), just notice that $\lim_\alpha \varphi \subseteq \lim_\alpha \psi \Leftrightarrow \lim_\alpha \psi = \lim_\alpha \varphi \cup \lim_\alpha \psi \Leftrightarrow \psi(n) = \varphi(n) \cup \psi(n)$ a.e. $\Leftrightarrow \varphi(n) \subseteq \psi(n)$ a.e.. □

We are now ready to show that sequences that are different *a.e.* must have different Alpha-limits.

PROPOSITION 4.15. *Let* $\varphi(n), \psi(n)$ *be arbitrary sequences. Then* $\lim_{n\uparrow\alpha} \varphi(n) = \lim_{n\uparrow\alpha} \psi(n) \implies \varphi(n) = \psi(n)$ a.e..

PROOF. Consider the following sequences of nonempty sets:

$$\sigma(n) = \{\varphi(n)\} \quad \text{and} \quad \tau(n) = \{\varphi(n), \psi(n)\}.$$

By Theorem 4.6 that we just proved above, we have that $\lim_\alpha \sigma = \{\lim_\alpha \varphi\} = \{\lim_\alpha \varphi, \lim_\alpha \psi\} = \lim_\alpha \tau$. Then, by the *Composition Axiom*, $\lim_\alpha(\text{Card} \circ \sigma) = \lim_\alpha(\text{Card} \circ \tau)$, where Card is the cardinality function. Since the last two sequences take real values, by Theorem 2.23 we can conclude that .

$$(\text{Card} \circ \sigma)(n) = |\{\varphi(n)\}| = |\{\varphi(n), \psi(n)\}| = (\text{Card} \circ \tau)(n) \ a.e.,$$

and hence $\varphi(n) = \psi(n)$ *a.e.*, as desired. $\square$

In order to show that the same result also holds for the membership relation, we need a useful "almost everywhere" reformulation of the *Internal Set Axiom*.

PROPOSITION 4.16 (Internal Set Axiom – *a.e.*-version).
If $\psi(n)$ is a sequence of nonempty sets a.e.*, then*

$$\lim_{n\uparrow\alpha} \psi(n) = \left\{ \lim_{n\uparrow\alpha} \varphi(n) \ \middle| \ \varphi(n) \in \psi(n) \ \text{a.e.} \right\}.$$

PROOF. Fix $\psi'(n)$ a sequence of nonempty sets such that $\psi'(n) = \psi(n)$ *a.e.*. If $\varphi(n) \in \psi(n)$ *a.e.* then also $\varphi(n) \in \psi'(n)$ *a.e.*; indeed,

$$\{n \mid \varphi(n) \in \psi'(n)\} \supseteq \{n \mid \varphi(n) \in \psi(n)\} \cap \{n \mid \psi(n) = \psi'(n)\},$$

where both sets of the intersection are qualified by the hypotheses. By the *Internal Set Axiom* and by Proposition 4.14, we conclude that $\lim_\alpha \varphi \in \lim_\alpha \psi' = \lim_\alpha \psi$. Conversely, let $x \in \lim_\alpha \psi = \lim_\alpha \psi'$. Then by the *Internal Set Axiom*, $x = \lim_\alpha \varphi$ for a suitable sequence such that $\varphi(n) \in \psi'(n)$ for all n. Since $\psi(n)$ and $\psi'(n)$ agree *a.e.*, it follows that $\varphi(n) \in \psi(n)$ *a.e.*. $\square$

PROPOSITION 4.17. *Let $\varphi(n), \psi(n)$ be arbitrary sequences. Then* $\lim_{n\uparrow\alpha} \varphi(n) \in \lim_{n\uparrow\alpha} \psi(n) \implies \varphi(n) \in \psi(n)$ *a.e.*.

PROOF. By the hypothesis, $\lim_{n\uparrow\alpha} \psi(n)$ is a nonempty set. So, by Proposition 4.3, $\psi(n)$ is a nonempty set *a.e.*. By the *a.e.*-version of the *Internal Set Axiom*, we obtain the existence of a sequence $\varphi'(n)$ such that $\varphi'(n) \in \psi(n)$ *a.e.* and $\lim_{n\uparrow\alpha} \varphi'(n) = \lim_{n\uparrow\alpha} \varphi(n)$. But then, by the previous proposition, $\varphi'(n) = \varphi(n)$ *a.e.*, and hence $\varphi(n) \in \psi(n)$ *a.e.*. $\square$

Notice that the proof of Theorem 4.4 is now completed. Indeed, the left-to-right implications are Proposition 4.14; and the converse implications of (1) and (2) are given by Proposition 4.15 and Proposition 4.17, respectively.

PROOF OF THEOREM 4.9. (1) follows from Theorem 4.6; indeed:

$$
\begin{aligned}
\lim_{n\uparrow\alpha}(\varphi(n),\psi(n)) &= \lim_{n\uparrow\alpha}\left\{\{\varphi(n)\},\{\varphi(n),\psi(n)\}\right\} \\
&= \left\{\lim_{n\uparrow\alpha}\{\varphi(n)\},\lim_{n\uparrow\alpha}\{\varphi(n),\psi(n)\}\right\} \\
&= \left\{\left\{\lim_{\alpha}\varphi\right\},\left\{\lim_{\alpha}\varphi,\lim_{\alpha}\psi\right\}\right\} = (\lim_{\alpha}\varphi,\lim_{\alpha}\psi).
\end{aligned}
$$

(2). Pick two sequences φ,ψ such that $\vartheta(n) = (\varphi(n),\psi(n))$ whenever $\vartheta(n)$ is an ordered pair. Then by the previous point we have that $\lim_{n\uparrow\alpha}\vartheta(n) = \lim_{n\uparrow\alpha}(\varphi(n),\psi(n)) = (\lim_{\alpha}\varphi,\lim_{\alpha}\psi)$ is an ordered pair. Conversely, assume that $\lim_{\alpha}\vartheta = (\xi,\eta) = \{\{\xi\},\{\xi,\eta\}\}$ is an ordered pair. Then, by the *Internal Set Axiom*, both ξ and η must be Alpha-limits because they are elements of elements of an Alpha-limit. If $\xi = \lim_{n\uparrow\alpha}\varphi(n)$ and $\eta = \lim_{n\uparrow\alpha}\psi(n)$ then $\lim_{\alpha}\vartheta = (\lim_{\alpha}\varphi,\lim_{\alpha}\psi) = \lim_{n\uparrow\alpha}(\varphi(n),\psi(n))$, and by *backward transfer* we conclude that $\vartheta(n) = (\varphi(n),\psi(n))$ *a.e.*.

(3) directly follows from (1) and (2).

The general case for k-tuples and for Cartesian products with k-many factors is then easily proved by induction. $\square$

PROOF OF THEOREM 4.10. Let us denote by $A = \lim_{n\uparrow\alpha}A_n$ and $B = \lim_{n\uparrow\alpha}B_n$. Without loss of generality, we can assume that $\mathcal{R}_n \neq \emptyset$, $A_n \neq \emptyset$ and $B_n \neq \emptyset$ for all n.

$\Longrightarrow$ Given an arbitrary sequence ϑ with $\vartheta(n) \in \mathcal{R}_n$ for all n, pick sequences φ and ψ such that $\vartheta(n) = (\varphi(n),\psi(n)) \in A_n \times B_n$ whenever $\mathcal{R}_n$ is a binary relation. By *transfer*, $\lim_{\alpha}\vartheta = (\lim_{\alpha}\varphi,\lim_{\alpha}\psi) \in A \times B$. This shows that $\mathcal{R}$ is a binary relation with $\mathrm{dom}(\mathcal{R}) \subseteq A$ and $\mathrm{ran}(\mathcal{R}) \subseteq B$. Now, given $\varphi(n) \in A_n$ for all n, pick a sequence ψ such that $(\varphi(n),\psi(n)) \in \mathcal{R}_n$ whenever $\mathcal{R}_n$ is a binary relation. Then

$$(\lim_{\alpha}\varphi,\lim_{\alpha}\psi) = \lim_{n\uparrow\alpha}(\varphi(n),\psi(n)) \in \mathcal{R}.$$

In particular, $\lim_{\alpha}\varphi \in \mathrm{dom}(\mathcal{R})$ and we can conclude that also the reverse inclusion $A \subseteq \mathrm{dom}(\mathcal{R})$ holds. The analogous result $B \subseteq \mathrm{ran}(\mathcal{R})$ is proved in the same manner.

$\Longleftarrow$ Let $X = \{n \mid \mathcal{R}_n \text{ is a binary relation}\}$. For every $n \notin X$, pick an element $\vartheta(n) \in \mathcal{R}_n$ which is *not* an ordered pair. If by contradiction

X was not qualified, then by (2) of Theorem 4.9, $\lim_\alpha \vartheta \in \mathcal{R}$ is *not* an ordered pair, and so $\mathcal{R}$ would not be a binary relation. This proves that X is qualified, that is, $\mathcal{R}_n$ is a binary relation *a.e.*.

Now fix sequences $\{A'_n\}, \{B'_n\}$ such that $A'_n = \mathrm{dom}(\mathcal{R}_n)$ and $B'_n = \mathrm{ran}(\mathcal{R}_n)$ whenever $n \in X$. By the above implication $\Longrightarrow$ we obtain that $\mathrm{dom}(\mathcal{R}) = \lim_{n \uparrow \alpha} A'_n$ and $\mathrm{ran}(\mathcal{R}) = \lim_{n \uparrow \alpha} B'_n$. By the hypotheses, $\mathrm{dom}(\mathcal{R}) = \lim_{n \uparrow \alpha} A_n$ and $\mathrm{ran}(\mathcal{R}) = \lim_{n \uparrow \alpha} B'_n$ and so, by *backward transfer*, $A'_n = A_n$ and $B'_n = B_n$ *a.e.*. We conclude that $\mathrm{dom}(\mathcal{R}_n) = A_n$ and $\mathrm{ran}(\mathcal{R}_n) = B_n$ *a.e.*, as desired. $\qquad \square$

PROOF OF THEOREM 4.11. By the previous theorem we know already that F is a binary relation with $\mathrm{dom}(F) = \lim_{n \uparrow \alpha} A_n$ and $\mathrm{ran}(F) = \lim_{n \uparrow \alpha} B_n$. In order to prove the "functionality" property, let us assume that $(\xi, \eta), (\xi, \eta') \in F$; we have to show that $\eta = \eta'$. Let $\vartheta(n) = (\varphi(n), \psi(n)) \in f_n$ and $\vartheta'(n) = (\varphi'(n), \psi'(n)) \in f_n$ be sequences such that $(\xi, \eta) = \lim_\alpha \vartheta$ and $(\xi, \eta') = \lim_\alpha \vartheta'$. Then $\xi = \lim_\alpha \varphi = \lim_\alpha \varphi'$, and so $\varphi(n) = \varphi'(n)$ *a.e.*. By the "functionality" property of the f_n, it follows that $\psi(n) = \psi'(n)$ *a.e.*, and hence $\eta = \lim_\alpha \psi = \lim_\alpha \psi' = \eta'$, as desired.

The last statement directly follows from the fact that Alpha-limits commute with ordered pairs. Indeed, for every n let $\psi(n) = f_n(\varphi(n))$, that is, $(\varphi(n), \psi(n)) \in f_n$. Then $\lim_{n \uparrow \alpha}(\varphi(n), \psi(n)) = (\lim_\alpha \varphi, \lim_\alpha \psi) \in F$, that is, $F(\lim_{n \uparrow \alpha} \varphi(n)) = \lim_{n \uparrow \alpha} \psi(n) = \lim_{n \uparrow \alpha} f_n(\varphi(n))$. $\qquad \square$

PROOF OF PROPOSITION 4.12. Given functions $f_n : A_n \to B_n$, let us denote $F = \lim_{n \uparrow \alpha} f_n$, $A = \lim_{n \uparrow \alpha} A_n$, and $B = \lim_{n \uparrow \alpha} B_n$.

F is *not* 1-1 $\Leftrightarrow$ there exist $\xi \neq \zeta$ in A with $F(\xi) \neq F(\zeta)$ $\Leftrightarrow$ there exist sequences φ, ψ such that $\varphi(n), \psi(n) \in A_n$ for all n and $\lim_{n \uparrow \alpha} \varphi(n) \neq \lim_{n \uparrow \alpha} \psi(n)$ and $F(\lim_{n \uparrow \alpha} \varphi(n)) \neq F(\lim_{n \uparrow \alpha} \psi(n))$ $\Leftrightarrow$ (by Theorem 4.11) there exist sequences φ, ψ such that $\varphi(n), \psi(n) \in A_n$ for all n and $\varphi(n) \neq \psi(n)$ *a.e.* and $f_n(\varphi(n)) \neq f_n(\psi(n))$ *a.e.* $\Leftrightarrow$ f_n is *not* 1-1 *a.e.*.

By Theorem 4.10, F is onto $\Leftrightarrow \mathrm{ran}(F) = \lim_{n \uparrow \alpha} \mathrm{ran}(f_n) = B \Leftrightarrow \mathrm{ran}(f_n) = B_n$ *a.e.* $\Leftrightarrow f_n$ is onto *a.e.*. $\qquad \square$

4. Hyper-images of sets

In Alpha-Calculus, we defined hyper-extensions A^* of sets A of real numbers, and hyper-extension f^* of real functions f. In the stronger axiomatic setting given by the Alpha-Theory, we have a more general notion of "idealisation" for *arbitrary* sets and functions, namely the *hyper-image*.

DEFINITION 4.18. *The* hyper-image *(or* star-transform*)* A^* *of an object A is the Alpha-limit of the constant sequence c_A with value A.*

$$A^* = \lim_{n\uparrow\alpha} c_A(n) = \left\{ \lim_{n\uparrow\alpha} \varphi(n) \,\middle|\, \varphi(n) \in A \text{ a.e.} \right\}.$$

By the *Real Number Axiom* and the *Internal Set Axiom*:

- If r is a real number, then its hyper-image $r^* = r$;
- The hyper-image of the empty set $\emptyset^* = \emptyset$.

Moreover, by Proposition 4.3,

- a is an atom if and only if a^* is an atom;
- A is a nonempty set if and only if A^* is a nonempty set.

Notice that for sets of real numbers the above definition of hyper-image coincides with that of hyper-extension as given in Alpha-Calculus (see Definition 2.12).

As straight consequences of the *transfer* properties seen in Section 2, hyper-images are coherent with respect to equality, membership and the basic set operations. In particular, what proved in Proposition 2.15 for hyper-extensions of sets of real numbers generalizes to arbitrary sets. All proofs are straightforward and are left as exercises.

THEOREM 4.19.
  (1) $A^* = B^* \Leftrightarrow A = B$;
  (2) $A^* \in B^* \Leftrightarrow A \in B$;
  (3) $A \subseteq B \Leftrightarrow A^* \subseteq B^*$;
  (4) $(A \cup B)^* = A^* \cap B^*$;
  (5) $(A \cup B)^* = A^* \cup B^*$;
  (6) $(A \setminus B)^* = A^* \setminus B^*$;
  (7) $\{a_1, \ldots, a_k\}^* = \{a_1^*, \ldots, a_k^*\}$;
  (8) $(a, b)^* = (a^*, b^*)$;
  (9) A *is an ordered pair* $\Longleftrightarrow A^*$ *is an ordered pair*;
  (10) $(A \times B)^* = A^* \times B^*$;
  (11) $\mathcal{R}$ *is a binary relation* $\Longleftrightarrow \mathcal{R}^*$ *is a binary relation*;
  (12) $\operatorname{dom}(\mathcal{R}^*) = \operatorname{dom}(\mathcal{R}^*)$ *and* $\operatorname{ran}(\mathcal{R}^*) = \operatorname{ran}(\mathcal{R}^*)$.

REMARK 4.20. *Any hyper-image A^* includes a canonical copy of the starting set, given by the hyper-images of its elements:*

$$\{a^* \mid a \in A\} \subseteq A^*.$$

Clearly, when $A \subseteq \mathbb{R}$ is a set of real numbers, then $A \subseteq A^*$ because $a^* = a$ for all $a \in A$. So, in this case, the hyper-image is actually an extension. The same property holds for subsets $A \subseteq \mathbb{R}^k$. However, notice that in general A^* is not a superset of A, because there may be elements $a \in A$ with $a \neq a^*$.

EXAMPLE 4.21. $\boldsymbol{\alpha}^* \neq \boldsymbol{\alpha}$.

PROOF. Since $\boldsymbol{\alpha} \notin \mathbb{N}$, trivially $\boldsymbol{\alpha} \neq n$ for all n, and so

$$\boldsymbol{\alpha}^* = \lim_{n\uparrow\alpha} \boldsymbol{\alpha} \neq \lim_{n\uparrow\alpha} n = \boldsymbol{\alpha}.$$

$\square$

DEFINITION 4.22. *The canonical image of a set A is defined as*

$$A^\sigma = \{a^* \mid a \in A\}.[6]$$

EXERCISE 4.23.

(1) $A \cap B = \emptyset \iff A^\sigma \cap B^* = \emptyset$.
(2) *If $A \subseteq B$ then $A^* \cap B^\sigma = A^\sigma$.*

PROPOSITION 4.24. *For every set A, $A^\sigma = A^*$ if and only if A is finite.*

PROOF. If $A = \{a_1, \ldots, a_k\}$ is a finite set, we have already seen that $A^* = \{a_1^*, \ldots, a_k^*\} = A^\sigma$. Conversely, let us assume that A is infinite, and pick a 1-1 sequence $\varphi : \mathbb{N} \to A$. Then $\lim_\alpha \varphi \in A^* \setminus A^\sigma$.

Indeed, for every $a \in A$, one has that $\varphi(n) \neq c_a(n)$ for all but at most one n, and so $\lim_\alpha \varphi \neq \lim_\alpha c_a = a^*$. $\square$

Let us now check that hyper-images also preserve *diagonals* of sets:

$$\mathrm{Diag}(A) = \{(a, a) \mid a \in A\}.$$

---

[6] Notation A^σ is borrowed from nonstandard analysis, where the Greek letter σ (corresponding to "s") stands for "standard".

PROPOSITION 4.25. *For every set A,*

$$Diag(A)^* = \{(x, x) \mid x \in A\}^* = \{(\xi, \xi) \mid \xi \in A^*\} = Diag(A^*).$$

PROOF. We can assume that $A \neq \emptyset$, otherwise trivially $\mathrm{Diag}(A) = \mathrm{Diag}(A^*) = \emptyset$. Given a sequence $F : \mathbb{N} \to \mathrm{Diag}(A)$, for every n let $\varphi(n) \in A$ be such that $F(n) = (\varphi(n), \varphi(n))$. Then $\lim_\alpha F = (\lim_\alpha \varphi, \lim_\alpha \varphi) \in \mathrm{Diag}(A^*)$ because $\lim_\alpha \varphi \in A^*$. Conversely, let $\xi \in A^*$, and pick a sequence $\varphi(n) \in A$ for all n such that $\lim_\alpha \varphi = \xi$. If $F(n) = (\varphi(n), \varphi(n))$ then $(\xi, \xi) = \lim_\alpha F \in \mathrm{Diag}(A)^*$.   □

EXERCISE 4.26. *For a given set A, denote by*

$$[A]^k = \{X \subseteq A \mid |X| = k\}$$

the family of all k-subsets of A. Then $([A]^k)^* = [A^*]^k$.

REMARK 4.27. *Our notion of hyper-image can be iterated. E.g., one can consider "second-level" hyper-images (or "double-stars") $\mathbb{N}^{**}$ and $\mathbb{R}^{**}$ of the natural and real numbers, respectively; or even "third level" hyper-images, and so forth. In §7.2 we will see an application of double-stars to give an alternative proof of Ramsey's Theorem.*

PROPOSITION 4.28. *The hyper-hypernatural numbers $\mathbb{N}^{**}$ are an end extension of the hypernatural numbers $\mathbb{N}^*$, that is,*

- *$\mathbb{N}^* \subset \mathbb{N}^{**}$, and*
- *$\xi < \zeta$ for every $\xi \in \mathbb{N}^*$ and for every $\zeta \in \mathbb{N}^{**} \setminus \mathbb{N}^*$.*

PROOF. Since the star-map preserves inclusions (see Theorem 4.19 (3)), from $\mathbb{N} \subset \mathbb{N}^*$ it directly follows that $\mathbb{N}^* \subset \mathbb{N}^{**}$. Given $\xi \in \mathbb{N}^*$ and $\zeta \in \mathbb{N}^{**} \setminus \mathbb{N}^* = (\mathbb{N}^* \setminus \mathbb{N})^*$, pick sequences $\varphi : \mathbb{N} \to \mathbb{N}$ and $\psi : \mathbb{N} \to \mathbb{N}^* \setminus \mathbb{N}$ such that $\lim_\alpha \varphi = \xi$ and $\lim_\alpha \psi = \zeta$. By Proposition 2.36 we know that $\varphi(n) < \psi(n)$ for every n and so, by passing to the Alpha-limit, we obtain the desired inequality $\xi < \zeta$.   □

5. Hyper-images of functions

To us, functions are sets (precisely, "functional" relations). Accordingly, the hyper-image f^* of a function f is the hyper-image of the set f, as defined in the previous section.

Recall that, in Alpha-Calculus, we defined the hyper-extensions of real functions (see Definition 2.26) by using the coherence property of Alpha-limits with respect to compositions (see Proposition 2.9), namely

$$\lim_{n\uparrow\alpha} \varphi(n) = \lim_{n\uparrow\alpha} \psi(n) \implies \lim_{\alpha} f(\varphi(n)) = \lim_{n\uparrow\alpha} f(\psi(n)).$$

Also in the general framework of Alpha-Theory, the notion of function is preserved under hyper-images, and the *Composition Axiom* (AT5) guarantees that the same coherence property as above holds for *all* functions.

THEOREM 4.29. *f is a function if and only if f^* is a function, and* $\mathrm{dom}(f^*) = \mathrm{dom}(f)^*$ *and* $\mathrm{ran}(f^*) = \mathrm{ran}(f)^*$; *so, $f : A \to B$ if and only if $f^* : A^* \to B^*$.*

Moreover, for every sequence $\varphi : \mathbb{N} \to A$,

$$f^* \left(\lim_{n\uparrow\alpha} \varphi(n) \right) = \lim_{n\uparrow\alpha} f(\varphi(n)).$$

PROOF. All properties are corollaries of Theorem 4.11 where one considers constant sequences. Indeed, notice that "$f^* : A^* \to B^*$" is the same as "$\lim_\alpha c_f : \lim_\alpha c_A \to \lim_\alpha c_B$" where c_f, c_A, and c_B are the constant sequences with value f, A, and B, respectively. Moreover, $f^*(\lim_{n\uparrow\alpha} \varphi(n)) = (\lim_{n\uparrow\alpha} c_f)(\lim_{n\uparrow\alpha} \varphi(n)) = \lim_{n\uparrow\alpha}(f(\varphi(n))$. □

PROPOSITION 4.30. *For every $a \in \mathrm{dom}(f)$, $f^*(a^*) = f(a)^*$.*

PROOF. We have the following chain of equalities:

$$f^*(a^*) = f^* \left(\lim_{n\uparrow\alpha} c_a(n) \right) = \lim_{n\uparrow\alpha} f(c_a(n)) = \lim_{n\uparrow\alpha} c_{f(a)}(n) = f(a)^*.$$

□

In consequence, one can see f^* as an actual extension of f *provided* the domain $\mathrm{dom}(f) = A$ be identified with its *canonical image* $A^\sigma = \{a^* \mid a \in A\} \subseteq A^*$.

By exactly the same arguments used in Alpha-Calculus for hyper-extensions of real functions, one proves the following list of general properties for hyper-images of functions (see Propositions 2.27 and 2.29).

PROPOSITION 4.31.

(1) *If $\imath_A : A \to A$ is the identity map on A, then $^*\imath_A : A^* \to A^*$ is the identity map on A^*;*

(2) $^*(f \circ g) = f^* \circ g^*$ *(provided the composition $f \circ g$ is defined);*

(3) *A function $f : A \to B$ is 1-1 if and only if its hyper-image $f^* : A^* \to B^*$ is 1-1;*

(4) *A function $f : A \to B$ is onto if and only if its hyper-image $f^* : A^* \to B^*$ is onto;*

(5) *If the function f is defined on X, then*

$$^*\{f(x) \mid x \in X\} = \{f^*(\xi) \mid \xi \in X^*\}.$$

In particular, $^*\mathrm{ran}(f) = \mathrm{ran}(f^*);$

(6) *If f and g are functions defined on X, then*

$$^*\{x \in X \mid f(x) = g(x)\} = \{\xi \in X^* \mid f^*(\xi) = g^*(\xi)\}.$$

The characterisations of "infinitely often" and "eventually" that were given for real sequences (see Proposition 3.19) can be extended to arbitrary sequences. The proofs are left as exercises.

PROPOSITION 4.32 (*Overspill*).
Let φ, ψ be two sequences, and let $\bowtie$ be any of the following relations: $=, \neq, \leq, <$. Then

(1) *$\varphi(n) \bowtie \psi(n)$ infinitely often (that is, for infinitely many $n \in \mathbb{N}$) $\iff \varphi^*(\nu) \bowtie \psi^*(\nu)$ for some infinite $\nu \in \mathbb{N}^* \setminus \mathbb{N}$;*

(2) *$\varphi(n) \bowtie \psi(n)$ eventually (that is, for all sufficiently large $n \in \mathbb{N}$) $\iff \varphi^*(\nu) \bowtie \psi^*(\nu)$ for all infinite $\nu \in \mathbb{N}^* \setminus \mathbb{N}$.*

EXERCISE 4.33. *Let f and g be functions defined on X. Prove that*

$$^*\{x \in X \mid f(x) \in g(x)\} = \{\xi \in X^* \mid f^*(\xi) \in g^*(\xi)\}.$$

SOLUTION. Denote by $\Gamma = \{x \in X \mid f(x) \in g(x)\}$. If $\varphi(n)$ is an arbitrary sequence taking values in Γ, then $f(\varphi(n)) \in g(\varphi(n))$ for all n, and so by the *Internal Set Axiom*:

$$f^*(\lim_{n \uparrow \alpha} \varphi(n)) = \lim_{n \uparrow \alpha} f(\varphi(n)) \in \lim_{n \uparrow \alpha} g(\varphi(n)) = g^*(\lim_{n \uparrow \alpha} \varphi(n)).$$

This proves the inclusion $^*\Gamma \subseteq \{\xi \in X^* \mid f^*(\xi) \in g^*(\xi)\}$.

For the other direction, pick an arbitrary sequence $\varphi : \mathbb{N} \to X$ and let $\xi = \lim_{n \uparrow \alpha} \varphi(n)$. If $f^*(\xi) \in g^*(\xi)$, that is, if $\lim_{n \uparrow \alpha} f(\varphi(n)) \in \lim_{n \uparrow \alpha} g(\varphi(n))$, then by *backward transfer* we have that $f(\varphi(n)) \in g(\varphi(n))$ a.e.. But then $\varphi(n) \in \Gamma$ a.e., and we can conclude that $\xi = \lim_{n \uparrow \alpha} \varphi(n) \in {}^*\Gamma$. $\square$

EXERCISE 4.34. *Let f and g be functions defined on A and B, respectively. Then:*

$$^*\{(x,y) \in A \times B \mid f(x) = g(y)\} = \{(\xi, \eta) \in A^* \times B^* \mid f^*(\xi) = g^*(\eta)\}.$$

SOLUTION. Let $(\xi, \eta) \in {}^*\{(x,y) \in A \times B \mid f(x) = g(y)\}$. Then $(\xi, \eta) = \lim_\alpha \Psi$ for some sequence

$$\Psi : \mathbb{N} \to \{(x,y) \in A \times B \mid f(x) = g(y)\}.$$

Now let $\varphi : \mathbb{N} \to A$ and $\psi : \mathbb{N} \to B$ be such that $\Psi(n) = (\varphi(n), \psi(n))$ for all n. Then $\lim_\alpha \Psi = (\lim_\alpha \varphi, \lim_\alpha \psi)$. As $f(\varphi(n)) = g(\psi(n))$ for all n, we have that

$$f^*(\xi) = f^*(\lim_\alpha \varphi) = \lim_\alpha(f \circ \varphi) = \lim_\alpha(g \circ \psi) = g^*(\lim_\alpha \psi) = g^*(\eta),$$

and this concludes the proof of one inclusion.

Conversely, let $\varphi : \mathbb{N} \to A$ and $\psi : \mathbb{N} \to B$ be such that $f^*(\lim_\alpha \varphi) = g^*(\lim_\alpha \psi)$, that is, $\lim_\alpha(f \circ \varphi) = \lim_\alpha(g \circ \psi)$. Then $f(\varphi(n)) = g(\psi(n))$ a.e., that is, $(\varphi(n), \psi(n)) \in \{(x,y) \in A \times B \mid f(x) = g(y)\}$ a.e.. We can finally reach the desired conclusion

$$(\lim_\alpha \varphi, \lim_\alpha \psi) = \lim_{n\uparrow\alpha}(\varphi(n), \psi(n)) \in {}^*\{(x,y) \in A \times A' \mid g(x) = h(y)\}.$$

$\square$

6. Functions of several variables

We already saw that Alpha-limits commute with ordered tuples (see Theorem 4.9):

$$\lim_{n\uparrow\alpha} (\varphi_1(n), \ldots, \varphi(k)) = \left(\lim_{n\uparrow\alpha} \varphi_1(n), \ldots, \lim_{n\uparrow\alpha} \varphi_k(n) \right).$$

In consequence, one can develop calculus in several variables along the same lines of Alpha-Calculus as treated in Chapter 3.

In this short section we present a few basic results of calculus on the spaces $\mathbb{R}^k$. We start with the fundamental definitions.

DEFINITION 4.35. *The* hyper-extension *of a set $\Omega \subseteq \mathbb{R}^k$ is:*

$$\Omega^* = \left\{ \left(\lim_\alpha \varphi_1, \ldots, \lim_\alpha \varphi_k \right) \in \mathbb{R}^{*k} \mid (\varphi_1(n), \ldots, \varphi_k(n)) \in \Omega \text{ for all } n \right\}.$$

DEFINITION 4.36. *Let $\Omega \subseteq \mathbb{R}^k$. If $f : \Omega \to \mathbb{R}$ is a function in k variables, then its* hyper-extension $f^* : \Omega^* \to \mathbb{R}^*$ *is defined by setting*

$$f^* : \left(\lim_{\alpha} \varphi_1, \ldots, \lim_{\alpha} \varphi_k \right) \longmapsto \lim_{n \uparrow \alpha} f(\varphi_1(n), \ldots, \varphi_k(n))$$

for all sequences $\varphi_i(n)$ such that $(\varphi_1(n), \ldots, \varphi_k(n)) \in \Omega$ for all n.

It can be verified in a straightforward manner that the above definitions are consistent with the general definitions of hyper-image of a set and hyper-image of a function.

A *multivariable function* $F : \Omega \to \mathbb{R}^h$ where $\Omega \subseteq \mathbb{R}^k$, can be seen as a h-tuple of functions $F = (f_1, \ldots, f_h)$ where each $f_i : \Omega \to \mathbb{R}$ and $F(r_1, \ldots, r_k) = (f_1(r_1, \ldots, r_k), \ldots, f_h(r_1, \ldots, r_k)) \in \mathbb{R}^h$ for all $(r_1, \ldots, r_k) \in \Omega$.

We now extend the previous definition to that more general setting as follows.

DEFINITION 4.37. *Let $F = (f_1, \ldots, f_h) : \Omega \to \mathbb{R}^h$ be a multivariable function where $\Omega \subseteq \mathbb{R}^k$. The* hyper-extension $F^* : \Omega^* \to \mathbb{R}^{*h}$ *is defined by setting $F^* = (f_1^*, \ldots, f_h^*)$, that is, for every $(\xi_1, \ldots, \xi_k) \in \Omega^*$:*

$$F^*(\xi_1, \ldots, \xi_k) = (f_1^*(\xi_1, \ldots, \xi_k), \ldots, f_h^*(\xi_1, \ldots, \xi_k)).$$

Again, it is easily seen that this definition is coherent with the general definition of hyper-images.

Properties of Proposition 4.31 also hold in the case of multivariable functions (the proofs are straightforward from the definitions).

PROPOSITION 4.38. *Let F and G be multivariable functions.*

(1) $(F \circ G)^* = F^* \circ G^*$ *(provided the composition $F \circ G$ is defined);*

(2) *If F is defined on $X \subseteq \mathbb{R}^k$, then*

$$\{F(\mathbf{x}) \mid \mathbf{x} \in X\}^* = \{F^*(\boldsymbol{\xi}) \mid \boldsymbol{\xi} \in X^*\}.$$

In particular, $\mathrm{ran}(F)^* = \mathrm{ran}(F^*)$;

(3) *If both F and G are defined on $X \subseteq \mathbb{R}^k$, then*

$$\{\mathbf{x} \in X \mid F(\mathbf{x}) = G(\mathbf{x})\}^* = \{\boldsymbol{\xi} \in X^* \mid F^*(\boldsymbol{\xi}) = G^*(\boldsymbol{\xi})\}.$$

7. Hyperfinite sets

In this section we introduce an important class of objects of Alpha-Theory, namely the *hyperfinite sets*. Roughly speaking, hyperfinite sets retain all nice "elementary properties" of finite sets, while they may contain infinitely many elements. In consequence, they will provide a useful bridge between the discrete and the continuum.

DEFINITION 4.39. *We call* hyperfinite set *the Alpha-limit of a sequence of finite sets.*

If we denote by $\mathfrak{Fin}(A)$ the set of all finite subsets of A then, by the *Internal Set Axiom*, it is readily verified that

$$\{X \subseteq A^* \mid X \text{ hyperfinite}\} = \mathfrak{Fin}(A)^*.$$

EXAMPLE 4.40. *The set* $\{1, \ldots, \boldsymbol{\alpha}\}$ *is hyperfinite, but* not *finite. Recall in fact that* $\{1, \ldots, \boldsymbol{\alpha}\} = \lim_{n\uparrow\alpha}\{1, \ldots, n\}$ *(see Proposition 4.5).*

Let us start with the simplest examples of hyperfinite sets.

PROPOSITION 4.41.

(1) *Every finite set of Alpha-limits is hyperfinite;*

(2) *A hyper-image* A^* *is hyperfinite if and only if* A *is finite;*

(3) *Every bounded interval of hyperintegers is hyperfinite.*

PROOF. (1). Let $A = \{\lim_\alpha \varphi_1, \ldots, \lim_\alpha \varphi_k\}$. Then $A = \lim_\alpha \vartheta$ where $\vartheta(n) = \{\varphi_1(n), \ldots, \varphi_k(n)\}$ is a sequence of finite sets.

(2). If A is finite, then trivially A^* is hyperfinite, because it is the Alpha-limit of the constant sequence with value the finite set A. Conversely, assume that A is infinite. Then A^* cannot be the Alpha-limit of any sequence $\varphi(n)$ of finite sets because $\varphi(n) \neq A$ for all n.

(3). Given hyperintegers $\nu < \mu$, pick sequences $\varphi, \psi : \mathbb{N} \to \mathbb{Z}$ such that $\nu = \lim_{n\uparrow\alpha}\varphi(n)$ and $\mu = \lim_{n\uparrow\alpha}\psi(n)$. Then it is easily verified that $[\nu, \mu]_{\mathbb{Z}^*} = \lim_{n\uparrow\alpha}[\varphi(n), \psi(n)]_{\mathbb{Z}}$ is the Alpha-limit of a sequence of finite sets. $\square$

DEFINITION 4.42. *The* hyperfinite cardinality $\|A\|$ *of a hyperfinite set* A *is the hypernatural number defined by putting:*

$$\|A\| = \lim_{n\uparrow\alpha} |A_n| \in \mathbb{N}^*$$

where $\langle A_n \mid n \in \mathbb{N} \rangle$ *is any sequence of finite sets with* $\lim_{n\uparrow\alpha} A_n = A$.

It is readily seen that the given definition is well-posed because it does not depend on the choice of the sequence $\langle A_n \mid n \in \mathbb{N} \rangle$. Moreover,

PROPOSITION 4.43.

(1) *If the hyperfinite set A is finite, then $\|A\| = |A|$;*
(2) *If $\nu < \mu$ are hyperintegers, then $\|[\nu, \mu]_{\mathbb{Z}^*}\| = \mu - \nu + 1$. In particular, initial segments $\|[1, \nu]_{\mathbb{N}^*}\| = \nu$ for every $\nu \in \mathbb{N}^*$.*

PROOF. (1). Since A is an Alpha-limit, all its elements are Alpha-limits. So, if $|A| = k$, then $A = \{\lim_\alpha \varphi_1, \ldots, \lim_\alpha \varphi_k\}$ for suitable pairwise distinct Alpha-limits $\lim_\alpha \varphi_i$. But then $\varphi_1(n), \ldots, \varphi_k(n)$ are pairwise distinct for almost all n, that is, $|\{\varphi_1(n), \ldots, \varphi_k(n)\}| = k$ for almost all n. We conclude that

$$\|A\| = \lim_{n\uparrow\alpha} |\{\varphi_1(n), \ldots, \varphi_k(n)\}| = k.$$

(2). If $\nu = \lim_{n\uparrow\alpha} \varphi(n)$ and $\mu = \lim_{n\uparrow\alpha} \psi(n)$, then

$$\|[\nu, \mu]_{\mathbb{Z}^*}\| = \lim_{n\uparrow\alpha} |[\varphi(n), \psi(n)]_{\mathbb{Z}}| = \lim_{n\uparrow\alpha}(\psi(n) - \varphi(n) + 1) = \mu - \nu + 1.$$

$\square$

EXERCISE 4.44. *If $f : X \to Y$ is a function, and $\Omega \subseteq X^*$ is a hyperfinite set, then the image $f^*[\Omega] = \{f^*(\xi) \mid \xi \in \Omega\}$ is hyperfinite as well.*

The following is a typical example of a property that hyperfinite sets inherit from finite sets.

PROPOSITION 4.45. *Every nonempty hyperfinite subset of $\mathbb{R}^*$ has a least and a greatest element.*

PROOF. Given a nonempty hyperfinite set $A \subset \mathbb{R}^*$, take a sequence of finite nonempty $A_n \subset \mathbb{R}$ such that $A = \lim_{n\uparrow\alpha} A_n$, and let

$$\xi = \lim_{n\uparrow\alpha}(\min A_n) \quad \text{and} \quad \zeta = \lim_{n\uparrow\alpha}(\max A_n).$$

We claim that $\xi = \min A$ and $\zeta = \max A$. First of all, both $\xi, \zeta \in A$ because trivially $\min A_n, \max A_n \in A_n$ for all n. Now, an arbitrary element $a \in A$ is of the form $a = \lim_{n\uparrow\alpha} \varphi(n)$ for a suitable sequence such that $\varphi(n) \in A_n$ a.e.. But then $\min A_n \leq \varphi(n) \leq \max A_n$ a.e. and, by taking the Alpha-limits, we obtain $\xi \leq a \leq \zeta$, as desired. $\square$

We close this section with two results about the cardinalities of hyperfinite sets. In particular, we show that no hyperfinite set can be countably infinite.

PROPOSITION 4.46. *Let A be a hyperfinite set of hyperfinite cardinality $\|A\| = \nu \in \mathbb{N}^*$. Then there exists a bijection $f : A \to [1, \nu]_{\mathbb{N}^*}$.*

PROOF. Fix a sequence of finite sets $\langle A_n \mid n \in \mathbb{N} \rangle$ with $\lim_{n \uparrow \alpha} A_n = A$, and let $\varphi(n) = |A_n| \in \mathbb{N}$. For every n, fix a bijection $f_n : A_n \to [1, \varphi(n)]_{\mathbb{N}}$ and consider the Alpha-limit $f = \lim_{n \uparrow \alpha} f_n$. By Proposition 4.12, the function f is a bijection between $\lim_{n \uparrow \alpha} A_n = A$ and $\lim_{n \uparrow \alpha} [1, \varphi(n)]_{\mathbb{N}} = [1, \lim_{n \uparrow \alpha} \varphi(n)]_{\mathbb{N}} = [1, \nu]_{\mathbb{N}^*}$, as desired. $\square$

PROPOSITION 4.47. *If A is hyperfinite, then either A is finite or $|A| = \mathfrak{c}$ has the cardinality of the* continuum.

PROOF. By the previous proposition, A is in bijection with an initial segment $[1, \nu]_{\mathbb{N}^*}$ of hypernatural numbers. If $\nu \in \mathbb{N}$ is finite, then trivially A is finite. We now have to show that $|[1, \nu]_{\mathbb{N}^*}| = \mathfrak{c}$ whenever ν is infinite.

On the one hand, trivially $|[1, \nu]_{\mathbb{N}^*}| \leq |\mathbb{N}^*| = \mathfrak{c}$ (see Proposition 2.38). Conversely, fix a sequence $\varphi : \mathbb{N} \to \mathbb{N}$ such that $\lim_{n \uparrow \alpha} \varphi(n) = \nu$. For every real number r with $0 \leq r \leq 1$, and for every $n \in \mathbb{N}$, pick a natural number $k_{r,n} \leq \varphi(n)$ such that $|k_{r,n}/\varphi(n) - r| \leq 1/\varphi(n)$. Then $\mu_r = \lim_{n \uparrow \alpha} k_{r,n} \leq \nu$ and $\mu_r/\nu \sim r$, because $|\mu_r/\nu - r| \leq 1/\nu \sim 0$. The function $f : [0, 1]_{\mathbb{R}} \to [1, \nu]_{\mathbb{N}^*}$ given by the correspondence $r \mapsto \mu_r$ is 1-1, because $\mu_r = \mu_s \Rightarrow r \sim s \Rightarrow r = s$ (two real numbers which are infinitely close are necessarily equal). Thus, we obtain the desired inequality $\mathfrak{c} = |[0, 1]_{\mathbb{R}}| \leq |[1, \nu]_{\mathbb{N}^*}|$. $\square$

In the sequel, whenever confusion is unlikely, we will simply write $|A|$ instead of $\|A\|$ to denote the internal cardinality.

8. Hyperfinite sums

Similarly to finite sums of real numbers, one can consider the sum of a *hyperfinite* set of hyperreal numbers.

DEFINITION 4.48. *For every hyperfinite subset $\Omega \subset \mathbb{R}^*$, the* hyperfinite sum *of the elements of Ω is defined by putting:*

$$\sum_{\xi \in \Omega} \xi = \lim_{n \uparrow \alpha} \left(\sum_{x \in \Omega_n} x \right)$$

where $\langle \Omega_n \mid n \in \mathbb{N} \rangle$ is any sequence of finite subsets of $\mathbb{R}$ whose Alpha-limit $\lim_{n \uparrow \alpha} \Omega_n = \Omega$.

It is easily verified that the above definition is well-posed.

Hyperfinite sums to be often found in the practice are those where the summands are values of a hyper-extension. In this case, for every real-valued function f defined on a set A, and for every hyperfinite subset $\Omega \subseteq A^*$, we denote

$$\sum_{\zeta \in \Omega} f^*(\zeta) = \sum_{\xi \in f^*(\Omega)} \xi,$$

where $f^*(\Omega) = \{f^*(\zeta) \mid \zeta \in \Omega\}$ is the image of Ω under f^*. Accordingly, one has that

$$\sum_{\zeta \in \Omega} f^*(\zeta) = \lim_{n \uparrow \alpha} \left(\sum_{x \in \Omega_n} f(x) \right)$$

where $\langle \Omega_n \mid n \in \mathbb{N} \rangle$ is any sequence of finite subsets of A whose Alpha-limit $\lim_{n \uparrow \alpha} \Omega_n = \Omega$.

REMARK 4.49. *Hyperfinitely long sums as introduced in Alpha-Calculus (see Definition 3.34) are particular cases of hyperfinite sums. Indeed, if $\langle a_n \mid n \in \mathbb{N} \rangle$ is a sequence of real numbers, and $\varphi(n)$ is a sequence of natural numbers with Alpha-limit $\nu = \lim_{n \uparrow \alpha} \varphi(n)$, then it is easily checked that*

$$\sum_{i=1}^{\nu} a_i = \sum_{\xi \in \Omega} \xi$$

where $\Omega = \lim_{n \uparrow \alpha} \{a_1, a_1, \ldots, a_{\varphi(n)}\} = \{a_1, a_1, \ldots, a_\nu\}$.

Also the grid integral was actually defined by means of a hyperfinite sum.

DEFINITION 4.50. *The Alpha-grid $\mathbb{H}_\alpha$ is the hyperfinite grid that determines a partition of the infinite interval $[-\alpha, \alpha]$ into α^2-many intervals of equal infinitesimal length $\eta = 1/\alpha$. Precisely:*

$$\mathbb{H}_\alpha = \left\{ \pm k \cdot \eta \,\middle|\, k = 0, 1, \ldots, \alpha^2 \right\}.$$

Notice that $\mathbb{H}_\alpha$ is the Alpha-limit of the n-grids (see Definition 3.39):

$$\mathbb{H}_\alpha = \lim_{n \uparrow \alpha} \mathbb{H}(n) = \lim_{n \uparrow \alpha} \left\{ \pm \frac{k}{n} \,\middle|\, k = 0, 1, \ldots, n^2 \right\}.$$

REMARK 4.51. *The grid integral as defined in Alpha-Calculus (see Definition 3.41) is the following standard part of a hyperfinite sum:*

$$\int_A f(x)\, d_\mathbb{H}x \;=\; \mathrm{st}\left(\sum_{\xi \in \mathbb{H}_\alpha \cap {}^*A} f^*(\xi) \right).$$

Indeed, it is easily verified that

$$Sum_A^f(\boldsymbol{\alpha}) \;=\; \lim_{n\uparrow\alpha} \sum_{x \in \mathbb{H}(n)\cap A} f(x) \;=\; \sum_{\xi \in \mathbb{H}_\alpha \cap A^*} f^*(\xi).$$

9. Internal objects

The class of those objects that are Alpha-limits is particularly relevant; indeed, as we will show in the sequel, they share the same "elementary properties" as the objects in the originating sequences.

DEFINITION 4.52. *An object is called* internal *if it is the Alpha-limit of a sequence.*[7]

EXAMPLE 4.53.

(1) *All hyperreal numbers $\xi \in \mathbb{R}^*$ are internal, as they are Alpha-limits of real sequences;*

(2) *All hyper-images A^* are internal, as they are Alpha-limits of constant sequences $c_A(n)$. In particular, the sets $\mathbb{N}^*, \mathbb{Z}^*, \mathbb{Q}^*, \mathbb{R}^*$ of hypernatural, hyperinteger, hyperrational, and hyperreal numbers, are all internal sets;*

(3) *All hyperfinite sets are internal, as they are Alpha-limits of sequences of finite sets;*

(4) *All intervals of hyperreal (or hyperinteger, or hyperrational) numbers are internal. E.g.,*

$$[\xi, \zeta]_{\mathbb{R}^*} = \{\eta \in \mathbb{R}^* \mid \xi \le \eta \le \zeta\}$$

is internal because $[\xi, \zeta]_{\mathbb{R}^*} = \lim_{n\uparrow\alpha}[\xi(n), \zeta(n)]_{\mathbb{R}}$, where the sequences $\xi(n)$ and $\zeta(n)$ are such that $\lim_{n\uparrow\alpha} \xi(n) = \xi$ and $\lim_{n\uparrow\alpha} \zeta(n) = \zeta$.

---

[7] We used the name "internal objects" because they precisely correspond to the *internal elements* of *nonstandard analysis* (see Theorem 13.4).

We remark right away that *not* all objects are internal; examples will be given further on in this section.

The first fundamental properties of internal objects are itemized below.

PROPOSITION 4.54.

(1) *The class of internal sets is* transitive, *that is, elements of internal sets are internal.*

(2) X *is internal if and only if* $X \in A^*$ *for some* A;

(3) *For every set* A,

$$\{X \subseteq A^* \mid X \text{ is internal}\} = \mathcal{P}(A)^*;$$

(4) *For all sets* A *and* B,

$$\{F : A^* \to B^* \mid F \text{ internal function}\} = \mathfrak{Fun}(A, B)^*.$$

PROOF. Property (1) directly follows from the *Internal Set Axiom*.

(2). Since any hyper-image A^* is internal, one implication follows from (1). Conversely, if $X = \lim_\alpha \varphi$ for some sequence $\varphi(n)$, let $A = \text{ran}(\varphi)$. Then trivially $\varphi(n) \in A$ for all n, and hence $X = \lim_\alpha \varphi \in A^*$.

(3). If $X \in \mathcal{P}(A)^*$ then X is internal by (2). Let us now show that $X \subseteq A^*$. Let us assume without loss of generality that $X \neq \emptyset$, and pick a sequence $\psi : \mathbb{N} \to \mathcal{P}(A)$ of nonempty sets such that $\lim_\alpha \psi = X$. For every $\xi \in X$, by the *Internal Set Axiom* there exists a sequence φ such that $\lim_\alpha \varphi = \xi$ and $\varphi(n) \in \psi(n)$ a.e.. But then $\varphi(n) \in A$ a.e., and so $\xi \in A^*$.

Conversely, assume that $X = \lim_\alpha \varphi \subseteq A^*$. Then, by *transfer* of the inclusion relation, $\lim_\alpha \varphi \subseteq \lim_\alpha c_A \Leftrightarrow \varphi(n) \subseteq A$ a.e., that is, $\varphi(n) \in \mathcal{P}(A)$ a.e., and hence $X = \lim_\alpha \varphi \in \mathcal{P}(A)^*$.

(4). Recall that, by Theorem 4.11, an Alpha-limit $F = \lim_{n\uparrow\alpha} f_n$ is a function $F : A^* \to B^*$ if and only if $f_n : A \to B$ for almost all n. So, if $F : A^* \to B^*$ is internal then $F \in \mathfrak{Fun}(A, B)^*$, because it is the Alpha-limit of a sequence f_n that belongs to $\mathfrak{Fun}(A, B)$ a.e.. Conversely, if $F \in \mathfrak{Fun}(A, B)^*$ then, by the *Internal Set Axiom*, F is the Alpha-limit of functions $f_n \in \mathfrak{Fun}(A, B)$, and so F is internal and $F : A^* \to B^*$. $\square$

We tell in advance that all "elementary properties" of subsets of a given set A *transfer* to the *internal* subsets of A^*; and similarly, all "elementary properties" of functions $f : A \to B$ *transfer* to the *internal* functions $F : A^* \to B^*$. Although in order to correctly formulate and prove that general *transfer* result one needs the formalism of first-order logic (see Theorem 5.8), a few relevant examples can be proved directly.

PROPOSITION 4.55.

(1) *Every nonempty internal subset of $\mathbb{N}^*$ has a least element.*

(2) *Every nonempty internal subset of $\mathbb{R}^*$ which is bounded above (or bounded below) has a least upper bound (a greatest lower bound, respectively).*

PROOF. (1). Let $A = \lim_{n \uparrow \alpha} A_n \subseteq \mathbb{N}^*$ be the Alpha-limit of a sequence of nonempty sets of reals. If $\varphi(n) = \min A_n$ then it is easily checked that the hypernatural number $\nu = \lim_{n \uparrow \alpha} \varphi(n)$ is the least element of A.

(2). If the nonempty set $A = \lim_{n \uparrow \alpha} A_n \subseteq \mathbb{R}^*$ is bounded above, then it is easily shown that $A_n \subset \mathbb{R}^*$ is bounded above for all n in a qualified set Q. By the *completeness* property of the real numbers, we can take a sequence $\psi : \mathbb{N} \to \mathbb{R}$ such that $\psi(n) = \sup A_n$ for all $n \in Q$. Then the hyperreal number $\zeta = \lim_{n \uparrow \alpha} \psi(n)$ is the least upper bound of A. □

The above results make it easy to produce relevant examples of *non*-internal sets.

EXAMPLE 4.56.

(1) *The set $\mathbb{N}_\infty = \mathbb{N}^* \setminus \mathbb{N}$ of infinite hypernatural numbers is not internal, because it does not admit a least element;*

(2) *The set $\mathbf{mon}(0)$ of infinitesimal numbers and the set $\mathbb{R}^*_{\mathrm{fin}}$ of finite hyperreal numbers are not internal. Indeed, they are both bounded while they do not admit a least upper bound (nor a greatest lower bound);*

(3) *More generally, all monads $\mathbf{mon}(\xi)$ and all galaxies $\mathbf{gal}(\xi)$ are non-internal.*

In the next propositions the basic closure properties of the class of internal sets are collected. Basically, internal sets are closed under all "elementary" set operations, with the only relevant exceptions of the *powerset* and the *function set*.

Closure under finite sets and tuples directly follow from Propositions 4.6 and 4.9, respectively.

PROPOSITION 4.57. *Let $a_1, \ldots, a_k$ be internal objects. Then also the k-set $\{a_1, \ldots, a_k\}$ and the k-tuple $(a_1, \ldots, a_k)$ are internal.*

As a straight consequence of Proposition 4.8, one also obtains:

PROPOSITION 4.58. *Let A, B be internal sets. Then also the following sets are internal:*

(1) *Union: $A \cup B$.*
(2) *Intersection: $A \cap B$.*
(3) *Set Difference: $A \setminus B$.*
(4) *Cartesian Product: $A \times B$.*

We can now easily obtain other examples of non-internal set.

EXAMPLE 4.59.

(1) *The set $\mathbb{N}$ of natural numbers is* not *internal, as otherwise also the set difference $\mathbb{N}_\infty = \mathbb{N}^* \setminus \mathbb{N}$ would be internal, a contradiction;*
(2) *The set $\mathbb{R}_\infty$ of infinite hyperreal numbers is* not *internal, as otherwise also the set difference $\mathbb{R}^*_{\text{fin}} = \mathbb{R}^* \setminus \mathbb{R}_\infty$ would be internal, a contradiction.*

PROPOSITION 4.60. *Let F be an internal function, and let A and B be internal sets. Then the following are internal sets as well:*

(1) *Image: $F(A) = \{F(a) \mid a \in A\}$.*
(2) *Pre-image: $F^{-1}(B) = \{x \mid F(x) \in B\}$.*

PROOF. Let $F = \lim_{n\uparrow\alpha} F_n$, $A = \lim_{n\uparrow\alpha} A_n$, and $B = \lim_{n\uparrow\alpha} B_n$. Then for almost all n, F_n is a function, and it is verified in a straightforward manner that:

$$F(A) = \lim_{n\uparrow\alpha} F_n(A_n) \quad \text{and} \quad F^{-1}(B) = \lim_{n\uparrow\alpha} F_n^{-1}(B_n).$$

$\square$

In Proposition 4.54 we saw that the set of all internal subsets of an hyper-image A^* is itself an hyper-image, namely $\mathcal{P}(A)^*$, and we also saw a similar properties for sets of functions. The same closure properties also holds if one replaces hyper-images with internal sets.

EXERCISE 4.61. *Prove the following properties:*

(1) *If A is an internal set, then also the family $\mathcal{P}_\mathcal{I}(A)$ of internal subsets of A is internal;*
(2) *If A and B are internal sets, then also the set $\mathfrak{Fun}_\mathcal{I}(A, B)$ of internal functions from A to B is internal.*

DEFINITION 4.62. *An* ideal object *is an internal object that is not an hyper-image.*

Important examples of ideal elements are given below.

PROPOSITION 4.63.

(1) *All infinite hypernatural numbers $\nu \in \mathbb{N}_\infty$ are ideal elements; in particular, the number α is ideal;*

(2) *All hyperreal numbers $\xi \in \mathbb{R}^* \setminus \mathbb{R}$ that are not real, are ideal elements;*

(3) *All infinite hyperfinite sets are ideal;*

(4) *All intervals of hyperreal (or hyperinteger, or hyperrational) numbers where at least one end-point is ideal, are ideal. E.g., if $\xi \in \mathbb{R}^*$ and $\zeta \in \mathbb{R}^* \setminus \mathbb{R}$ then the following intervals are ideal:*

$$[\xi, \zeta]_{\mathbb{R}^*}, \quad (-\infty, \zeta)_{\mathbb{R}^*}.$$

PROOF. (1) and (2). By the *Real Number Axiom*, a number $\xi \in \mathbb{R}^*$ is a hyper-image if and only if $\xi = r^* = r$ for some real number r.

(3). A hyperfinite set A is a hyper-image if and only if $A = \lim_{n \uparrow \alpha} B = B^*$ for some finite set $B = \{b_1, \ldots, b_k\}$, and in this case, $A = B^* = \{b_1^*, \ldots, b_k^*\}$ is finite.

(4). Notice that an interval $[\xi, \zeta]_{\mathbb{R}^*}$ is an hyper-image if and only if there exists a real interval $[a, b]_{\mathbb{R}}$ such that $[\xi, \zeta] = \lim_{n \uparrow \alpha}[a, b] = [a^*, b^*] = [a, b]$, and so both $\xi, \zeta \in \mathbb{R}$. The cases of other intervals are treated similarly. $\square$

As shown by the next proposition, "most" Alpha-limits produce ideal objects.

PROPOSITION 4.64. *If the sequence $\varphi(n)$ is finite-to-one[8] then the internal object $\lim_{n \uparrow \alpha} \varphi(n)$ is not an hyper-image.*

PROOF. By the hypothesis, for every object A one has that $\varphi(n) = A$ only for finitely many n. Then $\varphi(n) \neq c_A(n)$ a.e., and so the Alpha-limit $\lim_{n \uparrow \alpha} \varphi(n) \neq \lim_{n \uparrow \alpha} c_A(n) = A^*$. $\square$

As companions of the hyperfinite sets, one defines the hyperinfinite sets.

DEFINITION 4.65. *We call* hyperinfinite set *the Alpha-limit of a sequence of infinite sets.*

---

[8] Recall that a function is *finite-to-one* if each value is taken finitely many times, that is, if all preimages are finite.

Clearly, an internal set is hyperinfinite if and only if it is *not* hyperfinite. In fact, an Alpha-limit of sets $A = \lim_{n\uparrow\alpha} A_n$ is either hyperfinite or hyperinfinite, depending on whether the sets A_n are finite *a.e.* or infinite *a.e.*.

PROPOSITION 4.66. *If A is a hyperinfinite set, then there exists a 1-1 function $f : \mathbb{N}^* \to A$.*

PROOF. Let $A = \lim_{n\uparrow\alpha} A_n$ where the sets A_n are infinite. Then for every n we can pick a 1-1 function $f_n : \mathbb{N} \to A_n$. By taking the Alpha-limit $f = \lim_{n\uparrow\alpha} f_n$ we obtain the desired 1-1 function $f : \mathbb{N}^* \to A$. □

As a corollary, we get a general result on the cardinality of internal sets.

PROPOSITION 4.67. *If A is an internal set, then either A is finite or $|A| \geq \mathfrak{c}$ has at least the cardinality of the continuum. In particular, every infinite countable set is not internal.*

PROOF. If A is hyperfinite, we already showed that either A is finite or $|A| = \mathfrak{c}$ (see Proposition 4.47); and if A is hyperinfinite, then $|A| \geq |\mathbb{N}^*| = \mathfrak{c}$, by the above Proposition. □

Notice that, as a straight consequence of the above proposition, one obtains another proof that the set of natural numbers $\mathbb{N}$ is not internal.

As shown by items (3) and (4) of Proposition 4.54, in general one has the following inclusions:

$$\mathcal{P}(A)^* \subseteq \mathcal{P}(A^*) \quad \text{and} \quad \mathfrak{Fun}(A, B)^* \subseteq \mathfrak{Fun}(A^*, B^*).$$

We now show that equalities hold only in the case of finite sets.

PROPOSITION 4.68. *Let A and B be nonempty sets. Then:*

(1) $\mathcal{P}(A)^* = \mathcal{P}(A^*)$ *if and only if A is finite. In consequence, every infinite hyper-image A^* has ideal subsets;*

(2) $\mathfrak{Fun}(A, B)^* = \mathfrak{Fun}(A^*, B^*)$ *if and only if $\mathfrak{Fun}(A, B)$ is finite, that is, if both A and B are finite or if B is a singleton.*

PROOF. (1). If $A = \{a_1, \ldots, a_k\}$ is finite. Then every subset of $A^* = \{a_1^*, \ldots, a_k^*\}$ is internal, because a finite set of internal objects a_i^*. Conversely, if A is infinite then we can pick a countably infinite subset $B \subset A$. Such a set B is not internal, and hence $B \notin \mathcal{P}(A)^*$. If A is infinite, every object in $\mathcal{P}(A^*) \setminus \mathcal{P}(A)^*$ is an ideal subset of A^* (see Proposition 4.54 (3)).

(2). If $B = \{b\}$ is a singleton, then trivially $\mathfrak{Fun}(A^*, B^*) = \{c_{b^*}\} = \{(c_b)^*\} = \mathfrak{Fun}(A, B)^*$, where $c_{b^*} : A^* \to B^*$ and $c_b : A \to B$ are the constant functions with value b^* and b, respectively. If both A and B are finite then $A^* = \{a^* \mid a \in A\}$ and $B^* = \{b^* \mid b \in B\}$ only contains hyper-extensions. As a consequence, every $F : A^* \to B^*$ is a finite set of hyper-extensions of the form $(a^*, b^*) = (a, b)^*$, and it is readily verified that $F = f^*$ where $f(a) = b \Leftrightarrow F(a^*) = b^*$. Since $\mathfrak{Fun}(A, B)$ is finite, we obtain the desired conclusion:

$$\mathfrak{Fun}(A^*, B^*) \;=\; \{f^* \mid f \in \mathfrak{Fun}(A, B)\} \;=\; \mathfrak{Fun}(A, B)^*.$$

$\square$

By slightly generalizing the above arguments, also the following properties are proved.

EXERCISE 4.69.

(1) If the internal set A is infinite, then $\mathcal{P}_{\mathcal{I}}(A)$ is a proper subset of $\mathcal{P}(A)$;

(2) If A and B are internal sets, then $\mathfrak{Fun}_{\mathcal{I}}(A, B)$ is a proper subset of $\mathfrak{Fun}(A, B)$ whenever the latter set is infinite.

We close this section with an interesting extension property of internal functions.

THEOREM 4.70 (Comprehensiveness property).
Let A and B be arbitrary sets. For every $f : A \to B^*$ taking values into the hyper-image of B, there exists an internal $F : A^* \to B^*$ such that $F(a^*) = f(a)$ for every $a \in A$.

PROOF. For every $a \in A$, fix a sequence $\langle b_n^a \mid n \in \mathbb{N} \rangle$ of elements of B whose Alpha-limit $\lim_{n \uparrow \alpha} b_n^a = f(a)$. Then, for every sequence $\langle a_n \mid n \in \mathbb{N} \rangle$ of elements of A, let

$$F \left(\lim_{n \uparrow \alpha} a_n \right) \;=\; \lim_{n \uparrow \alpha} b_n^{a_n}.$$

We remark that the above definition is well-posed, because $\lim_{n \uparrow \alpha} a_n = \lim_{n \uparrow \alpha} a_n' \Leftrightarrow a_n = a_n'$ a.e. $\Leftrightarrow b_n^{a_n} = b_n^{a_n'}$ a.e. $\Leftrightarrow \lim_{n \uparrow \alpha} b_n^{a_n} = \lim_{n \uparrow \alpha} b_n^{a_n'}$.

For every $a \in A$, $F(a^*) = F(\lim_{n \uparrow \alpha} a) = \lim_{n \uparrow \alpha} b_n^a = f(a)$. Finally, F is an internal function because it is easily seen that $F = \lim_{n \uparrow \alpha} f_n$ where $f_n : A \to B$ is the function defined by $f_n(a) = b_n^a$. $\square$

10. Remarks and comments

1. As we will see further on in §13.1 and §13.2, our hyper-images correspond to the "star-extensions" (or "nonstandard extensions") as considered in *nonstandard analysis*. Typically, in *nonstandard analysis* one distinguishes between the "standard universe" and the "nonstandard universe". Only objects A that belong to the standard universe have a star-extension A^* and, with some ambiguity, both the objects in the standard universe and their star-extensions are named "standard".

In our context, the adjective "standard" would be misplaced; indeed, a crucial feature of the Alpha-Theory is that there is no "standard universe" to be opposed to a "nonstandard universe", and in fact there is a single universe, namely the universe of all mathematical objects. Some objects are Alpha-limits (the *internal* objects), and some of the Alpha-limits are limits of constant sequences (the *hyper-images*).

Our Alpha-limits correspond to the internal elements of *nonstandard analysis*, and this is the only reason why we adopted that same terminology, in spite of the fact that in the framework of Alpha-Theory the adjective "internal" does not make much sense. We avoided using the term *external* to indicate an object that is not internal, simply because also the non-internal objects are *inside* our universe, and can be used in Alpha-limits. Instead, we borrowed the adjective "ideal" from Leibniz, who used is to refer to numbers which are not real, such as the infinitesimals.

2. A nice feature of Alpha-Theory is that one can take iterated hyper-images, and consider objects such as the *hyper-hypernatural numbers* $(\mathbb{N}^*)^*$ (see §7.2 for a use of such numbers in Ramsey theory). Differently, in the usual framework of nonstandard analysis as given by the superstructure approach (see §13.1), one cannot consider the star-extension of a star-extension $(A^*)^*$; for this, one would need to construct a different standard universe to include the nonstandard universe of the previous stage. We remark that also in Nelson's Internal Set Theory IST, as well as in the related nonstandard set theories that have been elaborated upon that approach (see [**59**]), the notion of "iterated hyper-image" does make sense.

3. Historically, the idea of adding a new number to the system of natural numbers goes back to the pre-Robinsonian work [**85**] by C. Schmieden and D. Laugwitz. They considered a new symbol Ω and postulated that if a formula is true for all sufficiently large natural numbers, then the same formula is also true for Ω. The resulting theory has a constructive flavor, and in the later years, it inspired various "non-classic" presentations of nonstandard analysis.[9]

Even if there was no direct influence on nonstandard analysis, it is worth mentioning that in [82], A. Robinson himself acknowledged the work by C. Schmieden and D. Laugwitz as a "recent and rather successful effort of developing a calculus of infinitesimals."

Although it provides a sound foundation to infinitesimals and it allows to develop large parts of calculus, the Ω-approach revealed inadequate for advanced applications. The crucial drawback is that the nonstandard reals in that theory contain zero divisors, and they only have the structure of a partially ordered ring.

The Ω-approach has been developed by J.M. Henle in [53]. Related interesting considerations on the constructive viewpoint on nonstandard analysis can be found in the book [86].

---

[9] See, *e.g.*, E. Palmgren's [79, 80].

CHAPTER 5

Logic and Alpha-Theory

1. Some logic formalism

In this section we will use the formalism of first-order logic and make precise the notion of "elementary property".

Recall that we assumed that all our mathematical objects be sets or atoms. As a consequence, their properties can be described within the language of set theory, that is, by only appealing to the *equality* and the *membership* relations. (We will say more on this matter further on in this section.)

To begin with, let us specify the "alphabet" of symbols of the first-order *language of set theory*, which will be used to form our formulas.

- *Variables*: $x, y, z, \ldots, x_1, x_2, \ldots$
- *Logic Connectives*:
  - Negation: $\neg$ (not).
  - Conjunction: $\wedge$ (and).
  - Disjunction: $\vee$ (or).
  - Implication: $\Rightarrow$ (if ... then).
  - Double implication: $\Leftrightarrow$ (if and only if).
- *Quantifiers*:
  - Existential quantifier: $\exists$ (there exists).
  - Universal quantifier: $\forall$ (for all).
- *Equality*: $=$.
- *Membership*: $\in$.

To be precise, also parentheses " $($ " and " $)$ " must be included in the language.

DEFINITION 5.1. *An elementary formula is a finite string of symbols in the above language in which variables are distinguished into* free variables *and* bound variables, *and which are obtained according to the following rules.*

(EF1): Atomic formulas.

Let x, y be variables. Then

$$(x = y), \ (x \in y)$$

are elementary formulas, named atomic formulas, *where:*

- *the free variables are x and y;*
- *there are no bound variables.*

(EF2): Restricted quantifiers.

Let σ be an elementary formula and assume that:

(1) *the variable x is free in σ,*

(2) *the variable y is* not *bound in σ.*

Then also

$$(\forall x \in y) \ \sigma, \ (\exists x \in y) \ \sigma$$

are elementary formulas, where:

- *the free variables are y along with the free variables of σ,*
- *the bound variables are x along with the bound variables of σ.*

(EF3): Negation.

Let σ be an elementary formula. Then also

$$(\neg \sigma)$$

is an elementary formula, where:

- *the free variables are the same as in σ,*
- *the bound variables are the same as in σ.*

(EF4): Binary logic connectives.

Let σ and τ be elementary formulas, and assume that:

(1) *every free variable of σ is not bound in τ,*

(2) *every free variable of τ is not bound in σ.*

Then also

$$(\sigma \wedge \tau), \ (\sigma \vee \tau), \ (\sigma \Rightarrow \tau), \ (\sigma \Leftrightarrow \tau)$$

are elementary formulas, where:

- *the* free variables *are the union of the free variables of* σ *and free variables of* τ;
- *the* bound variables *are the union of the bound variables of* σ *and the bound variables of* τ.

A few comments are in order.

Clearly, in the inductive process of forming an elementary formula, the first rule to be applied is (**EF1**); in other words, every formula is built on *atomic formulas* (and this justifies the name "atomic"). An arbitrary formula is then obtained from atomic formulas by finitely many iterations of the other three rules, in whatever order.

The *restricted quantifiers rule* (**EF2**) is the only one that produces bound variables. The idea is that a variable is bound when it is quantified. Remark that a variable can be quantified only if it is free in the given formula, that is, only if it actually appears and it has been not quantified already.

It is worth remarking that quantifications are only permitted in the *restricted forms* $(\forall x \in y)$ or $(\exists x \in y)$, where the "scope" of the quantified variable x is "restricted" by another variable y. In order to avoid potential ambiguities, it is also required that the "bounding" variable y do not appear bound itself in the given formula.

The *negation rule* (**EF3**) is the simplest one, as it applies to all formulas without any restrictions, and it produces new formulas having the same free and bound variables as the starting ones.

Also the *binary connectives rule* (**EF4**) is quite simple, and it produces formulas where the free and the bound variables are simply obtained by putting together the free and the bound variables of the two given formulas, respectively.

The provisions in rule (**EF4**) are given to prevent "connecting" two formulas where a same variable appears free in one of them, and bound in the other.

For readability, we will follow the common use and adopt familiar short-hands to simplify notation. For instance:

- We write "$x \neq y$" to mean "$\neg(x = y)$".
- We write "$x \notin y$" to mean "$\neg(x \in y)$".
- We write "$\forall x_1, \ldots, x_k \in y \; \sigma$" to mean "$(\forall x_1 \in y) \ldots (\forall x_k \in y) \; \sigma$".
- We write "$\exists x_1, \ldots, x_k \in y \; \sigma$" to mean "$(\exists x_1 \in y) \ldots (\exists x_k \in y) \; \sigma$".

Moreover, we will take the freedom of using parentheses informally, and omit some of them whenever confusion is unlikely. So, we may write "$\forall x \in y\ \sigma$" instead of "$(\forall x \in y)\ \sigma$; or "$\sigma \wedge \tau$" instead of "$(\sigma \wedge \tau)$"; *etc.*

Another usual agreement is that negations $\neg$ bind more strongly than conjunctions $\wedge$ and disjunctions $\vee$, which in turn bind more strongly than implications $\Rightarrow$ and double implications $\Leftrightarrow$. So, *e.g.*,

- We may write "$\neg \sigma \wedge \tau$" to mean "$((\neg \sigma) \wedge \tau)$".
- We may write "$\neg \sigma \vee \tau \Rightarrow v$" to mean "$(((\neg \sigma) \vee \tau) \Rightarrow v)$".
- We may write "$\sigma \Rightarrow \tau \vee v$" to mean "$(\sigma \Rightarrow (\tau \vee v))$".

An important notation is the following.

NOTATION 5.2. *When writing*

$$\sigma(x_1, \ldots, x_k)$$

we will mean that $x_1, \ldots, x_k$ are all and only the free variables that appear in the formula σ.

The intuition is that the truth or falsity of a formula depends only on the values given to its *free variables*, whereas *bound variables* can be renamed without changing the meaning of a formula.

We are now ready for the fundamental definition.

DEFINITION 5.3. *A property of mathematical objects $A_1, \ldots, A_k$ is* expressed in elementary form *if it is written down by taking an elementary formula $\sigma(x_1, \ldots, x_k)$, and by replacing all occurrences of each free variable x_i by A_i. In this case we denote*

$$\sigma(A_1, \ldots, A_k),$$

and objects $A_1, \ldots, A_k$ are referred to as constants.

By a slight abuse, sometimes we will simply say elementary property *to mean "property expressed in elementary form".*

In order to make sense, it is implicitly assumed that in every quantification $(\forall x \in A_i)$ and $(\exists x \in A_i)$, the object A_i is a *set* (not an atom).

The motivation of our definition is the well-known fact that virtually *all* properties considered in mathematics can be formulated in elementary form. Below is a list of examples that include the fundamental ones. As an exercise, the reader can easily write down by him- or herself any other mathematical property that comes to his or her mind, in elementary form.

EXAMPLE 5.4. Each property is followed by one of its possible expressions in elementary form.[1]

(1) "$A \subseteq B$":
$(\forall x \in A)(x \in B)$.

(2) $C = A \cup B$:
$(A \subseteq C) \wedge (B \subseteq C) \wedge (\forall x \in C)(x \in A \vee x \in B)$.

(3) $C = A \cap B$:
$(C \subseteq A) \wedge (\forall x \in A)(x \in B \Leftrightarrow x \in C)$.

(4) $C = A \setminus B$:
$(C \subseteq A) \wedge (\forall x \in A)(x \in C \Leftrightarrow x \notin B)$.

(5) $C = \{a_1, \ldots, a_k\}$:
$(a_1 \in C) \wedge \ldots \wedge (a_k \in C) \wedge (\forall x \in C)(x = a_1 \vee \ldots \vee x = a_k)$.

(6) $\{a_1, \ldots, a_k\} \in C$:
$(\exists x \in C)(x = \{a_1, \ldots, a_k\})$.

(7) $C = (a, b)$: [2]
$C = \{\{a\}, \{a, b\}\}$

(8) $C = (a_1, \ldots, a_k)$:
The formula as obtained by induction from the previous item and the recursive definition of ordered tuple:
$(a_1, \ldots, a_n, a_{n+1}) = ((a_1, \ldots, a_n), a_{n+1})$.

(9) $(a_1, \ldots, a_k) \in C$:
$(\exists x \in C)(x = (a_1, \ldots, a_k))$.

(10) $C = A_1 \times \ldots \times A_k$:
$(\forall x_1 \in A_1) \ldots (\forall x_k \in A_k)((a_1, \ldots, a_k) \in C) \wedge$
$(\forall z \in C)(\exists x_1 \in A_1) \ldots (\exists x_k \in A_k)(z = (x_1, \ldots, x_k))$.

(11) R is a k-place relation on A: [3]
$(\forall z \in R)(\exists x_1, \ldots, x_k \in A)(z = (x_1, \ldots, x_k))$.

(12) $f : A \to B$: [4]
$(f \subseteq A \times B) \wedge (\forall a \in A)(\exists b \in B)((a, b) \in f) \wedge$
$(\forall a, a' \in A)(\forall b \in B)((a, b), (a', b) \in f \Rightarrow a = a')$.

(13) $f(a_1, \ldots, a_k) = b$:
$((a_1, \ldots, a_k), b) \in f$.

---

[1] For simplicity, in each item we use short-hands for properties that have been already expressed in previous items.

[2] Recall that $(a, b) = \{\{a\}, \{a, b\}\}$ was defined as the *Kuratowski pair*.

[3] Recall that a k-place relation was identified with the set of all k-tuples that satisfy it.

[4] Recall that a function $f : A \to B$ was defined as a functional binary relation, and hence it is identified with its graph.

(14) $x < y$ (in $\mathbb{R}$):

$(x, y) \in R$, where $R \subset \mathbb{R} \times \mathbb{R}$ is the order relation on $\mathbb{R}$.

REMARK 5.5. Recall when expressing a property in elementary form, quantifiers can only be used in the *restricted* forms:

$$\text{"}(\forall x \in A)\ \sigma(x, \dots)\text{"} \quad \text{and} \quad \text{"}(\exists x \in A)\ \sigma(x, \dots)\text{"}.$$

So, whenever quantifying on a variable x, one must always specify the set A where the variable x is ranging. Two crucial examples are the following:

- Quantifications ranging on *subsets*:

$$\text{"}\forall x \subseteq X \ \dots\text{"} \quad \text{or} \quad \text{"}\exists x \subseteq X \ \dots\text{"}$$

  must be reformulated in the forms

$$\text{"}\forall x \in \mathcal{P}(X) \ \dots\text{"} \quad \text{and} \quad \text{"}\exists x \in \mathcal{P}(X) \ \dots\text{"},$$

  respectively, where $\mathcal{P}(X)$ is the powerset of X.
- Quantifications ranging on *functions*:

$$\text{"}\forall f : A \to B \ \dots\text{"} \quad \text{or} \quad \text{"}\exists f : A \to B \ \dots\text{"}$$

  must be reformulated in the forms:

$$\text{"}\forall f \in \mathfrak{Fun}(A, B) \ \dots\text{"} \quad \text{and} \quad \text{"}\forall f \in \mathfrak{Fun}(A, B) \ \dots\text{"},$$

  respectively, where $\mathfrak{Fun}(A, B)$ is the set of all functions from A to B.

EXAMPLE 5.6. An elementary formulation of the *completeness property* of the reals requires the powerset $\mathcal{P}(\mathbb{R})$ as a constant:

$$\forall X \in \mathcal{P}(\mathbb{R}) \ (\text{"}X \text{ bounded above"} \Rightarrow (\exists x \in \mathbb{R}) \ \text{"}x = \sup A\text{"}).$$

Here, "X is bounded" and "$x = \sup A$" are short-hands for the corresponding expressions in elementary form, namely:

- $(\exists x \in \mathbb{R})(\forall y \in X)(x < y)$, and
- $(\forall y \in A)(y \leq x) \wedge (\forall z \in \mathbb{R}) \left((\forall t \in A)(t \leq z) \right) \Rightarrow x \leq z$,

respectively.

It is worth remarking that a same property may be expressed both in an elementary form and in a non-elementary form. The typical examples involve the powerset operation.

EXAMPLE 5.7. "$\mathcal{P}(A) = B$" is trivially an elementary property of constants $\mathcal{P}(A)$ and B, but *cannot* be formulated as an elementary property of constants A and B. In fact, while the inclusion "$B \subseteq \mathcal{P}(A)$" is formalized in elementary form by "$(\forall x \in B)(\forall y \in x)(y \in A)$", the other inclusion $\mathcal{P}(A) \subseteq B$ does not admit any elementary formulation with A and B as constants. The point here is that quantifications over subsets "$(\forall x \subseteq A)(x \in B)$" are not allowed by our rules.

2. Transfer principle

The next theorem is the fundamental result of Alpha-Theory. It states the strong property that *all* properties expressed in elementary form are preserved under Alpha-limits.

THEOREM 5.8 (Transfer Principle).
Let $\sigma(x_1, \ldots, x_k)$ be an elementary formula and let $\varphi_1, \ldots, \varphi_k$ be arbitrary sequences. Then

$$\sigma(\varphi_1(n), \ldots, \varphi_k(n)) \text{ a.e.} \iff \sigma(\lim_{n\uparrow\alpha}\varphi_1(n), \ldots, \lim_{n\uparrow\alpha}\varphi_k(n)).$$

PROOF. We proceed by induction on the complexity of the formulas. For atomic formulas $(x_1 = x_2)$ or $(x_1 \in x_2)$, the conclusion is given by Theorem 4.4.

The *negation* case $\neg\tau$ directly follows from the inductive hypothesis on τ and the property that a set *is not* qualified if and only if its complement *is* qualified. Precisely, we have the following chain of equivalences:

$$\neg\tau(\varphi_1(n), \ldots, \varphi_k(n)) \quad a.e.$$
$$\tau(\varphi_1(n), \ldots, \varphi_k(n)) \quad \underline{\text{not}} \ a.e.$$
$$\underline{\text{not}} \ \ \tau(\lim_{n\uparrow\alpha}\varphi_1(n), \ldots, \lim_{n\uparrow\alpha}\varphi_k(n))$$
$$\neg\tau(\lim_{n\uparrow\alpha}\varphi_1(n), \ldots, \lim_{n\uparrow\alpha}\varphi_k(n)).$$

The *conjunction* case $(\tau \wedge \tau')$ is proved similarly as above, by using the property that two sets are both qualified if and only if their intersection is qualified. The cases of the other binary connectives $(\tau \vee \tau')$, $(\tau \Rightarrow \tau')$ and $(\tau \Leftrightarrow \tau')$ then follow by using the equivalences:

- $(\tau \vee \tau') \Leftrightarrow \neg(\neg\tau \wedge \neg\tau')$;
- $(\tau \Rightarrow \tau') \Leftrightarrow \neg\tau \vee \tau'$;
- $(\tau \Leftrightarrow \tau') \Leftrightarrow (\tau \Rightarrow \tau') \wedge (\tau' \Rightarrow \tau)$.

We are left to consider the *existential quantifier*, as the *universal quantifier* is then treated by considering the equivalence

$$(\forall x \in y)\,\tau \iff \neg(\exists x \in y)\,\neg\tau.$$

Assume first that the following set is qualified:

$$\Lambda = \{n \mid \exists x \in \psi(n)\; \tau(x, \varphi_1(n), \ldots, \varphi_k(n))\}.$$

Pick a sequence ϑ of "witnesses" on Λ, that is, such that for all $n \in \Lambda$,

$$\vartheta(n) \in \psi(n) \quad \text{and} \quad \tau(\psi(n), \varphi_1(n), \ldots, \varphi_k(n)).$$

Then, by the inductive hypothesis, we have that $\lim_\alpha \vartheta \in \lim_\alpha \psi$ and $\tau(\lim_\alpha \psi, \lim_\alpha \varphi_1, \ldots, \lim_\alpha \varphi_k)$, and hence

$$(\exists x \in \lim_\alpha \psi)\; \tau(x, \lim_\alpha \varphi_1, \ldots, \lim_\alpha \varphi_k).$$

Conversely, if $(\exists x \in \lim_\alpha \psi)\; \tau(x, \lim_\alpha \varphi_1, \ldots, \lim_\alpha \varphi_k)$, then by the *Internal Set Axiom* there exists a sequence ϑ such that $\vartheta(n) \in \psi(n)$ a.e. and $\tau(\lim_\alpha \vartheta, \lim_\alpha \varphi_1, \ldots, \lim_\alpha \varphi_k)$. By the inductive hypothesis, we conclude that $\tau(\vartheta(n), \varphi_1(n), \ldots, \varphi_k(n))$ a.e., and the thesis follows. □

As a straight consequence, we obtain that every property expressed in elementary form holds for given objects $A_1, \ldots, A_k$ if and only if it holds for the corresponding hyper-images $A_1^*, \ldots, A_k^*$.

COROLLARY 5.9 (Transfer Principle for hyper-images).
Let $\sigma(x_1, \ldots, x_k)$ be an elementary formula. Then for all $A_1, \ldots, A_k$,

$$\sigma(A_1, \ldots, A_k) \iff \sigma(A_1^*, \ldots, A_k^*).$$

PROOF. It directly follows from the previous theorem by noticing that $\sigma(A_1, \ldots, A_k) \iff \sigma(c_{A_1}(n), \ldots, c_{A_k}(n))$ a.e.. □

We have seen in Example 5.4 that "$A \subseteq B$", "$C = A \cup B$", "$C = A \cap B$", and "$C = A \setminus B$" can all be expressed in elementary form. So, by *transfer*, one directly obtains the properties already shown in Theorem 4.8, namely:

- $\varphi(n) \subseteq \psi(n)$ a.e. $\iff$ $\lim_\alpha \varphi \subseteq \lim_\alpha \psi$.
- $\vartheta(n) = \varphi(n) \cup \psi(n)$ a.e. $\iff$ $\lim_\alpha \vartheta = \lim_\alpha \varphi \cup \lim_\alpha \psi$.
- $\vartheta(n) = \varphi(n) \cap \psi(n)$ a.e. $\iff$ $\lim_\alpha \vartheta = \lim_\alpha \varphi \cap \lim_\alpha \psi$.
- $\vartheta(n) = \varphi(n) \setminus \psi(n)$ a.e. $\iff$ $\lim_\alpha \vartheta = \lim_\alpha \varphi \setminus \lim_\alpha \psi$.

Similarly, the reader could re-prove all preservation properties under Alpha-limits presented in Chapter 4, by simply putting them in elementary form and then applying *transfer*. It is worth remarking that restricting to elementary sentences is not a limitation, because all mathematical properties can be rephrased in elementary terms.

Plenty of applications of *transfer* can be easily conceived by the reader. Some of them are given below as examples.

EXAMPLE 5.10. *If* $(A, <)$ *is a linearly ordered set, then by* transfer $(A^*, <^*)$ *is a linearly ordered set. Recall in fact that, by definition, a linear ordering* $<$ *on* A *is a set of ordered pairs of elements of* A *that satisfies the following three properties:*[5]

- $\forall x \in A \ (x \not< x)$;
- $\forall x, y, z \in A \ (x < y \wedge y < z) \Rightarrow x < z$;
- $\forall x, y \in A \ (x < y \vee y < x \vee x = y)$.

EXAMPLE 5.11. *Since* $\mathbb{N}$ *is an initial segment of* $\mathbb{N}^*$, *by* transfer *we obtain that* $\mathbb{N}^*$ *is an initial segment of the double hyper-image* $\mathbb{N}^{**}$, *that is,* $\mathbb{N}^* \subset \mathbb{N}^{**}$ *and* $\xi > \eta$ *whenever* $\xi \in \mathbb{N}^{**} \setminus \mathbb{N}^*$ *and* $\eta \in \mathbb{N}^*$.

EXAMPLE 5.12. *If* $(G, \cdot)$ *is a group with identity element* e, *then by* transfer $(G^*, \cdot^*)$ *is a group with identity element* e^*. *In fact, it is readily verified that all defining conditions of a group are expressed in elementary form.*

A delicate aspect that needs some caution, is the possibility of misreading a transferred sentence, once all asterisks $*$ have been put in the right places. A typical example is given by the Archimedean property.

EXAMPLE 5.13. *The* Archimedean property *of real numbers is expressed in elementary form as follows:*

- *"For all positive* $r \in \mathbb{R}$, *there exists* $n \in \mathbb{N}$ *such that* $n \cdot r > 1$".

By transfer, *one obtains:*

- *"For all positive* $\xi \in \mathbb{R}^*$, *there exists* $\nu \in \mathbb{N}^*$ *such that* $\nu \cdot \xi > 1$".

Notice that the above sentence does not express the Archimedean property of $\mathbb{R}^*$, *because the hypernatural number* ν *could be infinite.*

---

[5] As usual, we write $x < y$ to mean that the ordered pair (x, y) belongs to the binary relation $<$; and $x \not< y$ to mean that (x, y) does *not* belong to $<$.

3. Transfer and internal sets

Now that we have formalized the notion of elementary property, we can define the notion of definability.

DEFINITION 5.14. *We say that a set A is* definable *from constants* $C, B_1, \ldots, B_k$ *if*

$$A = \{x \in C \mid \sigma(x, B_1, \ldots, B_k)\}$$

for a suitable elementary formula $\sigma(x, x_1, \ldots, x_k)$.

For a set of constants $\mathcal{C}$, *we denote by* $Def(\mathcal{C})$ *the family of all sets that are definable from constant in* $\mathcal{C}$.

EXERCISE 5.15. *Taking definable sets is an idempotent operation, that is, for every set of constants* $\mathcal{C}$:

$$Def(Def(\mathcal{C})) = Def(\mathcal{C}).$$

EXERCISE 5.16. *For every set of constants* $\mathcal{C}$, *the cardinality*

$$|Def(\mathcal{C})| = \max\{|\mathcal{C}|, \aleph_0\}.$$

The following straightforward consequence of the *transfer principle* provides a general tool to generate internal sets.

THEOREM 5.17 (Internal Definition Principle).
The family of internal sets is closed under definable sets, that is, if $\sigma(x, x_1, \ldots, x_k)$ *is an elementary formula and* $C, B_1, \ldots, B_k$ *are internal objects then also the following set is internal:*

$$\{x \in C \mid \sigma(x, B_1, \ldots, B_k)\}.$$

PROOF. Pick a sequence of sets $\psi(n)$ and sequences $\varphi_i(n)$ such that $\lim_{n \uparrow \alpha} \psi(n) = C$ and $\lim_{n \uparrow \alpha} \varphi_i(n) = B_i$ for every $i = 1, \ldots, k$. Define the following sequence of sets:

$$\vartheta(n) = \{x \in \psi(n) \mid \sigma(x, \varphi_1(n), \ldots, \varphi_k(n))\}.$$

Then $\xi \in \lim_{n \uparrow \alpha} \vartheta(n) \Leftrightarrow \xi = \lim_{n \uparrow \alpha} \xi(n)$ for some sequence such that $\xi(n) \in \vartheta(n)$ a.e.. This last condition is equivalent to the conjunction of "$\xi(n) \in \psi(n)$ a.e." and "$\sigma(\xi(n), \varphi_1(n), \ldots, \varphi_k(n))$ a.e.", which in turn is equivalent to the conjunction of "$\xi \in C$" and "$\sigma(\xi, B_1, \ldots, B_k)$", *by transfer.* Thus we have proved that

$$\{x \in C \mid \sigma(x, B_1, \ldots, B_k)\} = \lim_{n \uparrow \alpha} \vartheta(n).$$

□

EXAMPLE 5.18. *By the* Internal Definition Principle, *the following sets are internal:*

(1) $A = \{\nu \in \mathbb{N}^* \mid \sin^* \nu > 0\}$.
(2) $A = \{\xi \in \mathbb{Q}^* \mid \xi^2 < 2\}$.
(3) $A = \{X \in \mathcal{P}(\mathbb{R})^* \mid X$ *is bounded*$\}$.

The notion of internal set plays a central role when applying the *transfer principle*. Recall the following facts (see Propositions 4.54):

- $\mathcal{P}(A)^*$ contains all and only the *internal* subsets of A^*;
- $\mathfrak{Fun}(A, B)^*$ contains all and only the *internal* functions $F : A^* \to B^*$.

In the practice, one can keep in mind the following general heuristic rule:

> **Rule of thumb for *transfer*.** Properties about *subsets* or about *functions* transfer to the corresponding properties about *internal subsets* or *internal functions*, respectively.

We already saw two relevant examples of this rule in Proposition 4.55, where it is shown that

- The *well-ordering* of $\mathbb{N}$ transfers to: "Every nonempty *internal* subset of $\mathbb{N}^*$ has a least element";
- The *completeness* of $\mathbb{R}$ transfers to: "Every nonempty *internal* subset of $\mathbb{R}^*$ which is bounded has a least upper bound and a greatest lower bound".

We warn the reader that getting familiar with the distinction between internal and non-internal (ideal) objects is probably the hardest step in learning to correctly use the *transfer principle*.

EXERCISE 5.19. *Prove that:*

(1) *All* internal *ideals of the commutative ring* $\mathbb{Z}^*$ *are principal, that is, they are generated by a single element.*
(2) $\mathbb{Z}^*$ *is* not *a principal ideal domain, that is, not all of its ideals are generated by a single element.*

SOLUTION. (1). Notice that the property "I is an ideal" is expressed in elementary form as follows:

$$I \neq \emptyset \;\wedge\; (\forall x \in I)\,(-x \in I) \;\wedge$$
$$(\forall\, x, y \in I)\,(x + y \in I) \;\wedge\; (\forall x \in I)\,(\forall y \in \mathbb{Z})\,(x \cdot y \in I).$$

Also the property "I is principal" is elementary:

$$(\exists x \in \mathbb{Z})\,(\forall z \in \mathbb{Z})\,(z \in I \Leftrightarrow (\exists y \in \mathbb{Z})\,(z = y \cdot x)).$$

Now, it is a well-known fact from algebra that $\mathbb{Z}$ is a principal ideal domain, and so we obtain the desired conclusion by transferring the following elementary property:

$$(\forall I \in \mathcal{P}(\mathbb{Z}))\,(\text{"I ideal"} \Rightarrow \text{"I principal"}).$$

(2). Let

$$I = \{\nu \in \mathbb{Z}^* \mid 2^a \text{ divides } \nu \text{ for some infinite } a \in \mathbb{N}\}.$$

It is readily verified that I is an ideal of $\mathbb{Z}^*$. Now assume by contradiction that $I = \langle \xi \rangle$ is generated by a single element $\xi \in \mathbb{Z}^*$. In particular, there exists an infinite hypernatural number a such that 2^a divides ξ. Then clearly $2^{a-1} \notin \langle \xi \rangle$, but $2^{a-1} \in I$ because also $a - 1$ is infinite. $\qquad\square$

EXERCISE 5.20. *Show that every non-decreasing internal function* $F : \mathbb{N}^* \to \mathbb{N}^*$ *is eventually constant (that is, there exists* $c, M \in \mathbb{N}^*$ *such that* $F(\nu) = c$ *for all* $\nu \geq M$*).*

SOLUTION. Apply *transfer* to the following property, which is expressed in elementary form:

$$\forall f \in \mathfrak{Fun}(\mathbb{N}, \mathbb{N}) \text{ "f non-increasing"} \Rightarrow$$

$$(\exists c, M \in \mathbb{N})(\forall n \in \mathbb{N})(n \geq M \Rightarrow f(n) = c).$$

$\qquad\square$

4. The transfer as a unifying principle

The *transfer principle* is a really powerful tool and can be seen as the primary feature of Alpha-Theory. Provided one has at hand the (non-elementary) notion of "elementary property", *transfer* allows for an equivalent shorter presentation of our theory. Indeed, the following result holds.

THEOREM 5.21. *Assume we are given a notion of Alpha-limit that satisfies axioms (AT0), (AT1), (AT2), and the condition* $\lim_\alpha c_\emptyset = \emptyset$. *Then the* transfer principle *implies (AT3), (AT4) and (AT5).*

PROOF. First of all, we already pointed out in Section 2 of this chapter that by using *transfer*, one can directly prove all preservation properties as presented in Chapter 4, by simply expressing them in elementary form. In particular, if we denote by $S : \mathbb{R} \times \mathbb{R} \to \mathbb{R}$ and by $P : \mathbb{R} \times \mathbb{R} \to \mathbb{R}$ the sum and product operations respectively, then $S^* : \mathbb{R} \times \mathbb{R}^* \to \mathbb{R}^*$ and $P^* : \mathbb{R}^* \times \mathbb{R}^* \to \mathbb{R}^*$ are binary functions such that for every $\varphi, \psi : \mathbb{N} \to \mathbb{R}$:

- $S^*(\lim_{n\uparrow\alpha} \varphi(n), \lim_{n\uparrow\alpha} \psi(n)) = \lim_{n\uparrow\alpha} S(\varphi(n), \psi(n))$;
- $P^*(\lim_{n\uparrow\alpha} \varphi(n), \lim_{n\uparrow\alpha} \psi(n)) = \lim_{n\uparrow\alpha} P(\varphi(n), \psi(n))$.

All the properties of a field are expressed in elementary terms, as one can readily verify. So, by *transfer*, we obtain that $\mathbb{R}^*$ is a field with operations S^* and P^*, and this concludes the proof of the *Field Axiom* (AT3).

The *Internal Set Axiom* (AT4) is obtained as a straightforward application of *transfer* to the membership relation $\varphi(n) \in \psi(n)$. Finally, also the *Composition Axiom* (AT5) is easily proved by *transfer*. In fact, $\lim_{n\uparrow\alpha} \varphi(n) = \lim_{n\uparrow\alpha} \psi(n) \Leftrightarrow \varphi(n) = \psi(n) \ a.e. \Rightarrow f(\varphi(n)) = f(\psi(n))$ $a.e. \Leftrightarrow \lim_{n\uparrow\alpha} f(\varphi(n)) = \lim_{n\uparrow\alpha} f(\psi(n))$. $\square$

Thus, one can give an equivalent alternative presentation of the Alpha-Theory as follows.

Alternative Alpha-Theory ALT

(ALT0) Existence Axiom. *Every sequence $\varphi(n)$ has a unique "Alpha-limit" denoted by $\lim_{n\uparrow\alpha} \varphi(n)$ (or more simply by $\lim_{\alpha} \varphi$).*

(ALT1) Atom-Set Axiom

(1) *If $\varphi(n)$ is a sequence of atoms, then also $\lim_{n\uparrow\alpha} \varphi(n)$ is an atom.*

(2) *If $c_r(n) = r$ is the constant sequence with value a real number r, then $\lim_{n\uparrow\alpha} c_r(n) = r$.*

(3) *If $c_\emptyset$ is the constant sequence with value the empty set, then $\lim_{n\uparrow\alpha} c_\emptyset(n) = \emptyset$.*

(ALT2) Alpha Number Axiom. *The Alpha-limit of the identity sequence $\imath(n) = n$ is a "new" number denoted by $\boldsymbol{\alpha}$, that is $\lim_{n\uparrow\alpha} n = \boldsymbol{\alpha} \notin \mathbb{N}$.*

(ALT3) Leibniz Transfer Axiom.   *Let $\sigma(x_1, \ldots, x_k)$ be an elementary formula and let $\varphi_1, \ldots, \varphi_k$ be arbitrary sequences. Then*

$$\sigma(\varphi_1(n), \ldots, \varphi_k(n)) \; eventually \; \Longrightarrow \; \sigma\left(\lim_{n \uparrow \alpha} \varphi_1(n), \ldots, \lim_{n \uparrow \alpha} \varphi_k(n)\right).$$

5. Remarks and comments

1. With respect to more common definitions of a *first-order formula*, in our definition of *elementary formula*:

- We can quantify over a variable *only* when that variable actually appears free in the formula under consideration;
- We *cannot* join two formulas by means of a binary connective when a same variable appears free in one formula and bound in the other.

As a consequence, we do not have mixed situations, and in every formula a given variable is either free in all its occurrences, or it is bound in all its occurrences.

Finally, let us remark that the provisions as required by rules (EF2) and (EF4) are not actual limitations. Indeed, one can always "rename" variables so as to fulfill those conditions, while the meaning of the formula is left unchanged.

2. In *nonstandard analysis*, the *transfer principle* is formulated as in Corollary 5.9. In that context, the stronger property of Theorem 5.8 is usually referred to by saying that "elementary properties" are preserved by internal sets.

CHAPTER 6

Complements of Alpha-Theory

In this chapter we collect a number of complementary results about the Alpha-Theory. Some of them are important topics also in the foundations of *nonstandard analysis*, and some others are closely related to the problem of consistency of Alpha-Theory, which will be fully addressed in the next chapter.

1. Overspill and underspill

If an internal set contains arbitrarily large natural numbers, then it necessarily contains also some infinite hypernatural number. Similarly, if an internal set contains arbitrarily small infinite hypernatural numbers, then it necessarily contains also some finite natural number. These peculiar phenomena are really useful in the practice when dealing with internal sets, and they are known as the *overspill* and the *underspill* principle, respectively.

THEOREM 6.1.

(1) Overspill Principle: *Let A be an internal set. If $A \cap \mathbb{N}$ is unbounded in $\mathbb{N}$ then $A \cap \mathbb{N}_\infty \neq \emptyset$;*

(2) Underspill Principle: *Let A be an internal set. If $A \cap \mathbb{N}_\infty$ is unbounded below in $\mathbb{N}_\infty$, then $A \cap \mathbb{N} \neq \emptyset$.*

PROOF. (1). Consider the following set:

$$B = \{\nu \in \mathbb{N}^* \mid \exists a \in A \text{ s.t. } a > \nu\}.$$

We remark that B is internal by the *Internal Definition Principle*. If by contradiction A does not contain any infinite natural number, while it contains arbitrarily large finite numbers, then we would have $B = \mathbb{N}$. This is not possible because $\mathbb{N}$ is not internal.

(2). We proceed similarly as above, and consider the following set

$$C = \{\nu \in \mathbb{N}^* \mid \exists a \in A \text{ s.t. } a < \nu\},$$

which is internal by the *Internal Definition Principle*. If by contradiction $A \cap \mathbb{N} = \emptyset$, while A contains arbitrarily small infinite hypernatural numbers, then $B = \mathbb{N}_\infty$, and this cannot be since $\mathbb{N}_\infty$ is not internal. $\qquad\square$

Two closely related properties are itemized below.

PROPOSITION 6.2. *Let A be an internal set. Then*

(1) Overspill Principle for Intervals: *$A \supseteq [n, +\infty]_\mathbb{N}$ contains all finite natural numbers from some point on if and only if $A \supseteq [n, \nu]_{\mathbb{N}^*}$ for some infinite ν.*

(2) Underspill Principle for Intervals: *$A \supseteq \mathbb{N}_\infty$ contains all infinite hypernatural numbers if and only if $A \supseteq [n, +\infty]_{\mathbb{N}^*}$ for some finite $n \in \mathbb{N}$.*

PROOF. (1). The following set is internal by the *Internal Definition Principle*:

$$B = \{\nu \in \mathbb{N}^* \mid [n, \nu]_{\mathbb{N}^*} \subseteq A\}.$$

By the hypothesis, B contains all finite natural numbers $k \geq n$ and so, by *overspill*, there exists an infinite $\nu \in B \cap \mathbb{N}_\infty$.

(2). Similarly as above, the set

$$C = \{\nu \in \mathbb{N}^* \mid [\nu, +\infty]_{\mathbb{N}^*} \subseteq A\}$$

is internal, and contains all infinite hypernatural numbers. By *underspill* there exists a finite $n \in \mathbb{N} \cap B$, and hence $[n, +\infty]_{\mathbb{N}^*} \subseteq A$, as desired. $\qquad\square$

Similar properties also hold with respect to real and hyperreal numbers. The proofs are left to the reader as an exercise.

EXERCISE 6.3. *Let A be an internal set. Prove the following:*

(1) *If $A \cap \mathbb{R}$ is unbounded in $\mathbb{R}$, then A also contains some infinite hyperreal number;*

(2) *If $A \supseteq \mathbb{R}^*_{\text{fin}}$, then $A \supseteq [-\xi, \xi]_{\mathbb{R}^*}$ for some positive $\xi \in \mathbb{R}_\infty$;*

(3) *If A contains arbitrarily large positive (or arbitrarily small negative) infinitesimal numbers, then it also contains some positive (negative, respectively) non-infinitesimal hyperreal number;*

(4) *If A contains all infinitesimal numbers, then $A \supseteq [-\epsilon, +\epsilon]_{\mathbb{R}^*}$ for some positive real number ϵ.*

PROOF. (1). By the hypothesis, the internal set

$$B = \{\nu \in \mathbb{N}^* \mid \exists a \in A \cap \mathbb{R}^* \text{ s.t. } |a| > \nu\}$$

contains all finite natural numbers. Then, by *overspill*, there exists an infinite $\nu \in B \cap \mathbb{N}_\infty$, and so B contains an infinite hyperreal number.

(2). By the hypothesis, the internal set

$$B = \{\nu \in \mathbb{N}^* \mid [-\nu, \nu]_{\mathbb{R}^*} \subseteq A\}$$

contains all finite natural numbers and hence, by *overspill*, it also contains an infinite $\nu \in \mathbb{N}_\infty$. So $[-\nu, +\nu]_{\mathbb{R}^*} \subseteq A$.

(3). By the hypothesis, the internal set

$$B = \{\nu \in \mathbb{N}^* \mid \exists a \in A \text{ s.t. } |a| > 1/\nu\}$$

contains all infinite hypernatural numbers. By *underspill*, there exists also a finite $n \in \mathbb{N} \cap B$, and so there exists an element $a \in A$ with $|a| > 1/n$, as desired.

(4). By the hypothesis, the internal set

$$B = \{\nu \in \mathbb{N}^* \mid [-1/\nu, 1/\nu]_{\mathbb{R}^*} \subseteq A\}$$

contains all infinite hypernatural numbers. By *underspill*, there exists also a finite $n \in \mathbb{N} \cap B$, and so $[-1/n, 1/n]_{\mathbb{R}^*} \subseteq A$, as desired. $\square$

EXERCISE 6.4. *Prove that a set $A \subseteq \mathbb{R}$ of real numbers is internal if and only if A is finite.*

PROOF. One direction is straightforward because any real number $r = \lim_{n \uparrow \alpha} c_r(n)$ is internal, and a finite set of internal objects is internal.

Conversely, assume by contradiction that the internal set A is infinite. If A is unbounded, then A contains infinite hypernatural numbers, and so $A \not\subseteq \mathbb{R}$ (see Exercise 6.3 (1)).

If A is bounded, then it has a cluster point r. So, the following set contains all finite natural numbers:

$$B = \{\nu \in \mathbb{N}^* \mid \exists a \in A \text{ s.t. } a \neq x \ \& \ |a - r| < 1/\nu\}.$$

We remark that B is internal by the *Internal Definition Principle*, and hence, by *overspill*, there exists an infinite $\nu \in \mathbb{N}_\infty \cap B$. Pick $a \in A$ with $a \neq r$ and $|a - r| < 1/\nu$. Finally, notice that $a \sim r$ and $a \neq r$ imply that $a \notin \mathbb{R}$, and we reach the contradiction $A \not\subseteq \mathbb{R}$. $\qquad \square$

EXERCISE 6.5 (*Overspill of infinitesimals*).
*Let $f, g : \mathbb{R} \to \mathbb{R}$ be a real function, and let $\bowtie$ be any of the relations:
$=, \neq, \leq, <$. Then the following two properties are equivalent:*

(1) *$f^*(\varepsilon) \bowtie g^*(\varepsilon)$ for all positive $\varepsilon \sim 0$;*

(2) *There exists $x_0 > 0$ such that $f(x) \bowtie g(x)$ for all $0 < x < x_0$.*

2. Countable saturation

DEFINITION 6.6. *A nonempty family of sets $\mathcal{F}$ satisfies the* finite intersection property *("FIP" for short) if all finite intersections of elements of $\mathcal{F}$ are nonempty.*

$$\forall A_1, \ldots, A_k \in \mathcal{F} \quad \bigcap_{i=1}^{k} A_i \neq \emptyset.$$

It is not hard to show that every countable family with the FIP has a nonempty intersection, *provided* one passes to hyper-images.

PROPOSITION 6.7. *If $\{A_n \mid n \in \mathbb{N}\}$ is a countable family with the finite intersection property then:*

$$\bigcap_{n \in \mathbb{N}} A_n^* \neq \emptyset.$$

PROOF. Let σ be the sequence where $\sigma(n) = A_1 \cap \ldots \cap A_n$, and consider its hyper-image σ^*. By the hypothesis, $\sigma(n) \neq \emptyset$ for every $n \in \mathbb{N}$ and so, by *transfer*, also $\sigma^*(\nu) \neq \emptyset$, where ν is any fixed infinite hypernatural number. Since $\sigma(n) \subseteq \sigma(m)$ whenever $n > m$, again by *transfer* we obtain that $\sigma^*(\nu) \subseteq [\sigma(n)]^* = A_1^* \cap \ldots \cap A_n^*$ for every $n \in \mathbb{N}$. But then $\emptyset \neq \sigma^*(\nu) \subseteq \bigcap_{n \in \mathbb{N}} A_n^*$, as desired. $\qquad \square$

An important tool used in *nonstandard analysis* is *countable saturation*, an intersection property for internal sets. The same property also holds in the Alpha-Theory.

THEOREM 6.8 (Countable Saturation Property).
*If $\{B_k \mid k \in \mathbb{N}\}$ is a countable family of internal sets with the FIP,
then:*

$$\bigcap_{k \in \mathbb{N}} B_k \neq \emptyset.$$

We give two different proofs of this important theorem. The first
one is obtained by an elementary purely combinatorial argument. The
second one is obtained by a combined application of the *Comprehensiveness Property* and of the *Overspill Principle*.

PROOF # 1. For every k, let φ_k be a sequence with $\lim_{n \uparrow \alpha} \varphi_k(n) = B_k$. For any fixed n, pick an element $\psi(n) \in \varphi_1(n) \cap \cdots \cap \varphi_n(n)$ if that
intersection is nonempty. Otherwise, pick $\psi(n) \in \varphi_1(n) \cap \cdots \cap \varphi_{n-1}(n)$
if that intersection is nonempty, and continue in this manner until
the element $\psi(n)$ is defined. (In case $\varphi_1(n) = \emptyset$ then let $\psi(n)$ be an
arbitrary element.) As a consequence of our definition, the following
property holds:

- If $\varphi_1(n) \cap \cdots \cap \varphi_k(n) \neq \emptyset$, then for every $n \geq k$:

$$\psi(n) \in \varphi_1(n) \cap \cdots \cap \varphi_k(n).$$

Now let k be fixed. By the hypothesis of FIP,

$$\emptyset \neq \bigcap_{i=1}^{k} \lim_{n \uparrow \alpha} \varphi_i(n) = \lim_{n \uparrow \alpha} \left(\bigcap_{i=1}^{k} \varphi_i(n) \right),$$

and hence, $\varphi_1(n) \cap \cdots \cap \varphi_k(n)$ is nonempty *a.e..* By the property
itemized above, it follows that $\psi(n) \in \varphi_1(n) \cap \cdots \cap \varphi_k(n)$ *a.e.*, and so
$\lim_\alpha \psi \in B_1 \cap \cdots \cap B_k$. As this holds for every k, we conclude that
$\lim_\alpha \psi \in \bigcap_k B_k$ and the proof is completed.   □

PROOF # 2. Pick a set C such that $B_n \subseteq C$ for all n, and define
the function $f : \mathbb{N} \to C^*$ by letting

$$f(n) = \bigcap_{i \leq n} B_i.$$

By the *comprehensiveness property* of Theorem 4.70, we can pick an
internal function $F : \mathbb{N}^* \to B^*$ such that $F(n) = f(n)$ for all $n \in \mathbb{N}$.
Now consider the following internal set

$$\Lambda = \{\nu \in \mathbb{N}^* \mid \forall \mu \leq \nu \;\; F(\mu) \supseteq F(\nu) \neq \emptyset\}.$$

By the hypothesis, Λ contains all finite natural numbers $n \in \mathbb{N}$,
and hence, by *overspill*, it also contains an infinite $\nu \in \mathbb{N}_\infty \cap \Lambda$. Then

$\emptyset \neq F(\nu) \supseteq F(n) = f(n) = \bigcap_{i \leq n} B_i$ for all $n \in \mathbb{N}$, and so we reach the desired conclusion $\bigcap_{n \in \mathbb{N}} B_n \supseteq F(\nu) \neq \emptyset$. $\square$

As a corollary, we obtain an alternative proof that no countable internal set can be internal (see Proposition 4.67).

PROPOSITION 6.9. *Any infinite countable set is not internal.*

PROOF. By contradiction, assume that the infinite countable set $A = \{a_n \mid n \in \mathbb{N}\}$ is internal. Then, for each n, also

$$A_n = A \setminus \{a_1, \ldots, a_n\}$$

is internal. The countable family $\{A_n \mid n \in \mathbb{N}\}$ satisfies the FIP and so, by *saturation*, its intersection is nonempty, a contradiction. $\square$

Other useful consequences of saturation are itemized below.

PROPOSITION 6.10.

(1) *Every sequence $f : \mathbb{N} \to \mathbb{R}^*$ of hyperreal numbers is bounded;*

(2) *Every sequence $f : \mathbb{N} \to \mathbf{mon}(0)$ of infinitesimal numbers is bounded in $\mathbf{mon}(0)$;*

(3) *Every sequence $f : \mathbb{N} \to \mathbb{N}_\infty$ of infinite hypernatural numbers is bounded below in $\mathbb{N}_\infty$, that is, there exists $N \in \mathbb{N}_\infty$ such that $N < f(k)$ for all $k \in \mathbb{N}$.*

PROOF. (1). For every $k \in \mathbb{N}$, let $A_k = \{\xi \in \mathbb{R}^* \mid \xi > f(k)\}$. It is easily verified that the family of internal sets $\{A_k\}$ satisfies the FIP. So, by *saturation*, we can pick $\xi \in \bigcap_k A_k$. Clearly $\xi > f(k)$ for all k.

(2). For every $k \in \mathbb{N}$, let $A_k = (f(k), 1/k)_{\mathbb{R}^*}$. It directly follows from the hypothesis that the family of internal intervals $\{A_k\}$ satisfies the FIP. Then, by *saturation* we can pick $\varepsilon \in \bigcap_k A_k$. Clearly ε is infinitesimal and $\varepsilon > f(k)$ for all k.

(3). Define the sequence $g : \mathbb{N} \to \mathbf{mon}(0)$ by letting $g(k) = 1/f(k)$. By (2) there exists an infinitesimal ε such that $g(k) < \varepsilon$ for all k. If $N \in \mathbb{N}_\infty$ is the integer part of $1/\varepsilon$, then $N < f(k)$ for all k. $\square$

3. Cauchy infinitesimal principle

A famous and much discussed historical viewpoint on the *continuum* was the one expressed by Augustin-Louis Cauchy, one of the fathers of modern calculus. In his celebrated textbooks published in the first half of the nineteenth century, Cauchy made use of infinitely small quantities, that he described as "variables converging to zero".

> *One says that a variable quantity becomes infinitely small when its numerical [that is, absolute] value decreases indefinitely in such a way as to converge toward the limit zero ...*
>
> *... an infinitely small quantity, that is, a variable whose numerical value decreases indefinitely ...*
>
> (Cauchy, *Cours d'Analyse de l'École Royale Polytechnique Sur le Calcul Infinitésimal*, 1821, pp. 26-27)[1]

What Cauchy really meant by those words is not made clear in his writings, and possible interpretations of his ideas of *variable* and of *infinitely small quantity* have been repeatedly disputed in the recent historical literature.[2] In this section we show how Alpha-Calculus provides a natural framework where one can formalize a possible interpretation to the above vague notions.

Inspired by Cauchy's view of an infinitesimal quantity as a variable when its value "decreases indefinitely in such a way as to converge toward the limit zero", we formulate:

(CIP) Cauchy Infinitesimal Principle
Every positive infinitesimal number is the Alpha-limit of some decreasing infinitesimal sequence.[3]

We remark right away that *Cauchy Infinitesimal Principle* is actually an *independent* property with respect to Alpha-Theory; that is to say, it cannot be proved nor disproved by only assuming axioms (AT0), ..., (AT5). On the other hand, (CIP) can be safely added to Alpha-Theory, because the resulting theory admits models. However, the existence of such models requires additional set-theoretic assumptions such as the *continuum hypothesis*, and so interesting foundational issues arise. A discussion of these topics is given in §14.3.

---

[1] As translated in [**48**]).

[2] See, *e.g.*, [**28, 66, 48, 31**].

[3] By "infinitesimal sequence" we mean a sequence converging to zero (in the sense of classic calculus).

At this point, the question naturally arises as to whether it is worthwhile to also include *Cauchy Infinitesimal Principle* in the list of our axioms. We believe so for at least three different reasons:

- First of all, (CIP) formalizes a natural intuition, namely the idea that any positive infinitesimal number is obtained as the result of a limit process under some decreasing infinitesimal sequence;
- Moreover, as we will see in Section 5 of this chapter, (CIP) has the pleasant consequence that our Alpha-limits are actually the limits with respect to a suitable topology (see Theorem 6.23);
- Finally, *Cauchy Infinitesimal Principle* is equivalent to *Zermelo's Principle* for *Alpha-numerosity* (see §16.8).

(CIP) is equivalent to the seemingly stronger property that every hyperreal number ξ is either the Alpha-limit of a constant sequence (when $\xi \in \mathbb{R}$) or the Alpha-limit of a *monotone* sequence that converges to its standard part $\mathrm{st}(\xi)$.

PROPOSITION 6.11. *Assume the axioms of Alpha-Theory. Then (CIP) holds if and only if for every $\xi \in \mathbb{R}^* \setminus \mathbb{R}$ there exists a monotone real sequence $\varphi(n)$ such that $\lim_{n\uparrow\alpha} \varphi(n) = \xi$ and $\lim_{n\to\infty} \varphi(n) = \mathrm{st}(\xi)$.*

PROOF. The "if" implication is trivial. Conversely, if $\varepsilon < 0$ is a negative infinitesimal number, by (CIP) we have that $-\varepsilon = \lim_{n\uparrow\alpha} \psi(n)$ for some decreasing infinitesimal sequence $\psi(n)$, and so $\varepsilon = \lim_{n\uparrow\alpha} \varphi(n)$, where $\varphi(n) = -\psi(n)$ is infinitesimal and increasing. This shows that the assertion is satisfied by every non-zero infinitesimal number.

Now assume that $\xi \in \mathbb{R}^* \setminus \mathbb{R}$ is finite. Then $\xi = r + \varepsilon$ for some real number $r \in \mathbb{R}$ and some infinitesimal $\varepsilon \neq 0$. Take a monotone infinitesimal sequence $\psi(n)$ such that $\lim_{n\uparrow\alpha} \psi(n) = \varepsilon$, and let $\varphi(n) = r + \psi(n)$. Then clearly $\lim_{n\uparrow\alpha} \varphi(n) = \xi$ and $\lim_{n\to\infty} \varphi(n) = r = \mathrm{st}(\xi)$.

Finally, assume that ξ is infinite. Since $1/\xi$ is a non-zero infinitesimal number, there exists a monotone infinitesimal sequence $\psi(n) \neq 0$ such that $\lim_{n\uparrow\alpha} \psi(n) = 1/\xi$. Then $\xi = \lim_{n\uparrow\alpha} \varphi(n)$ where $\varphi(n) = \frac{1}{\psi(n)}$ is monotone. Moreover, it is readily seen that $\lim_{n\to\infty} \varphi(n) = +\infty = \mathrm{st}(\xi)$ if ξ is positive infinite, and $\lim_{n\to\infty} \varphi(n) = -\infty = \mathrm{st}(\xi)$ if ξ is negative infinite. $\qquad\square$

Usually, the existence of several equivalent formulations of a given property is considered as a sign of its relevance. (CIP) fulfills this condition, as shown by the following theorem.

THEOREM 6.12. *(CIP) is equivalent to each of the following:*[4]

(1) *For every $\nu, \eta \in \mathbb{N}_\infty$ there exists a bijection $\varphi : \mathbb{N} \to \mathbb{N}$ such that $\varphi^*(\nu) = \eta$;*

(2) *For every $\nu \in \mathbb{N}_\infty$ there exists a bijection $\varphi : \mathbb{N} \to \mathbb{N}$ such that $\lim_{n\uparrow\alpha} \varphi(n) = \nu$;*

(3) *For every $\nu \in \mathbb{N}_\infty$ there exists a 1-1 sequence $\varphi : \mathbb{N} \to \mathbb{N}$ such that $\lim_{n\uparrow\alpha} \varphi(n) = \nu$;*

(4) *For every $\nu \in \mathbb{N}_\infty$ there exists a sequence $\varphi : \mathbb{N} \to \mathbb{N}$ and a qualified set X such that the restriction $\varphi_{|X}$ is 1-1 and $\lim_{n\uparrow\alpha} \varphi(n) = \nu$;*

(5) *For every $\nu \in \mathbb{N}_\infty$ there exists a sequence $\varphi : \mathbb{N} \to \mathbb{N}$ and a qualified set X such that the restriction $\varphi_{|X}$ is increasing and $\lim_{n\uparrow\alpha} \varphi(n) = \nu$;*

(6) *For every $\nu \in \mathbb{N}^*$ there exists a non-decreasing sequence $\varphi : \mathbb{N} \to \mathbb{N}$ such that $\lim_{n\uparrow\alpha} \varphi(n) = \nu$;*

PROOF. (1) $\Leftrightarrow$ (2). One implication is trivial, because $\varphi^*(\alpha) = \lim_{n\uparrow\alpha} \varphi(n)$ for every sequence φ. Conversely, pick bijections $\varphi, \psi : \mathbb{N} \to \mathbb{N}$ with $\varphi^*(\alpha) = \nu$ and $\psi^*(\alpha) = \eta$. Then $\vartheta = \psi \circ \varphi^{-1}$ is a bijection with $\vartheta^*(\nu) = \eta$.

(2) $\Rightarrow$ (3). Trivial.

(3) $\Rightarrow$ (5). Pick a 1-1 sequence $\varphi : \mathbb{N} \to \mathbb{N}$ such that $\lim_{n\uparrow\alpha} \varphi(n) = \nu$. Then define inductively:

$$n_1 = 1; \quad n_{k+1} = \min\{n \mid \varphi(n) > \varphi(n_1), \ldots, \varphi(n_k)\}.$$

Notice that this definition is well-posed by the injectivity of φ. Notice also that $\varphi(i) < \varphi(j)$ whenever $i \leq n_k$ and $j \geq n_{k+1}$ for some k.

Now consider the following sequence:

$$\vartheta(n) = k \iff n \in [n_k, n_{k+1}).$$

The Alpha-limit $\lim_{n\uparrow\alpha} \vartheta(n) = \eta \in \mathbb{N}_\infty$ is infinite, and so $\eta = \lim_{n\uparrow\alpha} \tau(n)$ for some 1-1 sequence τ. Clearly, the qualified set

$$A = \{n \in \mathbb{N} \mid \vartheta(n) = \tau(n)\}$$

meets each interval $[n_k, n_{k+1})$ in at most one point. The sets $B' = \bigcup_{s \in \mathbb{N}} [n_{2s-1}, n_{2s})$ and $B'' = \bigcup_{k \in \mathbb{N}} [n_{2s}, n_{2s+1})$ form a partition of $\mathbb{N}$ and so exactly one of them is qualified, say B' (the other case is entirely similar). Finally, consider the qualified set $X = A \cap B'$. We claim that the restriction $\varphi_{|X}$ is increasing.

---

[4] In §14.3, it is shown that this list of properties correspond to the properties of a *selective ultrafilter* on $\mathbb{N}$.

Let $x < y$ be in X. Then $n_{2s-1} \leq x < n_{2s}$ and $n_{2t-1} \leq y < n_{2t}$ for suitable $s < t$, and $x < n_{2s} < n_{2s+1} \leq n_{2t-1} \leq y$ implies that $\varphi(x) < \varphi(y)$, as desired.

(5) $\Rightarrow$ (4). Trivial.

(4) $\Rightarrow$ (2). Given $\nu \in \mathbb{N}_\infty$, fix a sequence $\psi : \mathbb{N} \to \mathbb{N}$ and a qualified set X such that $\lim_{n\uparrow\alpha} \psi(n) = \nu$ and the restriction $\psi_{|X}$ is 1-1. Then split $X = X' \cup X''$ into two disjoint infinite pieces. Exactly one of them is qualified, say X'. Since both $\mathbb{N} \setminus X' \supseteq X''$ and $\mathbb{N} \setminus \psi[X'] \supseteq \psi[X'']$ are infinite, we can pick a bijection $f : \mathbb{N} \setminus X' \to \mathbb{N} \setminus \psi[X']$. Finally, define the sequence $\varphi : \mathbb{N} \to \mathbb{N}$ by putting together $\psi_{|X'}$ and f (*i.e.* let $\varphi(n) = \psi(n)$ if $n \in X'$ and $\varphi(n) = f(n)$ if $n \notin X'$). It is easily verified that φ is a bijection. Moreover $\lim_{n\uparrow\alpha} \varphi(n) = \nu$, since ψ and φ agree on the qualified set X'.

Remark that we have now completed the proof that the first five conditions are equivalent.

(5) $\Rightarrow$ (CIP). Given a positive infinitesimal number $\varepsilon \sim 0$, pick a sequence $\psi(n)$ of positive real numbers such that $\lim_{n\uparrow\alpha} \psi(n) = \varepsilon$, and define the sequence $\vartheta : \mathbb{N} \to \mathbb{N}$ as follows.

$$\vartheta(n) = \begin{cases} 1 & \text{if } \psi(n) > 1 \\ i+1 & \text{if } \frac{1}{i+1} < \psi(n) \leq \frac{1}{i}. \end{cases}$$

Notice that $\vartheta(n) = k \in \mathbb{N}$ *a.e.* would imply that $\varepsilon < \frac{1}{k}$, against the hypothesis $\varepsilon \sim 0$. So, $\lim_{n\uparrow\alpha} \vartheta(n) = \nu \in \mathbb{N}_\infty$ is infinite, and by the hypothesis there exists a qualified set X such that the restriction $\vartheta_{|X}$ is increasing, and hence $\psi_{|X}$ is decreasing. Finally, pick any decreasing real sequence $\varphi : \mathbb{N} \to \mathbb{R}^+$ that agrees with ψ on X, so that $\lim_{n\uparrow\alpha} \varphi(n) = \lim_{n\uparrow\alpha} \psi(n) = \varepsilon$, as desired.

(CIP) $\Rightarrow$ (6). Given $\nu \in \mathbb{N}_\infty$, by Proposition 6.11 we can pick an increasing *real* sequence $\psi : \mathbb{N} \to \mathbb{R}$ such that $\lim_{n\uparrow\alpha} \psi(n) = \nu$. Let $\varphi(n) = \lfloor \psi(n) \rfloor$ be the integer part of $\psi(n)$. Clearly $\varphi : \mathbb{N} \to \mathbb{N}$ is non-decreasing. Moreover, since ν is a hypernatural number, the set $A = \{n \mid \varphi(n) = \psi(n)\} = \{n \mid \psi(n) \in \mathbb{N}\}$ is qualified and so $\lim_{n\uparrow\alpha} \varphi(n) = \lim_{n\uparrow\alpha} \psi(n) = \nu$.

(6) $\Rightarrow$ (4). Given $\nu \in \mathbb{N}_\infty$, let $\varphi : \mathbb{N} \to \mathbb{N}$ be a non-decreasing sequence such that $\lim_{n\uparrow\alpha} \varphi(n) = \nu$. Remark that $\varphi(n)$ is necessarily unbounded. Enumerate the elements of its range in increasing order:

$$\mathrm{ran}(\varphi) = \{m_1 < \ldots < m_k < m_{k+1} < \ldots\}.$$

Notice that the preimages $\varphi^{-1}(m_k) = [n_k, n_{k+1})$ for $k \in \mathbb{N}$ are consecutive finite intervals that partition $\mathbb{N}$. Now define $\vartheta : \mathbb{N} \to \mathbb{N}$ by putting:

$$\vartheta(n) = n_{k+1} + n_k - n - 1 \quad \text{when } n \in [n_k, n_{k+1}).$$

Notice that ϑ is a bijection that becomes a *decreasing* bijection when restricted to each interval $[n_k, n_{k+1})$. Moreover ϑ is "almost increasing" in the sense that $\vartheta(x) < \vartheta(y)$ whenever $x < y$ belong to different intervals. Clearly $\lim_{n \uparrow \alpha} \vartheta(n) = \eta \in \mathbb{N}_\infty$ is infinite and so, by the hypothesis, we can pick a non-decreasing sequence $\psi : \mathbb{N} \to \mathbb{N}$ such that $\lim_{n \uparrow \alpha} \psi(n) = \eta$. By the properties of ϑ, it is easily checked that the qualified set $X = \{n \mid \psi(n) = \vartheta(n)\}$ contains at most one point from each interval $[n_k, n_{k+1})$. We conclude that the restriction $\varphi_{|X}$ is increasing, as desired.   □

4. The S-topology

Every hyper-image X^* is naturally endowed with a topology, namely the topology generated by the hyper-images of subsets of X.

DEFINITION 6.13. *The* S-topology *(standard topology) on the hyper-image X^* is the topology having the family $\{A^* \mid A \subseteq X\}$ as a basis of open sets.*[5]

Every basic open set A^* is also closed because the complement $X^* \setminus A^* = {}^*(X \setminus A)$ is open. In particular, the S-topologies are totally disconnected.

Trivially, the canonical copy of X in its hyper-image, namely the *canonical image*

$$X^\sigma = \{x^* \mid x \in X\} \subseteq X^*,$$

is a discrete and dense subspace of X^*. Two more basic properties of the S-topology are the following.

PROPOSITION 6.14.

(1) *For every A, the closure $\overline{A^\sigma}$ is the hyper-image A^*;*

(2) *For every function $f : X \to Y$, its hyper-image $f^* : X^* \to Y^*$ is continuous with respect to the S-topologies on X^* and Y^*.*

---

[5] The notion of "S-topology" was introduced in *nonstandard analysis* by Abraham Robinson himself (see [**83**] §4.4). The letter "S" stands for "standard" (recall that, in *nonstandard analysis*, sets of the form A^* are named *standard*).

PROOF. (1). Since A^* is closed and $A^\sigma \subseteq A^*$, trivially $\overline{A^\sigma} \subseteq A^*$. Conversely, we need to show that every $\xi \in A^*$ is in the closure of A^σ. If B^* is an arbitrary basic open neighborhood of ξ, then $\xi \in B^* \cap A^* \neq \emptyset$, so by *transfer* we can pick an element $c \in B \cap A \neq \emptyset$, and clearly $c^* \in B^* \cap A^\sigma \neq \emptyset$. (See also Proposition 4.23.)

(2). For every $B \subseteq Y$, the preimage $(f^*)^{-1}(B^*) = (f^{-1}(B))^*$ is open.   □

EXERCISE 6.15. *If $U \subseteq X^*$ is open, then its closure*

$$\overline{U} = \{x \in X \mid x^* \in U\}^*.$$

SOLUTION. Denote by $\Gamma = \{x \in X \mid x^* \in U\}$. By the previous proposition $\Gamma^* = \overline{\Gamma^\sigma}$, so we have to prove that Γ^σ and U have the same closure.

Notice that $\Gamma^\sigma = U \cap X^\sigma \subseteq U$, and hence $\overline{\Gamma^\sigma} \subseteq \overline{U}$. Conversely, let $\xi \in \overline{U}$ and let A^* be a basic open neighborhood of ξ; then $A^* \cap U \neq \emptyset$. By the hypothesis on U, that intersection is open and so, by the density of X^σ in X^*, we have that $\emptyset \neq (A^* \cap U) \cap X^\sigma = A^* \cap (U \cap X^\sigma) = A^* \cap \Gamma^\sigma$. This shows that ξ is in the closure of Γ^σ, as desired.   □

EXERCISE 6.16. *Let X be any infinite set. Prove that the S-topology on X^* is not first countable, by showing that no point $\xi \in X^* \setminus X^\sigma$ has a countable basis of neighborhoods.*

SOLUTION. Let $\{A_n^* \mid n \in \mathbb{N}\}$ be a countable family of basic open neighborhoods of $\xi \in X^* \setminus X^\sigma$, and let $B_n = \bigcap_{i=1}^n A_i$. All sets B_n are infinite because $\xi \in \bigcap_{i=1}^n A_i^* = B_n^*$, and so we can inductively define a sequence $\langle b_n \mid n \in \mathbb{N} \rangle$ where $b_1 \in B_1 = A_1$ and $b_{n+1} \in B_{n+1} \setminus \{b_1, \ldots, b_n\}$. Then $B \subseteq B_n \cup \{b_1, \ldots, b_{n-1}\} \subseteq A_n \cup \{b_1, \ldots, b_{n-1}\}$. Notice that if $\eta \in B^* \setminus B^\sigma$ then $\eta \in A_n^*$ for all n. This shows that the whole infinite set $B^* \setminus B^\sigma$ is included in the intersection $\bigcap_{n \in \mathbb{N}} A_n^*$, and hence $\{A_n^* \mid n \in \mathbb{N}\}$ cannot be a basis of neighborhoods of ξ.   □

A first natural question that one may ask is whether the S-topologies are Hausdorff or not. With respect to this question, the next proposition shows that – without loss of generality – one can restrict to the S-topology on the hypernatural numbers.

PROPOSITION 6.17. *The S-topology on $\mathbb{N}^*$ is Hausdorff if and only the S-topology is Hausdorff on every hyper-image X^*.*

PROOF. Let ξ and η be two distinct elements of X^*, and pick sequences $\varphi, \psi : \mathbb{N} \to X$ such that $\xi = \lim_{n\uparrow\alpha} \varphi(n)$ and $\eta = \lim_{n\uparrow\alpha} \psi(n)$.

Assume first that the union $\mathrm{ran}(\varphi) \cup \mathrm{ran}(\psi)$ is finite. In this case both φ and ψ are constant a.e., that is, there exists elements $x, y \in X$ such that $\varphi(n) = x$ and $\psi(n) = y$ a.e.. Then $\xi = \lim_{n\uparrow\alpha} \varphi(n) = {}^*x$ and $\eta = \lim_{n\uparrow\alpha} \psi(n) = {}^*y$, and hence $x \neq y$. So, in this case $\xi = {}^*x \in \{{}^*x\} = {}^*\{x\}$ and $\eta = {}^*y \in {}^*\{y\}$, where ${}^*\{x\} \cap {}^*\{y\} = {}^*(\{x\} \cap \{y\}) = {}^*\emptyset = \emptyset$.

Now assume that $Z = \mathrm{ran}(\varphi) \cup \mathrm{ran}(\psi)$ is infinite and fix a 1-1 map $f : Z \to \mathbb{N}$. Then also $f^* : Z^* \to \mathbb{N}^*$ is 1-1 and so $f^*(\xi) \neq f^*(\eta)$. Since the S-topology on $\mathbb{N}^*$ is Hausdorff, there exists disjoint subsets $A, B \subset \mathbb{N}$ such that $f^*(\xi) \in A^*$ and $f^*(\eta) \in B^*$. If $A' = f^{-1}(A)$ and $B' = f^{-1}(B)$, then $(A')^* \cap (B')^* = \emptyset$ are disjoint and separates ξ from η. Indeed, $\xi \in (f^*)^{-1}(A^*) = (f^{-1}(A))^* = (A')^*$ and $\eta \in (f^*)^{-1}(B^*) = (B')^*$, as desired. $\square$

As a consequence of the above proposition, in the sequel we will simply say that the S-topology is Hausdorff to mean that the S-topology is Haudorff on *any* hyper-image X^*.

To further investigate the separation properties of the S-topology, we now need a purely combinatorial result on the natural numbers. Our proof is arranged into two steps: the first one is a simple induction, in the second one we take a suitable Alpha-limit.

PROPOSITION 6.18. *Let $f : \mathbb{N} \to \mathbb{N}$ be such that $f(n) \neq n$ for all n. Then there exists a 3-coloring of $\mathbb{N}$ such that n and $f(n)$ have different colors for all n.*[6]

We remark that 2 colors are not enough. *E.g.*, if $f(1) = 2$, $f(2) = 3$ and $f(3) = 1$, the numbers $1, 2, 3$ must have pairwise different colors.

PROOF. By induction on n, we prove that every finite $X \subset \mathbb{N}$ of cardinality n admits a 3-coloring with the property that n and $f(n)$ have different colors whenever both $f(n), n \in X$. Notice that no condition is required in case at least one of n and $f(n)$ does not belong to X.

The base cases $n = 1, 2, 3$ are trivial. Now let $X = \{x_1, \ldots, x_n\}$ contain n elements. By the *pigeon principle*, at least one (actually $\lceil \frac{n}{2} \rceil$ many) of its elements, say x_i, has at most one preimage that belongs to X. Now pick a 3-coloring of $X \setminus \{x_i\}$ as given by the inductive

---

[6] That is, there exists a partition $\mathbb{N} = A_1 \cup A_2 \cup A_3$ such that $n \in A_i \Rightarrow f(n) \notin A_i$ for all n and for all $i = 1, 2, 3$.

hypothesis, and complete it to a 3-coloring of X as follows: Give x_i a color that is different both from the color of its preimage (if any) and from its image (if in X). Clearly, this is always possible, as we have 3 colors at disposal.

We now have to find a way to "glue together" the various colorings of the finite subsets of $\mathbb{N}$ so as to get a 3-coloring of the whole set $\mathbb{N}$. Notice that the same element n may well have different colors when considered as a member of different finite subsets X. We can overcome this difficulty by considering a suitable Alpha-limit.

For all n, let $[1,n] = A_1^n \cup A_2^n \cup A_3^n$ be a 3-coloring of the set of the first n natural numbers as above, and take the Alpha-limits

$$\lim_{n\uparrow\alpha}[1,n] = [1,\alpha]; \quad \lim_{n\uparrow\alpha}A_i^n = A_i^\alpha \text{ for } i = 1,2,3.$$

It is readily verified by *transfer* that $[1,\alpha] = A_1^\alpha \cup A_2^\alpha \cup A_3^\alpha$ is a 3-coloring such that $\nu, f^*(\nu)$ have different colors whenever both $\nu, f^*(\nu) \in [1,\alpha]$.

Now, the finite hypernatural numbers $\mathbb{N}$ are a (proper) subset of $[1,\alpha]$, and so we can take the intersections $A_i = A_i^\alpha \cap \mathbb{N}$ and obtain a 3-coloring $\mathbb{N} = A_1 \cup A_2 \cup A_3$. Finally, by recalling that $f^*(n) = f(n)$ for all $n \in \mathbb{N}$, one concludes that the above 3-coloring of the natural numbers has the desired property. $\qquad\square$

Although the S-topology may not be Hausdorff, the number α can always be "separated" from any other hypernatural number ν.

THEOREM 6.19. *Let $\nu \in \mathbb{N}^*$ and assume $\alpha \neq \nu$. Then there exists a set $A \subseteq \mathbb{N}$ such that $\alpha \in A^*$ but $\nu \notin A^*$.*

PROOF. Since $\nu \neq \alpha = \lim_{n\uparrow\alpha} n$, we can pick a sequence $\varphi : \mathbb{N} \to \mathbb{N}$ such that $\nu = \lim_{n\uparrow\alpha} \varphi(n)$ and $\varphi(n) \neq n$ for all n. Now apply the previous proposition to φ, and get a 3-coloring $\mathbb{N} = A_1 \cup A_2 \cup A_3$ such that $\varphi[A_i] \cap A_i = \emptyset$ for $i = 1,2,3$. By considering the hyper-images, we obtain a 3-coloring of the hypernatural numbers $\mathbb{N}^* = A_1^* \cup A_2^* \cup A_3^*$ such that $\varphi^*[A_i^*] \cap A_i^* = \emptyset$ for $i = 1,2,3$. If j is such that $\alpha \in A_j^*$, then $\nu = \varphi^*(\alpha) \notin A_j^*$, as desired. $\qquad\square$

A possible way of interpreting this result is the following: The number α is characterized by its "properties" because for every hypernatural number $\nu \neq \alpha$ there exists a "property" $A \subseteq \mathbb{N}$ which is satisfied by α but not by ν (that is, $\alpha \in A^*$ but $\nu \notin A^*$).

An important consequence of the *Cauchy Infinitesimal Principle* is the following.

THEOREM 6.20. *Assume (CIP). Then the S-topology is Hausdorff.*

PROOF. Let $\nu, \eta \in \mathbb{N}^*$ be two distinct hypernatural numbers. If one of the two numbers is finite, say $\nu \in \mathbb{N}$, then trivially $\{\nu\} = {}^*\{\nu\}$ separates ν from η. So, let us assume that both $\nu, \eta \in \mathbb{N}_\infty$.

By (CIP), there exists a bijection $\varphi : \mathbb{N} \to \mathbb{N}$ such that $\varphi^*(\nu) = \boldsymbol{\alpha}$. As $\varphi^*(\eta) \neq \varphi^*(\nu) = \boldsymbol{\alpha}$, by Proposition 6.18 we can find $A \subset \mathbb{N}$ such that $\boldsymbol{\alpha} \in A^*$ but $\varphi^*(\eta) \notin A^*$. It follows that $\nu \in (\varphi^*)^{-1}(A^*)$ and $\eta \notin (\varphi^*)^{-1}(A^*)$, and hence the hyper-image of $B = \varphi^{-1}(A) \subset \mathbb{N}$ separates ν from η, as desired. □

We will see further on in §14.3 that the converse implication does not hold; indeed, there exist models of Alpha-Theory where the S-topology is Hausdorff but (CIP) fails.

5. The topology of Alpha-limits

The classic notion of limit originates from topology; on the contrary, our Alpha-limit was introduced axiomatically, with no reference to any underlying topology. So, it is natural to ask whether the Alpha-limit $\lim_{n\uparrow\alpha} f(n)$ of a real sequence is actually the limit of $f(n)$ as n approaches to $\boldsymbol{\alpha}$ with respect to suitable topologies $\tau_\mathbb{N}$ and $\tau_\mathbb{R}$ on the hypernatural numbers and on the hyperreal numbers, respectively. To this end, the following is a necessary condition:

- $\boldsymbol{\alpha}$ must be a *cluster point* of $\mathbb{N}$ in the topology $\tau_\mathbb{N}$, so to give sense to a notion of $\lim_{n\to\alpha}$.

The above required property suggests the S-topology as a good candidate. Indeed, the following holds:

PROPOSITION 6.21.

(1) $\boldsymbol{\alpha} \in \mathbb{N}^*$ *is a cluster point of $\mathbb{N}$ in the S-topology;*

(2) *For every real sequence $f : \mathbb{N} \to \mathbb{R}$, the point $f^*(\boldsymbol{\alpha})$ is a cluster point of the image set $\{f(n) \mid n \in \mathbb{N}\}$ in the S-topology.*

PROOF. (1). If A^* is a basic open neighborhood of $\boldsymbol{\alpha}$, then trivially $\boldsymbol{\alpha} \in A^* \cap \mathbb{N}^* \neq \emptyset$ and so, by *transfer*, $A \cap \mathbb{N} \neq \emptyset$. The conclusion $A^* \cap \mathbb{N} \neq \emptyset$ follows by recalling that $A^* \cap \mathbb{N} = A \cap \mathbb{N}$ (see Proposition 2.16).

(2). The hyper-image f^* is continuous in the S-topology (see Proposition 6.14). Since $\boldsymbol{\alpha}$ is a cluster point of $\mathbb{N}$, it follows that $f^*(\boldsymbol{\alpha})$ is a cluster point of $f^*(\mathbb{N}) = \{f(n) \mid n \in \mathbb{N}\}$. □

Another requirement for the Alpha-limit to be an actual topological limit is the following:

- The topology $\tau_{\mathbb{R}}$ must be such that there exists a *unique* limit $\lim_{n \to \alpha} f(n)$ for every real sequence $f : \mathbb{N} \to \mathbb{R}$.

THEOREM 6.22. *Assume the S-topology is Hausdorff. Then, in the S-topology, every sequence $f : \mathbb{N} \to \mathbb{R}$ has the point $f^*(\alpha)$ as its unique limit as n approaches to α.*

PROOF. It directly follows from the previous proposition; indeed, the limit is unique because the topology is Haudorff.   □

One of the most striking and interesting consequences of the *Cauchy Infinitesimal Principle* is the following result.

THEOREM 6.23. *Assume Cauchy Infinitesimal Principle (CIP). Then there exists a Hausdorff topology τ on $\mathbb{R}^*$, namely the S-topology, such that*

(1) *The closure $\overline{A}^\tau = A^*$ for every $A \subseteq \mathbb{R}$;*

(2) *α is a cluster point of $\mathbb{N}$;*

(3) *Every real sequence $f : \mathbb{N} \to \mathbb{R}$ has a unique limit as n approaches to α, namely the Alpha-limit:*

$$\tau\text{-}\lim_{n \to \alpha} f(n) \;=\; f^*(\alpha) \;=\; \lim_{n \uparrow \alpha} f(n).$$

PROOF. (1). Recall that we assumed that $r^* = r$ for every $r \in \mathbb{R}$, so $A^\sigma = A$, and the closure $\overline{A} = \overline{A^\sigma} = A^*$, by Proposition 6.14.

Recall also that (CIP) implies that the S-topology is Hausdorff (see Theorem 6.20). Then (2) and (3) directly follow from above Proposition 6.21 and Theorem 6.22, respectively.   □

6. Superstructures

In order to go beyond the basics of calculus and develop advanced real analysis, one needs a universe of mathematical objects that is larger than just the family of real numbers, sets of real numbers, and real functions. Our intuitive idea of a *universe V* requires the following properties:

- V includes all objects of calculus, namely, the real numbers, the sets of real numbers, all real functions, and so forth;

- V is closed under all usual mathematical operations, namely unions, intersections, set differences, tuples, ordered tuples, Cartesian products, powersets, sets of functions, and so forth;
- V is *transitive*, that is, if a set belongs to V then also its elements belong to V (in formulas, $a \in A \in V \Rightarrow a \in V$).

Notice that the closure of V under subsets follows from the above properties. Indeed, if $A \in V$ and $A' \subseteq A$ then $A' \in V$ by transitivity, since $A' \in \mathcal{P}(A) \in V$.

The definition below aims to capture the intuitive idea of the "smallest universe" containing $\mathbb{R}$, or more generally, a given set of atoms X. In other words, it formalizes the notion of "universe generated" by X.

DEFINITION 6.24. *The* superstructure *over a given set of atoms X is the union $V(X) = \bigcup_{k \geq 0} V_k(X)$ where the k-th levels $V_k(X)$ are inductively defined by putting:*

$$\begin{cases} V_0(X) & = X \\ V_{k+1}(X) & = V_k(X) \cup \mathcal{P}(V_k(X)). \end{cases}$$

The set X is referred to as the ground set *of $V(X)$, and $V_k(X)$ as the k-th level of $V(X)$.*

Roughly speaking, a superstructure is simply obtained from a given set of atoms by iterating the powerset operation infinitely many times.

We will see next that the superstructure $V(X)$ is suitable to formalize our intuitive notion of "universe grounded on X," because it contains *all* mathematical objects that are potentially involved in the study of X. For this reason, superstructures provide a convenient framework where Alpha-Theory can operate in its full generality.

PROPOSITION 6.25. *Let $V(X)$ be the superstructure over a set of atoms X. Then:*

(1) *Levels $V_k(X)$ are increasing, that is, $V_k(X) \subseteq V_{k+1}$ for all k;*

(2) *If $A \in V_k(X)$ then $A \subseteq V_{k-1}(X)$;*

(3) *If $A \in V_k(X)$ is a set and $B \subseteq A$ then $B \in V_k(X)$;*

(4) *If $A \in V_k(X)$ is a set then $\mathcal{P}(A) \in V_{k+1}(X)$;*

(5) *If $A, B \in V_k(X)$ are sets then $A \cup B \in V_k(X)$;*

(6) *If $A, B \in V_k(X)$ are sets then $A \cap B \in V_k(X)$;*

(7) *If $A, B \in V_k(X)$ are sets then $A \setminus B \in V_k(X)$;*

(8) *If $a, b \in V_k(X)$ then $(a, b) \in V_{k+2}(X)$;*

(9) *If $A, B \in V_k(X)$ are sets then $A \times B \in V_{k+2}(X)$;*

(10) *If $\mathcal{R} \in V_k(X)$ is a binary relation then $\mathrm{dom}(\mathcal{R}), \mathrm{ran}(\mathcal{R}) \in V_{k-2}(X)$;*

(11) *If $A, B \in V_k(X)$ are sets and $f : A \to B$ then $f \in V_{k+2}(X)$;*

(12) *If $A, B \in V_k(X)$ are sets then $\mathfrak{Fun}(A, B) \in V_{k+3}(X)$.*

PROOF. (1) is trivial from the definition.

(2). Let $h \le k$ be the least level such that $A \in V_h(X)$. Notice that $h > 0$ because $V_0(X) = X$ contains only atoms. By the inductive definition of $V_h(X)$ and the minimality of h, it must be $A \in \mathcal{P}(V_{h-1}(X))$, and so $A \subseteq V_{h-1}(X) \subseteq V_k(X)$.

(3). By (2), if $B \subseteq A \in V_k(X)$ then $B \subseteq V_{k-1}(X)$, and hence $B \in \mathcal{P}(V_{k-1}(X)) \subseteq V_k(X)$.

(4). By (3), every subset of A belongs to $V_k(X)$, and so $\mathcal{P}(A) \subseteq V_k(X) \in \mathcal{P}(V_k(X)) \subseteq V_{k+1}(X)$. Again by (3), we conclude that $\mathcal{P}(A) \in V_{k+1}(X)$.

(5). By (2), both A and B are subsets of $V_{k-1}(X)$; so $A \cup B \subseteq V_{k-1}(X)$, that is, $A \cup B \in \mathcal{P}(V_{k-1}(X)) \subseteq V_k(X)$.

(6) and (7). Both $A \cap B$ and $A \setminus B$ are subsets of A, and so the desired properties follow from (3).

(8). By the hypothesis $\{a\}, \{b\} \subseteq V_k(X)$, so $\{a\}, \{b\} \in V_{k+1}(X)$. Then $(a, b) = \{\{a\}, \{a, b\}\} \subseteq V_{k+1}(X)$ and $(a, b) \in V_{k+2}(X)$.

(9). For every $a \in A$ and $b \in B$, by (2) we have that $a, b \in V_{k-1}(X)$. So, by (8), $(a, b) \in V_{k+1}(X)$. Then $A \times B \in V_{k+2}(X)$ because $A \times B = \{(a, b) \mid a \in A, b \in B\} \subseteq V_{k+1}(X)$.

(10). If $(a, b) \in \mathcal{R} \in V_k(X)$, then $(a, b) \in V_{k-1}(X)$. Now, $\{a\} \in (a, b)$ implies that $\{a\} \in V_{k-2}$, and $a \in \{a\}$ implies that $a \in V_{k-3}(X)$. This shows that $\mathrm{dom}(\mathcal{R}) = \{a \mid \exists b \, (a, b) \in \mathcal{R}\} \subseteq V_{k-3}(x)$, and hence $\mathrm{dom}(\mathcal{R}) \in V_{k-2}(X)$. The proof that also $\mathrm{ran}(\mathcal{R}) \in V_{k-2}(X)$ is entirely similar.

(11). Any function $f : A \to B$ is a subset of $A \times B$. Since $A \times B \in V_{k+2}(X)$ by (9), then also $f \in V_{k+2}(X)$ by (3).

(12). By (11), the set of all functions $\mathfrak{Fun}(A, B) \subseteq V_{k+2}(X)$, and hence $\mathfrak{Fun}(A, B) \in V_{k+3}(X)$. $\square$

EXERCISE 6.26. *A function f belongs to a superstructure $V(X)$ if and only if there exists a level k such that both $\mathrm{dom}(f), \mathrm{ran}(f) \in V_k(X)$.*

We are now ready to make precise our intuition that $V(X)$ is the universe generated by X.

THEOREM 6.27. *The* superstructure $V(X)$ *is the smallest family* $\mathcal{V}$ *that satisfies the following closure properties:*

(1) $X \in \mathcal{V}$;

(2) *If* $A \in \mathcal{V}$ *is a set and* $a \in A$, *then* $a \in \mathcal{V}$ *(transitivity)*;

(3) *If* $A \in \mathcal{V}$ *is a set and* $B \subseteq A$, *then* $B \in \mathcal{V}$;

(4) *If* $A \in \mathcal{V}$ *is a set, then its powerset* $\mathcal{P}(A) \in \mathcal{V}$;

(5) *If* $A, B \in \mathcal{V}$ *are sets, then* $A \cup B \in \mathcal{V}$;

(6) *If* $A, B \in \mathcal{V}$ *are sets, then* $A \cap B \in \mathcal{V}$;

(7) *If* $A, B \in \mathcal{V}$ *are sets, then* $A \setminus B \in \mathcal{V}$;

(8) *If* $a, b \in \mathcal{V}$, *then the ordered pair* $(a, b) \in \mathcal{V}$;

(9) *If* $A, B \in \mathcal{V}$ *are sets, then* $A \times B \in \mathcal{V}$;

(10) *If* $\mathcal{R} \in \mathcal{V}$ *is a binary relation then* $\mathrm{dom}(\mathcal{R}), \mathrm{ran}(\mathcal{R}) \in \mathcal{V}$;

(11) *If* $A, B \in \mathcal{V}$ *are sets and* $f : A \to B$ *then* $f \in \mathcal{V}$;

(12) *If* $A, B \in \mathcal{V}$ *are sets then* $\mathfrak{Fun}(A, B) \in \mathcal{V}$.

PROOF. As seen in the previous proposition, the superstructure $V(X)$ satisfies all properties above and so $\mathcal{V} \subseteq V(X)$.

Conversely, we proceed by induction and prove that each superstructure level $V_k(X) \in \mathcal{V}$. Then, by the *transitivity* property (2), it follows that $V_k(X) \subseteq \mathcal{V}$, and also the other inclusion $V(X) \subseteq \mathcal{V}$ is obtained.

The basis $V_0(X) \in \mathcal{V}$ is given by (1). Now assume $V_k(X) \in \mathcal{V}$. By the closure property (4), we have $\mathcal{P}(V_k(X)) \in \mathcal{V}$ and so, by (5), $V_{k+1}(X) = V_k(X) \cup \mathcal{P}(V_k(X)) \in \mathcal{V}$. □

Particularly relevant is the universe generated by the real numbers.

DEFINITION 6.28. *The* real-universe *is the superstructure* $V(\mathbb{R})$ *over the real numbers* $\mathbb{R}$.

We remark that *real analysis* can actually be fully operated within $V(\mathbb{R})$. Indeed, real functions, spaces of real functions, functionals, as well as norms and the involved topologies, and so forth, they *all* belong to $V(\mathbb{R})$. Besides, by identifying complex numbers $\mathbb{C}$ with ordered pairs of reals, also *complex analysis* lives within $V(\mathbb{R})$.

EXERCISE 6.29. *Prove that all the mathematical objects below belong to the real-universe* $V(\mathbb{R})$:

(1) *The sum and product operations* $+, \cdot : \mathbb{R} \times \mathbb{R} \to \mathbb{R}$;

(2) *The* ordered field $(\mathbb{R}, +, \cdot, 0, 1, \leq)$ *(as an ordered tuple)*;

(3) *The complex field* $(\mathbb{C}, 0, 1, +, \cdot)$;

(4) *The Banach spaces of real functions $L^p(\mathbb{R})$.*

Having found an appropriate universe for real calculus, we now look for a larger universe that contains all objects needed to perform Alpha-Theory and (many of) its applications. To this end, we need a family of sets $\mathcal{V}$ that satisfies the following requirements:

- $\mathcal{V}$ has all the closure properties of a *superstructure* (as itemized in Theorem 6.27);
- $\mathcal{V}$ contains all mathematical objects that are needed for Alpha-Calculus;
- $\mathcal{V}$ is closed under hyper-images of sets and functions and, more generally, under Alpha-limits.

As we have seen above, the *real-universe* as given by the superstructure $V(\mathbb{R})$ is a well-behaved family of mathematical objects, which is large enough for almost all traditional mathematics. However, it does not directly contain the hyperreal numbers $\mathbb{R}^*$, and so it is not suitable for Alpha-Calculus. To this end, one could enlarge $V(\mathbb{R})$ and consider the superstructure $V(\mathbb{R}^*)$ over the *hyperreal numbers*. However, we remark that the mathematical universe as given by $V(\mathbb{R}^*)$ is still too small to be closed under Alpha-limits, and so one cannot fully operate Alpha-Theory within.[7]

As it will be shown in the next section, this problem can be overcome by considering the superstructure $V(\mathfrak{R})$ grounded on a suitable field extension $\mathfrak{R} \supset \mathbb{R}^*$ of the hyperreal numbers.

7. Models of Alpha-Theory

In §2.11 we characterized the models of Alpha-Calculus. Precisely, we showed that such models can be identified with *Alpha-morphisms*, that are ring homomorphisms $J : \mathfrak{Fun}(\mathbb{N}, \mathbb{R}) \twoheadrightarrow \mathbb{R}^*$ onto fields $\mathbb{R}^* \supset \mathbb{R}$ that properly extend the reals.

Aiming to construct models of the more general Alpha-Theory, in this section we introduce the notion of "reflexive Alpha-morphism", a modified notion of Alpha-morphism $\mathfrak{J} : \mathfrak{Fun}(\mathbb{N}, \mathfrak{R}) \twoheadrightarrow \mathfrak{R}$ where one considers a single field $\mathfrak{R}$ in place of the two fields $\mathbb{R}$ and $\mathbb{R}^*$. The underlying idea is that the field extension $\mathfrak{R} \supset \mathbb{R}^*$ be closed under the notion of Alpha-limit as determined by $\mathfrak{J}$.

---

[7] *E.g.*, if $\varphi : \mathbb{N} \to \mathbb{R}^* \setminus \mathbb{R}$ is any sequence of hyperreal numbers that are not real, then it is readily seen that the Alpha-limit $\lim_{n \uparrow \alpha} \varphi(n) \notin \mathbb{R}^*$. In consequence, the *hyper-hyperreal numbers* $\mathbb{R}^{**}$ are not included in $V(\mathbb{R}^*)$.

DEFINITION 6.30. *A reflexive Alpha-morphism (RAM for short)*

$$\mathfrak{J} : \mathfrak{Fun}(\mathbb{N}, \mathfrak{R}) \twoheadrightarrow \mathfrak{R}$$

is a map from the set of $\mathfrak{R}$-valued sequences $\mathfrak{Fun}(\mathbb{N}, \mathfrak{R})$ onto $\mathfrak{R}$ such that:

(RAM1) $\mathfrak{J}(c_r) = r$ *for every $r \in \mathbb{R}$;*

(RAM2) $\mathfrak{R} \supseteq \mathbb{R}$ *is a field that properly extends the reals;*

(RAM3) $\mathfrak{J}$ *is a ring homomorphism, that is, for every $\varphi, \psi : \mathbb{N} \to \mathfrak{R}$,*

- $\mathfrak{J}(\varphi + \psi) = \mathfrak{J}(\varphi) + \mathfrak{J}(\psi)$;
- $\mathfrak{J}(\varphi \cdot \psi) = \mathfrak{J}(\varphi) \cdot \mathfrak{J}(\psi)$.

Reflexive Alpha-morphisms are extensions of Alpha-morphisms.

PROPOSITION 6.31. *Let $\mathfrak{J} : \mathfrak{Fun}(\mathbb{N}, \mathfrak{R}) \to \mathfrak{R}$ be a reflexive Alpha-morphism. If $\mathbb{R}^* = \{\mathfrak{J}(\varphi) \mid \varphi : \mathbb{N} \to \mathbb{R}\}$ then the restriction of $\mathfrak{J}$ to the real sequences is an Alpha-morphism:*

$$J = \mathfrak{J}|_{\mathfrak{Fun}(\mathbb{N}, \mathbb{R})} : \mathfrak{Fun}(\mathbb{N}, \mathbb{R}) \twoheadrightarrow \mathbb{R}^*.$$

We will refer to the above J as the Alpha-morphism induced by $\mathfrak{J}$.

PROOF. The only non-trivial fact we have to check is that $\mathbb{R}^*$ is a field that properly extends the reals. Since $\mathfrak{R} \cong \mathfrak{Fun}(\mathbb{N}, \mathfrak{R})/\ker(\mathfrak{J})$ is a field, $\ker(\mathfrak{J})$ is a maximal ideal of $\mathfrak{Fun}(\mathbb{N}, K)$, and hence its *contraction* $\ker(J) = \ker(\mathfrak{J}) \cap \mathfrak{Fun}(\mathbb{N}, \mathbb{R})$ is a prime ideal.[8] But then $\ker(J)$ is maximal by Proposition 2.50, and so $\mathbb{R}^* \cong \mathfrak{Fun}(\mathbb{N}, \mathbb{R})/\ker(J)$ is a field. Finally, $\mathbb{R}^*$ properly extends $\mathbb{R}$ because the maximal ideal $\ker(J)$ is not principal (see Proposition 2.49).[9]   □

The existence problem is addressed by the following:

THEOREM 6.32 (Existence of reflexive Alpha-morphisms).
Every Alpha-morphism $J : \mathfrak{Fun}(\mathbb{N}, \mathbb{R}) \to \mathbb{R}^*$ can be extended to a reflexive Alpha-morphism $\mathfrak{J} : \mathfrak{Fun}(\mathbb{N}, \mathfrak{R}) \to \mathfrak{R}$.

The proof of this result is not easy and quite technical. So, we temporarily take it as a known fact, and put off its proof to the next section.

---

[8] Recall the following well-known property from algebra: "Let $\varphi : R \to S$ be a ring homomorphism between commutative rings with unit. If $\mathfrak{p}$ is a prime ideal of S, then also its *contraction* $\varphi^{-1}(\mathfrak{p}) = \{x \in R \mid \varphi(x) \in S\}$ is a prime ideal."

[9] Notice that Proposition 2.49 also holds for the ring $\mathfrak{Fun}(\mathbb{N}, \mathfrak{R})$.

Starting from a given RAM $\mathfrak{J} : \mathfrak{Fun}(\mathbb{N}, \mathfrak{R}) \twoheadrightarrow \mathfrak{R}$, we will show below that the superstructure $V(\mathfrak{R})$ "almost" yields a model of Alpha-Theory. Precisely, what we will obtain is a weakened model of Alpha-Theory where the existence of Alpha-limits is restricted to *bounded level sequences*, that is, to sequences whose range is included in some level $V_k(\mathfrak{R})$. However, it is safe to say that – in the practice – considering only bounded level sequences is not an actual limitation.[10]

THEOREM 6.33. *Let a RAM $\mathfrak{J} : \mathfrak{Fun}(\mathbb{N}, \mathfrak{R}) \twoheadrightarrow \mathfrak{R}$ be given. Then there exists a unique notion of Alpha-limit for bounded level sequences in the superstructure $V(\mathfrak{R})$ such that:*

(1) $\lim_{n \uparrow \alpha} \varphi(n) = \mathfrak{J}(\varphi)$ *for every $\varphi : \mathbb{N} \to \mathfrak{R}$;*

(2) *All axioms of the Alpha-Theory* **AT** *are satisfied.*

PROOF. Let $\mathcal{Q}$ be the family of *qualified sets* with respect to the Alpha-morphism $J = \mathfrak{J} \restriction_{\mathfrak{Fun}(\mathbb{N}, \mathbb{R})}$ induced by $\mathfrak{J}$. As a consequence of condition (1), every candidate notion of Alpha-limit for bounded level sequences in $V(\mathfrak{R})$ must come with the same family $\mathcal{Q}$ of qualified sets. In particular, by the *transfer principle* for equality and membership as proved by Alpha-Theory (see Theorem 4.4), every candidate notion of Alpha-limit must satisfy:

(a) $\lim_{n \uparrow \alpha} \varphi(n) = \lim_{n \uparrow \alpha} \psi(n) \iff \{n \mid \varphi(n) = \psi(n)\} \in \mathcal{Q}$;

(b) $\lim_{n \uparrow \alpha} \varphi(n) \in \lim_{n \uparrow \alpha} \psi(n) \iff \{n \mid \varphi(n) \in \psi(n)\} \in \mathcal{Q}$.

The conclusion will be reached by showing that conditions (a) and (b) "force" a well-posed definition of a (unique) notion of Alpha-limit for bounded level sequences in $V(\mathfrak{R})$, and that such a notion satisfies all axioms of Alpha-Theory.

As a preliminary observation, notice that a sequence $\varphi : \mathbb{N} \to V(\mathfrak{R})$ belongs to $V(\mathfrak{R})$ if and only if it a bounded level sequence. The proof of this fact is straightforward and is left to the reader (see Exercise 6.26). Proceeding by induction on k, we now define the Alpha-limit of the sequences $\varphi \in V(\mathfrak{R})$ such that $\varphi(n) \in V_k(\mathfrak{R})$ a.e..

- $k = 0$. Let $\varphi(n) \in V_0(\mathfrak{R}) = \mathfrak{R}$ a.e.. Then pick any sequence $\varphi' \in \mathfrak{Fun}(\mathbb{N}, \mathfrak{R})$ such that $\varphi'(n) = \varphi(n)$ a.e., and put

$$\lim_{n \uparrow \alpha} \varphi(n) = \mathfrak{J}(\varphi').$$

---

[10] In §14.2, we will outline the construction of a set-theoretic model of the Alpha-Theory in its full generality.

- $k \geq 1$. If $\varphi(n) \in V_{k-1}(\mathfrak{R})$ *a.e.*, the Alpha-limit of φ has already been defined, so let us assume $\varphi(n) \in V_k(\mathfrak{R}) \setminus V_{k-1}(\mathfrak{R})$ *a.e.*, that is, $\varphi(n) \subseteq V_{k-1}(\mathfrak{R})$ *a.e.*.

  – If $\varphi(n) = \emptyset$ *a.e.*, then put $\lim_{n \uparrow \alpha} \varphi(n) = \emptyset$;
  – If $\varphi(n)$ is a nonempty set *a.e.*, then put

$$\lim_{n \uparrow \alpha} \varphi(n) \;=\; \left\{ \lim_{n \uparrow \alpha} \psi(n) \,\middle|\, \psi(n) \in \varphi(n) \ \textit{a.e.} \right\}$$

Notice that the above inductive definition is well-posed. Indeed, if $\varphi(n) \in V_k(\mathfrak{R})$ *a.e.* and $\psi(n) \in \varphi(n)$ *a.e.* then $\psi(n) \in V_{k-1}(\mathfrak{R})$ *a.e.*, and by the inductive hypothesis, the Alpha-limit $\lim_{n \uparrow \alpha} \psi(n)$ has already been defined.

By the base step $k = 0$, condition (1) is trivially satisfied. Since the restriction $J = \mathfrak{J}|_{\mathfrak{Fun}(\mathbb{N},\mathbb{R})}$ is an Alpha-morphism, we know already that the notion of Alpha-limit as obtained by putting $\lim_{n \uparrow \alpha} \varphi(n) = J(\varphi)$ for all real sequences $\varphi : \mathbb{N} \to \mathbb{R}$ is a model of Alpha-Calculus (see Theorem 2.54). In consequence, the first three axioms (AT1), (AT2), (AT3) of Alpha-Theory are realized. Moreover, it directly follows from the inductive step of the definition of Alpha-limit that the *Internal Set Axiom* (AT4) is satisfied as well.

We are left to verify the validity of the *Composition Axiom* (AT5). Notice that the given definitions imply the *transfer* properties (a) and (b) in a straightforward manner. As a consequence, for all sequences $\varphi, \varphi' \in V(\mathfrak{R})$ and for every function $f \in V(\mathfrak{R})$ such that the compositions $f \circ \varphi, f \circ \psi$ make sense, the following implications hold:

$$\lim_{n \uparrow \alpha} \varphi(n) = \lim_{n \uparrow \alpha} \varphi'(n) \iff \varphi(n) = \psi(n) \ \textit{a.e.} \implies$$

$$f(\varphi(n)) = f(\psi(n)) \ \textit{a.e.} \iff \lim_{n \uparrow \alpha} f(\varphi(n)) = \lim_{n \uparrow \alpha} f(\psi(n)).$$

$\square$

8. Existence of reflexive Alpha-morphisms

By the Characterization Theorem 2.55, the Alpha-morphisms can be put in a 1-1 correspondence with the non-principal maximal ideals of the ring of sequences $\mathfrak{Fun}(\mathbb{N}, \mathbb{R})$. Aiming to address the existence problem of Reflexive Alpha-morphisms (RAM), we now need a little preliminary study of the properties of such ideals. The first observation is that they naturally yield an equivalence relation.

Denote by $Z(\varphi) = \{n \mid \varphi(n) = 0\}$ the *zero-set* φ.

PROPOSITION 6.34. *Every maximal ideal* $\mathfrak{m}$ *of* $\mathfrak{Fun}(\mathbb{N}, \mathbb{R})$ *induces the following equivalence relation on the family of* all *sequences:*

$$\varphi \equiv_{\mathfrak{m}} \psi \iff Z(\varphi - \psi) = Z(\vartheta) \text{ for some } \vartheta \in \mathfrak{m}.^{[11]}$$

PROOF. Reflexivity of $\equiv_{\mathfrak{m}}$ holds because $Z(\varphi - \varphi) = Z(c_0)$ and $c_0 \in \mathfrak{m}$. The symmetric property is trivial. As for transitivity, assume that $Z(\varphi - \psi) = Z(\vartheta_1)$ and $Z(\psi - \tau) = Z(\vartheta_2)$ for suitable real sequences $\vartheta_1, \vartheta_2 \in \mathfrak{m}$. Let $\vartheta_3(n) = 0$ if $\varphi(n) = \tau(n)$ and $\vartheta_3(n) = 1$ otherwise. Clearly $Z(\varphi - \tau) = Z(\vartheta_3)$. We will reach the conclusion $\varphi \equiv_{\mathfrak{m}} \tau$ by showing that $\vartheta_3 \in \mathfrak{m}$. If by contradiction $\vartheta_3 \notin \mathfrak{m}$ then, by the maximality of $\mathfrak{m}$, $\vartheta_3 \cdot (1 - \vartheta_3) = c_0 \in \mathfrak{m}$ would imply $1 - \vartheta_3 \in \mathfrak{m}$. But then also

$$\zeta = \vartheta_1^2 + \vartheta_2^2 + (1 - \vartheta_3)^2 \in \mathfrak{m}.$$

This is not possible because $\zeta(n) \neq 0$ for all n, and no invertible element can belong to any ideal. □

The equivalence relation $\equiv_{\mathfrak{m}}$ is significant in that it determines the extensions of $\mathfrak{m}$ in all rings of sequences $\mathfrak{Fun}(\mathbb{N}, K)$ where K is a superfield of $\mathbb{R}$.[12]

PROPOSITION 6.35. *Let* $\mathfrak{m}$ *be a maximal ideal of* $\mathfrak{Fun}(\mathbb{N}, \mathbb{R})$ *and let* $K \supseteq \mathbb{R}$ *be any superfield of the reals. If* $\mathfrak{m}_K \subset \mathfrak{Fun}(\mathbb{N}, K)$ *is the extended ideal, then*

$$\varphi - \psi \in \mathfrak{m}_K \iff \varphi \equiv_{\mathfrak{m}} \psi.$$

In other words, the quotient modulo $\mathfrak{m}_K$ *coincides with the quotient modulo the equivalence relation* $\equiv_{\mathfrak{m}}$:

$$\mathfrak{Fun}(\mathbb{N}, K)/\mathfrak{m}_K = \mathfrak{Fun}(\mathbb{N}, K)/\equiv_{\mathfrak{m}}.$$

Moreover, $\mathfrak{m}_K$ *is maximal.*

---

[11] Recall that $Z(\tau) = \{n \in \mathbb{N} \mid \vartheta(n) = 0\}$ denotes the *zero-set* of a sequence τ.

[12] Recall from algebra that if $f : R \to S$ is a ring homomorphism, the extension of an ideal $\mathfrak{i} \subset R$ is the ideal generated by the image $f(\mathfrak{i})$.

PROOF. $\mathfrak{m}_K$ is the ideal generated by $\mathfrak{m}$ in $\mathfrak{Fun}(\mathbb{N}, K)$, and so

$$\mathfrak{m}_K = \{\tau_1 \cdot \vartheta_1 + \ldots + \tau_k \cdot \vartheta_k \mid \tau_i \in \mathfrak{Fun}(\mathbb{N}, K) \text{ and } \vartheta_i \in \mathfrak{m}\}.$$

Assume first that $\varphi \equiv_\mathfrak{m} \psi$, that is $Z(\varphi - \psi) = Z(\vartheta)$ for a suitable real sequence $\vartheta \in \mathfrak{m}$. Let τ be the sequence defined by putting

$$\tau(n) = \begin{cases} 0 & \text{if } \vartheta(n) = 0 \\ \frac{\varphi(n) - \psi(n)}{\vartheta(n)} & \text{if } \vartheta(n) \neq 0. \end{cases}$$

Then $\varphi - \psi = \tau \cdot \vartheta \in \mathfrak{m}_K$.

Conversely, given $\varphi - \psi = \sum_{i=1}^k \tau_i \cdot \vartheta_i \in \mathfrak{m}_K$ where $\vartheta_i \in \mathfrak{m}$, pick the sequence ϑ_0 defined by

$$\vartheta_0(n) = \begin{cases} 0 & \text{if } \varphi(n) = \psi(n) \\ 1 & \text{if } \varphi(n) \neq \psi(n). \end{cases}$$

We claim that $\vartheta_0 \in \mathfrak{m}$, and hence $\varphi \equiv_\mathfrak{m} \psi$ since $Z(\vartheta_0) = Z(\varphi - \psi)$. We argue similarly as in the proof of the previous proposition.

If by contradiction $\vartheta_0 \notin \mathfrak{m}$ then, by the maximality of $\mathfrak{m}$, from $\vartheta_0 \cdot (1 - \vartheta_0) = c_0 \in \mathfrak{m}$ would follow that $1 - \vartheta_0 \in \mathfrak{m}$. But then

$$\zeta = (1 - \vartheta_0)^2 + \vartheta_1^2 + \ldots + \vartheta_k^2 \in \mathfrak{m}.$$

This is impossible because $\zeta(n) \neq 0$ for all n, as one can easily verify.

Also the maximality of $\mathfrak{m}_K$ is proved by the same argument. In fact, assume by contradiction that there exists an ideal $\mathfrak{m}'$ that properly extends $\mathfrak{m}_K$, and pick a witness $\varphi \in \mathfrak{m}' \setminus \mathfrak{m}_K$. Let ϑ be the sequence such that $\vartheta(n) = 0$ if $\varphi(n) = 0$, and $\vartheta(n) = 1$ is $\varphi(n) \neq 0$. Since $Z(\vartheta) = Z(\varphi)$ and $\varphi \notin \mathfrak{m}_K$, it must be $\vartheta \notin \mathfrak{m}$. But then, by the maximality of $\mathfrak{m}$, it follows that $1 - \vartheta \in \mathfrak{m} \subset \mathfrak{m}'$ and so the product $\zeta = \varphi \cdot (1 - \vartheta) \in \mathfrak{m}'$. This is impossible because $\zeta(n) \neq 0$ for all n. $\square$

EXERCISE 6.36. *Prove that $\mathfrak{m}_K$ is principal if and only if $\mathfrak{m}$ is principal.*

In the sequel we will need the following (easy) result on cardinalities:

LEMMA 6.37. *Let X be any set cardinality $|X| = \mathfrak{c}$.[13] Then for every maximal ideal $\mathfrak{m}$ of $\mathfrak{Fun}(\mathbb{N}, \mathbb{R})$ also the quotient $|\mathfrak{Fun}(\mathbb{N}, X)/\equiv_\mathfrak{m}| = \mathfrak{c}$.*

---

[13] Recall that $\mathfrak{c}$ denotes the cardinality of the *continuum*, that is, $\mathfrak{c} = |\mathbb{R}|$.

PROOF. The diagonal embedding $d : X \to \mathfrak{Fun}(\mathbb{N}, X)/\equiv_m$ defined by putting

$$d : \xi \longmapsto [c_\xi]$$

is 1-1. (As usual, we denoted by $c_\xi : n \mapsto \xi$ the constant sequence with value ξ.) Indeed, if by contradiction $c_\xi \equiv_m c_\eta$ for some $\xi \neq \eta$, there would be a sequence $\vartheta \in m$ whose zero-set $Z(\vartheta) = Z(c_\xi - c_\eta) = \emptyset$; but then ϑ would be invertible and this is not possible. The conclusion is obtained by the following inequalities:

$$\mathfrak{c} = |X| \leq |\mathfrak{Fun}(\mathbb{N}, X)/\equiv_m| \leq |\mathfrak{Fun}(\mathbb{N}, X)| = \mathfrak{c}^{\aleph_0} = \mathfrak{c}.$$

$\square$

We are finally ready to prove the existence result stated in the previous section.

PROOF OF THEOREM 6.32. By the hypothesis, $\mathbb{R}^* \cong \mathfrak{Fun}(\mathbb{N}, \mathbb{R})/m$ where $m = \ker(J)$ is a maximal ideal. Let $\equiv_m$ be the equivalence relation on the class of all sequences induced by m. By Lemma 6.37, $\mathbb{R}^* \cong \mathfrak{Fun}(\mathbb{N}, \mathbb{R})/m = \mathfrak{Fun}(\mathbb{N}, \mathbb{R})/\equiv_m$ has cardinality $\mathfrak{c}$. Now pick a set of atoms X of cardinality $\mathfrak{c}$ and disjoint from $\mathbb{R}^*$. Since $|\mathbb{R}^* \cup X| = \mathfrak{c}$, again by Lemma 6.37, $|\mathfrak{Fun}(\mathbb{N}, \mathbb{R}^* \cup X)/\equiv_m| = \mathfrak{c}$. In consequence, we can pick a bijection

$$\Psi : \mathfrak{Fun}(\mathbb{N}, \mathbb{R}^* \cup X)/\equiv_m \longrightarrow \mathbb{R}^* \cup X$$

such that $\Psi([\varphi]) = J(\varphi)$ for every real sequence $\varphi \in \mathfrak{Fun}(\mathbb{N}, \mathbb{R})$.

By using the ideal m and the bijection Ψ, we now proceed by transfinite induction on the first uncountable ordinal ω_1 and define fields $\mathfrak{R}_\beta$ in such a way that the following properties are satisfied:

(1) $\mathfrak{R}_\beta \subset \mathbb{R}^* \cup X$;

(2) $\mathfrak{R}_{\beta'}$ is a subfield of $\mathfrak{R}_\beta$ for every $\beta' < \beta$;

(3) $\mathfrak{R}_{\beta+1} = \{\Psi([\varphi]) \mid \varphi \in \mathfrak{Fun}(\mathbb{N}, \mathfrak{R}_\beta)\}$;

(4) For all $\varphi, \psi : \mathfrak{Fun}(\mathbb{N}, \mathfrak{R}_\beta)$:[14]

- $\Psi([\varphi]) + \Psi([\psi]) = \Psi([\varphi + \psi])$;
- $\Psi([\varphi]) \cdot \Psi([\psi]) = \Psi([\varphi \cdot \psi])$;

(5) $\mathfrak{R}_\lambda = \bigcup_{\beta < \lambda} \mathfrak{R}_\beta$ for limit λ.

---

[14] With obvious notation, by $\varphi + \psi$ we indicate the sequence $n \mapsto \varphi(n) + \psi(n)$, and similarly for the product $\varphi \cdot \psi$.

Notice that in item (4), operations on the left side are in $\mathfrak{R}_{\beta+1}$ and operations on the right side are in $\mathfrak{R}_\beta$.

As inductive basis, let $\mathfrak{R}_0 = \mathbb{R}$ be the real field, and let $\mathfrak{R}_1 = \mathbb{R}^*$ be the hyperreal field given by Alpha-morphism J. By our choice of the map Ψ, we have

$$\mathfrak{R}_1 \;=\; \{J(\varphi) \mid \varphi \in \mathfrak{Fun}(\mathbb{N}, \mathbb{R})\} \;=\; \{\Psi([\varphi]) \mid \varphi \in \mathfrak{Fun}(\mathbb{N}, \mathfrak{R}_0)\},$$

and the required properties are satisfied.

Now let $\beta > 1$ be given, and assume that the sequence of fields $\langle \mathfrak{R}_\gamma \mid \gamma < \beta \rangle$ has been defined so that the five conditions above are fulfilled. For every $\gamma < \beta$, denote by $\mathfrak{m}_\gamma = \mathfrak{m}_{\mathfrak{R}_\gamma}$ the extended ideal of $\mathfrak{m}$ in the ring $\mathfrak{Fun}(\mathbb{N}, \mathfrak{R}_\gamma)$. By Proposition 6.35, the quotient

$$\mathfrak{R}'_\gamma \;=\; \mathfrak{Fun}(\mathbb{N}, \mathfrak{R}_\gamma)/\!\equiv_\mathfrak{m} \;=\; \mathfrak{Fun}(\mathbb{N}, \mathfrak{R}_\gamma)/\mathfrak{m}_\gamma$$

is a field. Moreover, if $\delta < \gamma < \beta$, then $\mathfrak{R}'_\delta$ is a subfield of $\mathfrak{R}'_\gamma$ because $\mathfrak{R}_\delta$ is a subfield of $\mathfrak{R}_\gamma$ by the inductive hypothesis. We now distinguish two cases.

<u>Case 1</u>. β is a limit. In this case, we put:

$$\mathfrak{R}_\beta \;=\; \bigcup_{\gamma < \beta} \mathfrak{R}_\gamma.$$

By the inductive hypothesis, $\mathfrak{R}_\delta$ is a subfield of $\mathfrak{R}_\gamma$ for all $\delta < \gamma < \beta$. As a consequence, the defined $\mathfrak{R}_\beta$ is a field that satisfies all the required properties.

<u>Case 2</u>. $\beta = \gamma + 1$ is a successor. In this case, define $\mathfrak{R}_{\gamma+1}$ as the isomorphic copy of $\mathfrak{R}'_\gamma$ under the 1-1 map Ψ. That is, we put

$$\mathfrak{R}_{\gamma+1} \;=\; \Psi(\mathfrak{R}'_\gamma) \;=\; \{\Psi([\varphi]) \mid \varphi \in \mathfrak{Fun}(\mathbb{N}, \mathfrak{R}_\gamma)\} \;\subseteq\; \mathbb{R}^* \cup X$$

and we define operations by letting for all $\varphi, \psi \in \mathfrak{Fun}(\mathbb{N}, \mathfrak{R}_\gamma)$:

$$\Psi([\varphi]) + \Psi([\psi]) = \Psi([\varphi + \psi]) \quad \text{and} \quad \Psi([\varphi]) \cdot \Psi([\psi]) = \Psi([\varphi \cdot \psi]).$$

We have to verify that $\mathfrak{R}_{\gamma+1}$ is a field extension of $\mathfrak{R}_\gamma$. If $\gamma = \delta + 1$ is a successor, by the inductive hypothesis $\mathfrak{R}_\delta$ is a subfield of $\mathfrak{R}_{\delta+1}$, and so $\mathfrak{R}_\gamma = \Psi(\mathfrak{R}'_\delta)$ is a subfield of $\mathfrak{R}_{\gamma+1} = \Psi(\mathfrak{R}'_{\delta+1})$. If γ is a limit, again by inductive hypothesis, it follows that $\mathfrak{R}_{\delta+1} = \Psi(\mathfrak{R}'_\delta)$ is a subfield of $\mathfrak{R}_{\gamma+1} = \Psi(\mathfrak{R}'_\gamma)$ for every $\delta < \gamma$. Since $\mathfrak{R}_\gamma = \bigcup_{\delta < \gamma} \mathfrak{R}_\delta = \bigcup_{\delta < \gamma} \mathfrak{R}_{\delta+1}$, we conclude that also $\mathfrak{R}_\gamma$ is a subfield of $\mathfrak{R}_{\gamma+1}$.

Finally, we define

- $\mathfrak{R} = \bigcup_{\beta < \omega_1} \mathfrak{R}_\beta \subseteq \mathbb{R}^* \cup X$;
- $\mathfrak{J} : \varphi \mapsto \Psi([\varphi])$ for all $\varphi \in \mathfrak{Fun}(\mathbb{N}, \mathfrak{R})$.

Clearly $\mathfrak{J}$ extends J because the bijection Ψ was chosen in such a way to satisfy $\Psi([\varphi]) = J(\varphi)$ for all $\varphi \in \mathfrak{Fun}(\mathbb{N}, \mathbb{R})$.

Notice that

$$\mathfrak{Fun}(\mathbb{N}, \mathfrak{R}) = \bigcup_{\beta < \omega_1} \mathfrak{Fun}(\mathbb{N}, \mathfrak{R}_\beta).$$

In fact, for every $\xi \in \mathfrak{R}$, define its rank $\rho(\xi) = \min\{\beta \mid \xi \in \mathfrak{R}_\beta\}$. For every $\varphi \in \mathfrak{Fun}(\mathbb{N}, \mathfrak{R})$, the set $\{\rho(\varphi(n)) \mid n \in \mathbb{N}\}$ is a countable subset of ω_1, so it is bounded and it admits supremum $\beta \in \omega_1$.[15] Then clearly $\varphi \in \mathfrak{Fun}(\mathbb{N}, \mathfrak{R}_\beta)$.

Now let $\varphi, \psi \in \mathfrak{Fun}(\mathbb{N}, \mathfrak{R})$. By the above, we can pick β such that $\varphi, \psi : \mathbb{N} \to \mathfrak{R}_\beta$. Then condition (4) guarantees that

- $\mathfrak{J}(\varphi + \psi) = \Psi([\varphi + \psi]) = \Psi([\varphi]) + \Psi([\psi]) = \mathfrak{J}(\varphi) + \mathfrak{J}(\psi)$;
- $\mathfrak{J}(\varphi \cdot \psi) = \Psi([\varphi \cdot \psi]) = \Psi([\varphi]) \cdot \Psi([\psi]) = \mathfrak{J}(\varphi) \cdot \mathfrak{J}(\psi)$.

This completes the proof that $\mathfrak{J}$ is a ring homomorphism. Finally, the map $\mathfrak{J}$ is onto $\mathfrak{R}$; indeed:

$$\begin{aligned}
\mathrm{ran}(\mathfrak{J}) &= \bigcup_{\beta < \omega_1} \{\mathfrak{J}(\varphi) \mid \varphi \in \mathfrak{Fun}(\mathbb{N}, \mathfrak{R}_\beta)\} \\
&= \bigcup_{\beta < \omega_1} \{\Psi([\varphi]) \mid \varphi \in \mathfrak{Fun}(\mathbb{N}, \mathfrak{R}_\beta)\} = \bigcup_{\beta < \omega_1} \mathfrak{R}_{\beta+1} = \mathfrak{R}.
\end{aligned}$$

$\square$

EXERCISE 6.38. *Prove that the RAM $\mathfrak{J}$ as constructed in the proof of Theorem 6.32 is "minimal" in the following sense: For every RAM $\mathfrak{J}' : \mathfrak{Fun}(\mathbb{N}, \mathfrak{R}') \twoheadrightarrow \mathfrak{R}'$ that extends J there exists a unique 1-1 homomorphism $\theta : \mathfrak{R} \to \mathfrak{R}'$ that makes the following diagram commute:*

$$\begin{array}{ccc}
\mathfrak{Fun}(\mathbb{N}, \mathfrak{R}) & \xrightarrow{\ \mathfrak{J}\ } & \mathfrak{R} \\
\widehat{\theta} \downarrow & & \downarrow \theta \\
\mathfrak{Fun}(\mathbb{N}, \mathfrak{R}') & \xrightarrow[\ \mathfrak{J}'\ ]{} & \mathfrak{R}'
\end{array}$$

where $\widehat{\theta} : \varphi \mapsto \theta \circ \varphi$ for every $\varphi \in \mathfrak{Fun}(\mathbb{N}, \mathfrak{R})$. As a consequence, if two RAM $\mathfrak{J}, \mathfrak{J}'$ are minimal extensions of the same Alpha-morphism, then the corresponding fields $\mathfrak{R} \cong \mathfrak{R}'$ are isomorphic.

---

[15] This is the very point that justifies our inductive construction over the *uncountable* ordinal ω_1.

9. Remarks and comments

1. *Countable saturation* is crucial to carry out the *Loeb measure* construction, currently one of the major sources of applications of nonstandard methods (see, *e.g.*, the book [**33**]). Stronger forms of saturation, namely κ-*saturation* for uncountable cardinals κ (see Definition 13.2) are required for a nonstandard study of topological spaces with uncountable bases. In this respect, we remark that also κ-saturation could be embedded in Alpha-Theory, provided one postulates the existence of Alpha-limits for κ-sequences $\langle a_i \mid i \in \kappa \rangle$, and modify the axioms accordingly.

2. In the definition of superstructure $V(X)$, the elements of the ground set X were assumed to be *atoms*, that is, mathematical entities that are not reducible to sets (and in particular, they have no elements). We remark that this assumption simplifies matters, but it is not strictly necessary. Indeed, superstructures $V(X)$ over sets of atoms can be "simulated" also in a pure set theory without atoms. This can be done by taking X a family of nonempty sets x such that $x \cap (V(X) \setminus X) = \emptyset$. (See [**29**] §4.4, where such sets X are called *base sets*.)

Part 3

Applications

CHAPTER 7

First Applications

1. The real line as a quotient of hyperrationals

In this section we show that there is a uniform procedure to recover the real line from the set $\mathbb{Q}^*$ of hyperrational numbers. In short, this is done by restricting to finite numbers, and then identifying numbers that are infinitely close.

THEOREM 7.1. *The quotient of the finite hyperrational numbers* $\mathbb{Q}^*_{\text{fin}} = \{\xi \in \mathbb{Q}^* \mid |\xi| < n \text{ for some } n\}$ *modulo the infinitesimal numbers* $\mu(\mathbb{Q}^*) = \mathfrak{mon}(0) \cap \mathbb{Q}^* = \{\xi \in \mathbb{Q}^* \mid |\xi| < 1/n \text{ for all } n \in \mathbb{N}\}$ *is an ordered field that is isomorphic to the real line:*

$$\mathbb{Q}^*_{\text{fin}}/\mu(\mathbb{Q}^*) \cong \mathbb{R}.$$

PROOF. Notice first that both $\mathbb{Q}^*_{\text{fin}}$ and $\mu(\mathbb{Q}^*)$ are subrings of $\mathbb{Q}^*$, because they are closed under sums, differences and products (see Proposition 1.6). Since the product of a finite number by an infinitesimal number is infinitesimal, $\mu(\mathbb{Q}^*)$ is an ideal of $\mathbb{Q}^*_{\text{fin}}$. Moreover, $\mu(\mathbb{Q}^*)$ is *maximal* because it consists of all non-invertible elements of $\mathbb{Q}^*_{\text{fin}}$. So, the quotient $\mathbb{K} = \mathbb{Q}^*_{\text{fin}}/\mu(\mathbb{Q}^*)$ is a field.

Since $\mu(\mathbb{Q}^*)$ is *convex* (that is, $\xi, \eta \in \mu(\mathbb{Q}^*)$ and $\xi < x < \eta$ imply that $x \in \mu(\mathbb{Q}^*)$), such a field $\mathbb{K}$ is an ordered field.[1] Indeed, denote by $\bar{\xi} \in \mathbb{K}$ the coset of $\xi \in \mathbb{Q}^*_{\text{fin}}$ modulo $\mu(\mathbb{Q}^*)$. Then it is easily verified that if $\mathbb{K}^+ = \{\bar{\xi} \neq \bar{0} \mid \xi \in \mathbb{Q}^*_{\text{fin}}, \xi > 0\}$, then the pair $(\mathbb{K}, \mathbb{K}^+)$ satisfies the properties of ordered field (see Definition 1.1).

---

[1] Recall the following general fact from algebra: The quotient R/I of an ordered ring R modulo an ideal I is an ordered field (with the ordering inherited from R) if and only if I is maximal and convex.

To complete the proof, we are left to show that $\mathbb{K}$ satisfies the *completeness property*.[2] Clearly $\mathbb{K}$ does not contain non-zero infinitesimals, as it directly follows from its very definition as a quotient modulo $\mathfrak{mon}(\mathbb{Q}^*)$. So, we can use the following well-known characterization:

- An Archimedean field $\mathbb{F}$ is *complete* if and only if every non-increasing countable chain $\{[a_n, b_n] \mid n \in \mathbb{N}\}$ of closed intervals (that is, $[a_n, b_n] \supseteq [a_{n+1}, b_{n+1}]$ for all n), has a nonempty intersection $\bigcap_{n \in \mathbb{N}}[a_n, b_n] \neq \emptyset$.

Let the non-increasing chain $\{[\overline{\xi_n}, \overline{\zeta_n}] \mid n \in \mathbb{N}\}$ of nonempty closed intervals of $\mathbb{K}$ be given. Then the family of (internal) closed intervals $\{[\xi_n, \zeta_n] \mid n \in \mathbb{N}\}$ of $\mathbb{Q}^*$ has the finite intersection property. Indeed, it is easily follows from the hypotheses that for every n, the mid-point $\frac{\xi_n + \zeta_n}{2} \in \bigcap_{i=1}^{n}[\xi_i, \zeta_i]$.[3] By *countable saturation*, there exists an element $\eta \in \bigcap_{n \in \mathbb{N}}[\xi_n, \zeta_n]$. It is readily verified that its equivalence class $\overline{\eta} \in \bigcap_{n \in \mathbb{N}}[\overline{\xi_n}, \overline{\zeta_n}]$. $\qquad\square$

2. Ramsey's Theorem

Ramsey's Theorem is one the fundamental results of combinatorics, and the starting point of a whole area of research known as *Ramsey theory*. It can be used in virtually any field of mathematics where combinatorial reasoning can be applied.

Probably the most elementary Ramsey-type property is the well-known *pigeon principle* stating the following: "If n items are put into m pigeonholes, and $m < n$, then at least one pigeonhole must contain more than one item." The analogous property of infinite sets is the following: "If an infinite set is partitioned into finitely many pieces, then at least one piece must be infinite."

One nice result that is usually mentioned in introductions to Ramsey theory is the following fact, that can be proved by a simple counting argument:

- In every 6 people party, one can always find 3 people such that either they mutually know each other or they mutually do not know each other.

More generally, the following result holds.

---

[2] Recall that all complete ordered fields are isomorphic to $\mathbb{R}$.

[3] Notice that we cannot assume $[\xi_{n+1}, \zeta_{n+1}] \subseteq [\xi_n, \zeta_n]$; in fact, it could be $\overline{\xi_n} = \overline{\xi_{n+1}}$ and $\xi_n > \xi_{n+1}$.

THEOREM 7.2 (Party Theorem). *For every m, one can find a large enough n such that in every n people party there are always m people that either know each other or do not know each other.*

In the theorem above, one consider the set of pairs of people, and partition it into two classes, depending on whether they know or they do not know each other. The full generality of that combinatorial result is given by the next theorem.

Let us first fix terminology. As usual in combinatorics, one thinks of a partition as a coloring that assigns different colors to elements that belong to different pieces of the partition. So, an *r-coloring* means a partition into *r* pieces, and a set is called *monochromatic* when it is included in one piece of the partition.

The fundamental Ramsey's Theorem deals with coloring of pairs (and more generally of *k*-tuples) of a given set.

THEOREM 7.3 (Finite Ramsey's Theorem). *Given r and m, one can always find a large enough n such that for every r-coloring of the pairs $[\{1, \ldots, n\}]^2 = C_1 \cup \ldots \cup C_r$, there exists a homogeneous set $|H| = m$, that is, a set H such that the set of pairs $[H]^2 \subseteq C_i$ is monochromatic.*[4]

A Ramsey result usually has a "finite version" and an "infinite version."

THEOREM 7.4 (Infinite Ramsey's Theorem). *Every finite colouring $[\mathbb{N}]^2 = C_1 \cup \ldots \cup C_r$ admits a homogeneous infinite set H, that is an infinite set such that the set of pairs $[H]^2 \subseteq C_i$ is monochromatic.*[5]

We remark that often the infinite versions are simpler to formulate, simpler to be proved, and imply the finite versions. In this case, it is fair to say that "infinite" is better than "finite".

A typical result in infinite Ramsey theory has the following form: Given a finite coloring $X = C_1 \cup \ldots \cup C_r$, if X has some "structural property" P, then one of the colors C_i preserves the "structural property" P.

---

[4] Recall that $[A]^k = \{X \subseteq A \mid |X| = k\}$ denotes the family of all k-tuples of elements of A.

[5] More generally, both the finite and the infinite version of Ramsey's Theorem also hold for k-tuples, with k any natural number.

In this section we limit ourselves to present alternative proofs of the two versions of Ramsey's Theorem, by using the tools of Alpha-Theory[6]. Let us start with the infinite version.

PROOF OF INFINITE RAMSEY'S THEOREM. Let us apply the star-map twice, and consider the hyper-hypernatural numbers $\mathbb{N}^{**} = (\mathbb{N}^*)^*$. It is easily verified that the set of pairs of $\mathbb{N}^{**}$ is the same as the hyper-hyper-image (double-star) of the set of pairs of natural numbers, that is, $[\mathbb{N}^{**}]^2 = ([\mathbb{N}]^2)^{**}$.[7] By taking double-stars of the partition $[\mathbb{N}]^2 = C_1 \cup \ldots \cup C_r$, we get a partition of the hyper-hypernatural numbers:

$$[\mathbb{N}^{**}]^2 = C_1^{**} \cup \ldots \cup C_r^{**}.$$

Now, $\alpha \in \mathbb{N}^* \setminus \mathbb{N} \Rightarrow \alpha^* \in \mathbb{N}^{**} \setminus \mathbb{N}^*$. Since $\mathbb{N}^{**}$ is an end extension of $\mathbb{N}^*$ (see Proposition 4.28), it is $\alpha < \alpha^*$, and so $\{\alpha, \alpha^*\} \in [\mathbb{N}^{**}]^2$. Pick i such that $\{\alpha, {}^*\alpha\} \in C_i^{**}$. We will reach the desired conclusion by constructing an infinite set H such that $[H]^2 \subseteq C_i$. Since

$$\lim_{n\uparrow\alpha} \{n, \alpha\} = \left\{\lim_{n\uparrow\alpha} n, \lim_{n\uparrow\alpha} \alpha\right\} = \{\alpha, \alpha^*\} \in C_i^{**} = \lim_{n\uparrow\alpha} C_i^*,$$

the set $A_0 = \{n \mid \{n, \alpha\} \in C_i^*\}$ is qualified. Now pick $a_1 \in A_0$, so that $\{a_1, \alpha\} \in C_i^*$. Since

$$\lim_{n\uparrow\alpha}\{a_1, n\} = \{a_1, \alpha\} \in C_i^* = \lim_{n\uparrow\alpha} C_i,$$

also the set $A_1 = \{n \mid \{a_1, n\} \in C_i\}$ is qualified. Pick an element $a_2 \in A_0 \cap A_1$ with $a_2 > a_1$ (this is possible because $A_0 \cap A_1$ is qualified, and hence infinite). So, we have that $\{a_2, \alpha\} \in C_i^*$ and $\{a_1, a_2\} \in C_i$.

By proceeding similarly as above, it is proved that $\{a_2, \alpha\} \in C_i^*$ implies that the set $A_2 = \{n \mid \{a_2, n\} \in C_i\}$ is qualified, and so we can pick an element $a_3 > a_2$ in the intersection $A_0 \cap A_1 \cap A_2$. In this way, $\{a_3, \alpha\} \in C_i^*$ and $\{a_1, a_3\}, \{a_2, a_3\} \in C_i$. By iterating this procedure, we obtain an increasing sequence $\langle a_n \mid n \in \mathbb{N}\rangle$ such that $\{a_n, \alpha\} \in {}^*C_i$ for all n, and $\{a_m, a_n\} \in C_i$ for all $m < n$. Then $H = \{a_n \mid n \in \mathbb{N}\}$ is the infinite homogeneous set we were looking for. □

INFINITE RAMSEY'S THEOREM $\Rightarrow$ FINITE RAMSEY'S THEOREM. By contradiction, assume that there exist r and m such that for every n one can find a counter-example to the *Finite Ramsey's Theorem*, that

---

[6] See the book [**39**] for a comprehensive presentation of recent applications of nonstandard methods in Ramsey Theory and related areas of combinatorics.

[7] See Exercise 4.26.

is an r-coloring of the pairs from $\{1, \ldots, n\}$ with no homogeneous sets of size m:

$$[\{1, \ldots, n\}]^2 = C_n^1 \cup \ldots \cup C_n^r.$$

Since $\lim_{n \uparrow \alpha} \{1, \ldots, n\} = \{1, \ldots, \alpha\}$ (see Exercise 4.5), by taking the Alpha-limits of the above objects, one gets the r-coloring

$$[\{1, \ldots, \alpha\}]^2 = C_1^\alpha \cup \ldots \cup C_r^\alpha,$$

where $C_i^\alpha = \lim_{n \uparrow \alpha} C_n^i$. By restricting to the subset $\mathbb{N} \subset \{1, \ldots, \alpha\}$, one obtains the r-coloring

$$[\mathbb{N}]^2 = C_1 \cup \ldots \cup C_r,$$

where $C_i = C_i^\alpha \cap [\mathbb{N}]^2$. Then, by the *Infinite Ramsey's Theorem*, there exists an infinite homogeneous set H, say $[H]^2 \subseteq C_i$. Pick a finite set $X \subset H$ containing m-many elements. As a finite set of natural numbers, $X^* = X$, and so

$$\lim_{n \uparrow \alpha} [X]^2 = [X^*]^2 = [X]^2 \subset [H]^2 \subseteq C_i \subseteq C_i^\alpha = \lim_{n \uparrow \alpha} C_n^i.$$

But then $[X]^2 \subset C_n^i$ for at least one n (actually, for almost all n). This means that X is a homogeneous set of cardinality m for the r-coloring $[\{1, \ldots, n\}]^2 = C_n^1 \cup \ldots \cup C_n^r$, contradicting our assumptions. $\qquad \square$

3. Grid functions

A grid function is a function whose argument ranges on a suitable hyperfinite "grid". Since the grid is hyperfinite, these functions are easy to handle and, from many points of view, they behave similarly to functions defined on finite sets; actually, they form a hyperfinite vector space. We will see that this simple kind of functions turn out to be very useful in applications.

Recall that in Section 3.10 we introduced the notion of *grid integral* by using the following notion of "grid" (see Definition 3.39):

DEFINITION. *Let $n \in \mathbb{N}$. The n-grid is the finite set*

$$\mathbb{H}(n) = \left\{ \pm \frac{i}{n} \mid i = 0, 1, \ldots, n^2 \right\} \subset [-n, n].$$

DEFINITION 7.5. *The* Alpha-grid *or the* hyperfinite grid *is the hyperfinite set $\mathbb{H}$ obtained as the Alpha-limit of the n-grids, namely*

$$\mathbb{H} = \lim_{n \uparrow \alpha} \mathbb{H}(n) = \left\{ \pm i \cdot \eta \mid i = 0, 1, \ldots, \alpha^2 \right\} \subset [-\alpha, \alpha].^8$$

---

[8] Recall that $\eta = 1/\alpha$.

EXERCISE 7.6. *Assume the **Divisibility Property** that α is a multiple of every $k \in \mathbb{N}$ (see §2.9). Show that a real number belongs to the hyperfinite grid if and only if it is rational, that is, $\mathbb{H} \cap \mathbb{R} = \mathbb{Q}$.*

For real numbers $a < b$, we denote by

- $[a, b]_{\mathbb{H}} = [a, b]^* \cap \mathbb{H}$, the hyperfinite grid of $[a, b]^*$.

DEFINITION 7.7. *A grid function is an internal function*

$$\sigma : \mathbb{H} \to \mathbb{R}^*.$$

The family of grid functions will be denoted by $\mathfrak{G}$.

More generally, for real numbers $a < b$, a grid function on $[a, b]^*$ is an internal function $\sigma : [a, b]_{\mathbb{H}} \to \mathbb{R}^*$. The family of grid functions on $[a, b]^*$ will be denoted by $\mathfrak{G}([a, b])$.

Notice that given a real function $f : [a, b] \to \mathbb{R}$ one canonically obtains a grid function $f_{\mathbb{H}} : [a, b]_{\mathbb{H}} \to \mathbb{R}^*$ by restricting the hyper-extension f^* to $[a, b]_{\mathbb{H}}$.

With abuse of terminology, we will say that a grid function σ is continuous or differentiable if $\sigma = f_{\mathbb{H}}$ for a continuous function f, or a differentiable function f, respectively.

When there is no risk of ambiguity, we will omit the subscript and write f instead of $f_{\mathbb{H}}$.

DEFINITION 7.8. *Let $\sigma : \mathbb{H} \to \mathbb{R}^*$ be a grid function. The* right grid derivative $D_{\mathfrak{G}}^+ \sigma$ *and the* left grid derivative $D_{\mathfrak{G}}^- \sigma$ *are defined by setting for every $t \in \mathbb{H}$, respectively:*

$$D_{\mathfrak{G}}^+\sigma(t) = \frac{\sigma(t + \eta) - \sigma(t)}{\eta} ; \quad D_{\mathfrak{G}}^-\sigma(t) = \frac{\sigma(t) - \sigma(t - \eta)}{\eta}.$$

The grid derivative $D_{\mathfrak{G}}\sigma$ *is defined by setting for every $t \in \mathbb{H}$:*

$$D_{\mathfrak{G}}\sigma(t) = \frac{D_{\mathfrak{G}}^+\sigma(t) + D_{\mathfrak{G}}^-\sigma(t)}{2} = \frac{\sigma(t + \eta) - \sigma(t - \eta)}{2\eta}.$$

Grid functions are also suitable for a simple definition of Laplacian.

DEFINITION 7.9. *The* grid Laplacian $\Delta_{\mathfrak{G}}\sigma$ *is defined by setting for every $t \in \mathbb{H}$:*

$$\Delta_{\mathfrak{G}}\sigma(t) = D_{\mathfrak{G}}^+ D_{\mathfrak{G}}^- \sigma(t) = D_{\mathfrak{G}}^- D_{\mathfrak{G}}^+ \sigma(t) = \frac{\sigma(t + \eta) + \sigma(t - \eta) - 2\sigma(t)}{\eta^2}.$$

In order to make $D_{\mathfrak{G}}^+\sigma(t), D_{\mathfrak{G}}^-\sigma(t), D_{\mathfrak{G}}\sigma(t), \Delta_{\mathfrak{G}}\sigma(t)$ defined for every $t \in \mathbb{H}$, in the above definitions we implicitly agreed that $\sigma(\alpha + \eta) = \sigma(-\alpha - \eta) = 0$. Notice that in this way the grid derivative of a grid function is again a grid function.

For simplicity, in the remainder of the book, whenever confusion is unlikely, we will omit the subscript $\mathfrak{G}$ and write D^+, D^-, D instead of $D_{\mathfrak{G}}^+, D_{\mathfrak{G}}^-, D_{\mathfrak{G}}$, respectively.

It is readily checked that if $f \in \mathscr{C}^1(\mathbb{R})$ then for every $t \in \mathbb{R} \cap \mathbb{H} = \mathbb{Q}$ we have:
$$D^+ f(t) \ \sim \ D^- f(t) \ \sim \ Df(t) \ \sim \ f'(t).$$
Moreover, if $f \in \mathscr{C}^2(\mathbb{R})$, then we also have:
$$\Delta f(t) \ \sim \ f''(t).$$

Most of the properties of the usual derivative also hold for the grid derivative with some approximation. For example,

$$
\begin{aligned}
D^+(\sigma \cdot \tau) &= \frac{\sigma(t+\eta)\tau(t+\eta) - \sigma(t)\tau(t)}{\eta} \\
&= \frac{\sigma(t+\eta)\tau(t+\eta) - \sigma(t+\eta)\tau(t) + \sigma(t+\eta)\tau(t) - \sigma(t)\tau(t)}{\eta} \\
&= \sigma(t+\eta)D^+\tau(t) + D^+\sigma(t)\tau(t).
\end{aligned}
$$

The analogous of integral for grid functions is the following.

DEFINITION 7.10. *The grid sum of a grid function σ over the internal set $\Gamma \subseteq \mathbb{H}$ is the hyperfinite sum:*
$$\eta \cdot \sum_{t \in \Gamma} \sigma(t).$$
When the above grid sum is finite, we say that σ is grid integrable *and we define its* grid integral *over Γ as:*
$$\int_\Gamma \sigma(x)\, d_\mathbb{H}x \ = \ \mathrm{st}\left(\eta \cdot \sum_{t \in \Gamma} \sigma(t)\right).$$
We say that σ is absolutely grid integrable *if $|\sigma|$ is grid integrable.*

We remark that the above definition of grid integral is consistent with Definition 3.41. Indeed, for every real function $f : A \to \mathbb{R}$ we have
$$\int_a^b f(x)\, d_\mathbb{H}x \ = \ \int_{[a,b]^* \cap \mathbb{H}} f_\mathbb{H}(x)\, d_\mathbb{H}x.$$

In particular, if f is Riemann-integrable then by Theorem 3.54 we have that:

$$\int_{[a,b]^*\cap\mathbb{H}} f(x)\,d_\mathbb{H}x \;=\; \int_a^b f(x)\,dx.$$

We have the following relationships between the grid derivative and the grid sum, that correspond to the fundamental theorem of calculus.

THEOREM 7.11. *If* $\sigma : \mathbb{H} \to \mathbb{R}^*$ *is a grid function, then for every* $a < b$ *in* $\mathbb{H}$:

(1) $\eta \cdot \sum_{s\in[a,b)_\mathbb{H}} D^+\sigma(s) = \sigma(b) - \sigma(a)$;

(2) $\eta \cdot \sum_{s\in(a,b]_\mathbb{H}} D^-\sigma(s) = \sigma(b) - \sigma(a)$.

Moreover, if $S_\sigma^+(t) = \eta \cdot \sum_{s\in[a,x)_\mathbb{H}} \sigma(s)$ *and* $S_\sigma^-(t) = \eta \cdot \sum_{s\in[a,x)_\mathbb{H}} \sigma(s)$ *are the grid sum functions of* σ *starting at* a, *then*

(3) $D^+S_\sigma^+(b) = \sigma(b)$;

(4) $D^-S_\sigma^-(b) = \sigma(b)$.

PROOF. We have

$$\eta \cdot \sum_{s\in[a,b)} D^+\sigma(s) = \eta \cdot \left(\frac{\sigma(a+\eta)-\sigma(a)}{\eta} + \frac{\sigma(a+2\eta)-\sigma(a+\eta)}{\eta} + \dots \right.$$

$$\left. \dots + \frac{\sigma(b)-\sigma(b-\eta)}{\eta} \right) = \sigma(b) - \sigma(a).$$

Furthermore,

$$D^+S_\sigma^+(b) \;=\; \frac{\eta \cdot \sum_{s\in[a,b+\eta)} \sigma(s) - \eta \cdot \sum_{s\in[a,b)} \sigma(s)}{\eta}$$

$$=\; \sum_{s\in[a,b+\eta)} \sigma(s) - \sum_{s\in[a,b)} \sigma(s) \;=\; \sigma(b).$$

The proofs of the other properties about D^- are entirely similar.   □

The above notions can be easily extended to functions of several variables. If $\Omega \subseteq \mathbb{R}^N$, then a grid function on Ω is an internal function

$$\rho : \mathbb{H}^N \cap \Omega^* \to \mathbb{R}^*.$$

The family of such functions will be denoted by $\mathfrak{G}(\Omega)$. For functions $\rho \in \mathfrak{G}(\Omega)$, one defines the *grid partial derivatives* in the obvious way;

for example, if $\rho : \mathbb{H} \times \mathbb{H} \to \mathbb{R}^*$ we set:

$$D_x^+ \rho(x, y) \;=\; \frac{\rho(x + \eta, y) - \rho(x, y)}{\eta}$$

$$D_y^+ \rho(x, y) \;=\; \frac{\rho(x, y + \eta) - \rho(x, y)}{\eta}.$$

It is interesting to see the *grid Laplacian* $\Delta_{\mathfrak{G}}$. If $\rho : \mathbb{H}^N \to \mathbb{R}^*$, then for every $x = (x_1, \ldots, x_N) \in \mathbb{H}^N$ we set:

$$\Delta_{\mathfrak{G}} \rho(x) \;=\; \sum_{j=1}^{N} D_{x_j}^+ D_{x_j}^- \rho(x) \;=\; \sum_{j=1}^{N} D_{x_j}^- D_{x_j}^+ \rho(x)$$

$$=\; \frac{1}{\eta^2} \left[\left(\sum_{\xi \in N(x)} \rho(\xi) \right) - 2N \rho(x) \right]$$

where $N(x)$ is the set of the points of the grid which are the nearest to x, namely,

$$N(x) \;=\; \{ \xi \in \mathbb{H}^N \mid |\xi - x| = \eta \}.$$

Notice that $N(x)$ contains $2N$-many points and so, if we write

$$\Delta_{\mathfrak{G}} \rho(x) \;=\; \frac{2N}{\eta^2} \left[\left(\frac{1}{2N} \sum_{\xi \in N(x)} \rho(\xi) \right) - \rho(x) \right]$$

we see that $\Delta_{\mathfrak{G}} \rho(x)$ represents how much $\rho(x)$ differs form the average of the neighboring points, normalized by the factor $\eta^2 / 2N$.

4. Grid differential equations

A grid ordinary differential equation is a differential equation whose time step ranges on the hyperfinite grid. To a great extent, such equations behave like discrete time objects, and this simplifies many formal aspects.

DEFINITION 7.12. *A grid ordinary differential equation, grid ODE for short, is an equation of the form*

$$D^{\pm} x(t) \;=\; g(t, x(t))$$

where $x(t)$ is a grid function and $g : \mathbb{H} \times \mathbb{R}^* \to \mathbb{R}^*$ is an internal function. A grid function $x(t)$ is a solution if it satisfies the above equation at every point of the grid.

The following result shows that the Cauchy problem associated to a grid ODE has a unique solution, without assuming any regularity assumption on g.

THEOREM 7.13. *Given an initial time $t_0 \in \mathbb{H}$ and an initial data $x_0 \in \mathbb{R}^*$, for every internal $g : [t_0, +\infty)_{\mathbb{H}} \times \mathbb{R}^* \to \mathbb{R}^*$ the grid Cauchy problem*

(1)
$$\begin{cases} x(t_0) = x_0 \\ D^+ x(t) = g(t, x(t)) \quad t \in \mathbb{H}, \ t \geq t_0 \end{cases}$$

admits a unique grid solution $x : [t_0, +\infty)_{\mathbb{H}} \to \mathbb{R}^*$. Moreover, if $g(t, x)$ is bounded by $M \in \mathbb{R}^+$, then the solution $x(t)$ satisfies the inequality

(2)
$$|x(t_2) - x(t_1)| \leq M \cdot |t_2 - t_1|$$

for every $t_1, t_2 \in [t_0, +\infty)_{\mathbb{H}}$. In this case, if x_0 is finite then also the values $x(t)$ are finite whenever $t - t_0$ is finite.

PROOF. Pick sequences of numbers $t_{0,n} \in \mathbb{H}(n)$ and $x_{0,n} \in \mathbb{R}$ such that $t_0 = \lim_{n \uparrow \alpha} t_{0,n}$ and $x_0 = \lim_{n \uparrow \alpha} x_{0,n}$, and pick a sequence of functions $g_n : \mathbb{H}(n) \times \mathbb{R} \to \mathbb{R}$ such that $g = \lim_{n \uparrow \alpha} g_n$. For every $n \in \mathbb{N}$, inductively define a function $x_n : \mathbb{H}(n) \cap [t_{0,n}, +\infty) \to \mathbb{R}$ by letting for every $m \in \mathbb{N}_0$ such that $t_{0,n} + \frac{m+1}{n} \in \mathbb{H}(n)$:

$$\begin{cases} x_n(t_{0,n}) = x_{0,n} \\ x_n\left(t_{0,n} + \frac{m+1}{n}\right) = x_n\left(t_{0,n} + \frac{m}{n}\right) + \frac{1}{n} \cdot g_n\left(t_{0,n} + \frac{m}{n}, \ x_n\left(t_{0,n} + \frac{m}{n}\right)\right). \end{cases}$$

Let $x = \lim_{n \uparrow \alpha} x_n$. Notice that $x : [t_0, +\infty)_{\mathbb{H}} \to \mathbb{R}^*$ is internal, and hence it is a grid function. By passing the above equations to the Alpha-limits, we obtain that $x(t_0) = x_0$, and for every $m \in \mathbb{N}_0^*$ such that $t_0 + (m+1)\boldsymbol{\eta} \in \mathbb{H}$:

$$x\left(t_0 + (m+1)\boldsymbol{\eta}\right) = x\left(t_0 + m\boldsymbol{\eta}\right) + \boldsymbol{\eta} \cdot g\left(t_0 + m\boldsymbol{\eta}, \ x(t_0 + m\boldsymbol{\eta})\right).$$

This shows that $D^+ x(t_0 + m\boldsymbol{\eta}) = g(t_0 + m\boldsymbol{\eta}, x(t_0 + m\boldsymbol{\eta}))$ for every $m \in \mathbb{N}_0^*$ such that $t_0 + (m+1)\boldsymbol{\eta} \in \mathbb{H}$, and so x is the desired solution. It is then easily checked that the solution x is also unique.

Now assume that g is bounded by M, and let $t_1 < t_2$ be numbers in $[t_0, +\infty)_{\mathbb{H}}$. Take sequences $t_{1,n}, t_{2,n} \in \mathbb{H}(n)$ such that $t_1 = \lim_{n \uparrow \alpha} t_{1,n}$ and $t_2 = \lim_{n \uparrow \alpha} t_{2,n}$, and let $t_{0,n}, x_{0,n}, g_n$ be defined as above. For every n, let us assume without loss of generality that g_n is bounded by M and that $t_{2,n} > t_{1,n} \geq t_{0,n}$; and let $m \in \mathbb{N}$ be such that $t_{2,n} = t_{1,n} + m/n$.

Then we have:

$$
\begin{aligned}
|x_n(t_{2,n}) - x_n(t_{1,n})| &= \left| x_n\left(t_{1,n} + \frac{m}{n}\right) - x_n(t_{1,n}) \right| \\
&= \sum_{i=0}^{m-1} \left| x_n\left(t_{1,n} + \frac{i+1}{n}\right) - x_n\left(t_{1,n} + \frac{i}{n}\right) \right| \\
&= \sum_{i=0}^{m-1} \frac{1}{n} \cdot \left| D^+ x_n\left(t_{1,n} + \frac{i}{n}\right) \right| \\
&= \sum_{i=0}^{m-1} \frac{1}{n} \cdot \left| g_n\left(t_{1,n} + \frac{i}{n}, \; x_n\left(t_{1,n} + \frac{i}{n}\right)\right) \right| \\
&\leq \sum_{i=0}^{m-1} \frac{1}{n} \cdot M = M \cdot \frac{m}{n} = M \cdot (t_{2,n} - t_{1,n}).
\end{aligned}
$$

By taking the Alpha-limit, we obtain the desired inequality

$$
|x(t_2) - x(t_1)| \leq M \cdot (t_2 - t_1).
$$

The last statement directly follows from the above inequality where $t_1 = t_0$ and $t_2 = t$. $\qquad\square$

5. Peano's theorem

In this section we focus on a "classic" Cauchy problem.

THEOREM 7.14 (Peano). *Let $f : [t_0, t_1] \times \mathbb{R} \to \mathbb{R}$ be a bounded continuous function. Then there exists a solution $u \in \mathscr{C}^1([t_0, t_1])$ to the following problem:*[9]

$$
(3) \qquad \begin{cases} u(t_0) = u_0 \\ u'(t) = f(t, u(t)) \quad t \in [t_0, t_1]. \end{cases}
$$

PROOF. For every $t \in \mathbb{R}$ denote by

$$
t^+ = \min\{\xi \in \mathbb{H} \mid t \leq \xi\}.
$$

Clearly, $t^+ \sim t$. Let $\xi_0 = t_0^+$ and $\xi_1 = \max\{\xi \in \mathbb{H} \mid \xi \leq t_1\}$, so that $[t_0, t_1]_\mathbb{H} = [\xi_0, \xi_1]_\mathbb{H}$, and let $g : [\xi_0, \xi_1]_\mathbb{H} \times \mathbb{R}^* \to \mathbb{R}^*$ be the grid function obtained by restricting the hyper-extension f^* to the internal

---

[9] $u'(t)$ denotes the derivative of u with respect to t.

set $[\xi_0, \xi_1]_\mathbb{H} \times \mathbb{R}^* \subseteq \mathbb{H} \times \mathbb{R}^*$. Pick $x(\xi)$ the solution to the following grid Cauchy problem, as given by Theorem 7.13:

$$(4) \qquad \begin{cases} x(\xi_0) = u_0 \\ D^+x(\xi) = g(s, x(\xi)) & \xi \in [\xi_0, \xi_1]_\mathbb{H}. \end{cases}$$

Since f is bounded, also g is bounded, and so all values $x(\xi)$ for $\xi \in [\xi_0, \xi_1]_\mathbb{H}$ are finite. Therefore, we can define

$$u(t) = \mathrm{st}(x(t^+)) \quad \text{for } t \in [t_0, t_1)_\mathbb{R}.$$

Let us also put $u(t_1) = \mathrm{st}(x(\xi_1))$. We claim that the function $u : [t_0, t_1] \to \mathbb{R}$ is the desired solution.

Let $M \in \mathbb{R}^+$ be a bound for g. Then for every $t \in [t_0, t_1)_\mathbb{R}$ and for every $\varepsilon \sim 0$, we have that

$$\begin{aligned} |u(t+\varepsilon) - u(t)| &= |\mathrm{st}(x((t+\varepsilon)^+)) - \mathrm{st}(t^+)| \\ &\sim |x((t+\varepsilon)^+) - x(t^+)| \leq M \cdot |(t+\varepsilon)^+ - t^+| \sim 0. \end{aligned}$$

This shows that u is continuous on $[t_0, t_1)$. Similarly, it is also readily checked that u is continuous at t_1. We have seen in Theorem 7.11 that for every $\xi \in [\xi_0, \xi_1]_\mathbb{H}$ one has the following equality:

$$x(\xi) = x(\xi_0) + \eta \cdot \sum_{s \in [\xi_0, \xi)_\mathbb{H}} D^+x(s).$$

Since $x(\xi)$ is a solution of (4), it follows that

$$x(\xi) = x(\xi_0) + \eta \cdot \sum_{s \in [\xi_0, \xi)_\mathbb{H}} g(s, x(s)),$$

and so, for every $t \in [t_0, t_1]_\mathbb{R}$, we have that

$$u(t) = \mathrm{st}\left(x(t^+)\right) = u_0 + \mathrm{st}\left[\eta \cdot \sum_{s \in [\xi_0, t^+)_\mathbb{H}} f^*(s, x(s))\right].$$

Now notice the following fact:

- $f^*(s, x(s)) \sim f^*(s, u^*(s))$ for every $s \in [t_0, t_1]_\mathbb{H}$.

Indeed, let $t = \mathrm{st}(s)$. Since u is continuous, $u^*(s) \sim u(t) = x(t^+) \sim x(s)$; then, by the continuity of f, we have that both $f^*(s, x(s))$ and $f^*(s, u^*(s))$ are infinitely close to the real number $f(t, u(t))$. Now, for every $s \in [t_0, t_1]_\mathbb{H}$, pick the number $\theta_s \sim 0$ such that $f^*(s, x(s)) = f^*(s, u^*(s)) + \theta_s$. Then:

$$u(t) = u_0 + \mathrm{st}\left[\eta \cdot \sum_{s \in [\xi_0, t^+)_\mathbb{H}} f^*(s, u^*(s))\right] + \mathrm{st}\left[\eta \cdot \sum_{s \in [\xi_0, t^+)_\mathbb{H}} \theta_s\right].$$

Notice that $\mathrm{st}(\boldsymbol{\eta} \cdot \sum_{s \in [\xi_0, t^+)_\mathbb{H}} f^*(s, u^*(s)))$ equals the grid integral $\int_{t_0}^t f(s, u(s))d_\alpha s$ (see Definition 3.41). Since $f(s, u(s))$ is continuous, and hence Riemann integrable, that quantity coincides with the usual Riemann integral (see Theorem 3.54.) Finally, the last term equals 0: indeed, if $\theta = \max\{|\theta_s| \mid s \in [\xi_0, t^+]_\mathbb{H}\}$, then

$$\left| \boldsymbol{\eta} \cdot \sum_{s \in [\xi_0, t^+)_\mathbb{H}} \theta_s \right| \leq \boldsymbol{\eta} \cdot \theta \cdot |[\xi_0, t^+)_\mathbb{H}| \leq (t^+ - t_0) \cdot \theta \sim 0.$$

Thus we have shown that

$$u(t) = u_0 + \int_{t_0}^t f(s, u(s))ds.$$

Clearly $u(t_0) = u_0$. Moreover, by the fundamental theorem of calculus, we also obtain the desired equality $u'(t) = f(t, u(t))$. $\qquad\square$

REMARK 7.15. *The "nonstandard" proof of Peano's Theorem seen above is simpler than the "standard" one. In the usual proof, one approximates the solution u with a suitable sequence of functions $\{x_n\}$; the key point is to have the continuity of the function u, and this is shown by applying Ascoli-Arzelà's theorem, a quite involved result. Notice that in our proof, this is a trivial consequence of inequality 8.39 in Theorem 7.13.*

It is well known that, in general, problem (3) does not have a unique solution. On the other hand, we have seen that problem (4) has a unique solution. How is it possible? The point is that for every choice of the initial data $\xi_0 \sim t_0$ and $x(\xi_0) \sim u_0$, t_0, one has a solution of (4). When uniqueness fails, it happens that different choices of $\xi_0 \sim t_0$ and $x(\xi_0) \sim u_0$ produce different solutions $u(t) = \mathrm{st}(x(t^+))$. So, the Alpha-Theory approach to ordinary differential equations provides an interesting interpretation of the lack of uniqueness.

EXERCISE 7.16. *Consider the following Cauchy problem:*

$$\begin{cases} u(0) = 0 \\ u'(t) = \sqrt{|u(t)|} \quad t \in \mathbb{R}_{\geq 0} \end{cases}$$

Show that by considering all different choices of the initial data $\xi_0 \sim 0$ and $x(\xi_0) \sim 0$ for the approximating grid differential equation, one obtains all the solutions of that Cauchy problem.

CHAPTER 8

Gauge Spaces

The notion of topological space was introduced as a convenient framework where the Weierstrass limit and, consequently, the other basic notions of calculus, could be accommodated. As we have seen in Chapter 3, one can develop a calculus which is not grounded on the notion of limit, but rather on the notion of infinitesimal number. By keeping this in mind, one may ask whether the generalization of calculus must necessarily lead to the notion of topological space. Actually, as we will see in this chapter, one can extend the notion of *"infinitely close"* from the hyperreal numbers $\mathbb{R}^*$ to the hyper-extensions of more general spaces (such as infinite dimensional vector spaces), and obtain a new notion, namely the notion of *gauge space*.

The name "gauge space" is already present in the mathematical literature to denote a space equipped with a family of pseudometrics, and our notion of gauge space is indeed a generalization of it.[1]

1. Main definitions

DEFINITION 8.1. *A gauge space is a pair* $(X, \approx)$ *where* X *is a set and* $\approx$ *is an equivalence relation, called* gauge, *on the hyper-extension* X^*. *If* $x \approx y$ *we say that* x *is* $\approx$-close *to* y, *or* x *is* infinitely close *to* y.

Equivalence classes of a gauge $\approx$ will be denoted by $[\,\cdot\,]_\approx$, or simply by $[\,\cdot\,]$ when the gauge is clear from the context.

We have decided to consider gauge spaces rather than general topological spaces for at least three reasons:

---

[1] See Remark 8.5.

- The notion of gauge is a really natural extension of the equivalence relation of "infinitesimal distance" between hyperreal numbers;

- The gauge spaces are much easier to deal with than general topological spaces, since the notion of equivalence relation is simpler than the notion of topology;

- The nonstandard approach to topology is already present in the literature, and we think that it is now worth exploring new directions.

Let us remark that, in any case, the concept of a gauge space is not too far from that of a topological space, and indeed in the simplest cases, such as metric spaces, the two notions coincide (see Section 5 of this chapter).

DEFINITION 8.2. *The* canonical gauge space *on $\mathbb{R}$ is the pair $(\mathbb{R}, \sim)$, where $\sim$ is the "infinitesimal distance" relation.*

Primary examples of gauge spaces are obtained by considering metric spaces.

DEFINITION 8.3. *Let (X, d) be a metric space. The gauge $\approx_d$ induced by the metric d is defined as follows:*

$$\xi \approx_d \eta \iff d^*(\xi, \eta) \sim 0.$$

In this case we say that $(X, \approx_d)$ is a metric gauge.

Probably, the second simpler way to construct a gauge is to use a family of pseudometrics.[2]

DEFINITION 8.4. *Given a family $\mathfrak{M} = \{d_j \mid j \in J\}$ of pseudometrics on the set X, the gauge $\approx_{\mathfrak{M}}$ induced by $\mathfrak{M}$ is defined by setting:*

$$\xi \approx_{\mathfrak{M}} \eta \iff d_j^*(\xi, \eta) \sim 0 \text{ for all } j \in J.$$

REMARK 8.5. *In the mathematical literature, a "gauge space" is a topological space whose topology is defined by a family of pseudometrics. This fact and the above example motivate our using the same name "gauge space" in Definition 8.1. However, we remark that our notion of a gauge space is more general.*

---

[2] Recall that a function $d : X \times X \to [0, +\infty)_{\mathbb{R}}$ is called *pseudometric* if it satisfies the following three assumptions: (1) $d(x, x) = 0$; (2) $d(x, y) = d(y, x)$; (3) $d(x, z) \leq d(x, y) + d(y, z)$. So, a pseudometric d is a metric if and only if $d(x, y) = 0 \Rightarrow x = y$.

As further examples, let us see some gauges that may be considered on spaces of real functions.

EXAMPLE 8.6. *Let $\mathscr{C}^0(\Omega)$ be the set of continuous real-valued functions defined on $\Omega \subseteq \mathbb{R}^n$, and let f and g denote generic elements of $\mathscr{C}^0(\Omega)^*$ (which are internal functions $f, g : \Omega^* \to \mathbb{R}^*$).*

(1) *The gauge $\approx_p$ of* pointwise convergence *is defined by:*

$$f \approx_p g \iff \forall x \in \Omega \ f(x^*) \sim g(x^*).$$

(2) *The gauge $\approx_u$ of* uniform convergence *is defined by:*

$$f \approx_u g \iff \forall \xi \in \Omega^* \ f(\xi) \sim g(\xi).$$

(3) *Let $1 \le p < +\infty$. The L^p-gauge $\approx_{L^p}$ is defined by:*

$$f \approx_{L^p} g \iff \int^* |f(\xi) - g(\xi)|^p d\xi \sim 0.$$

(4) *Let $\mathscr{D} = \mathscr{D}(\Omega) \subset \mathscr{C}^0(\Omega)$ be the set of infinitely differentiable functions on Ω with compact support. The* distribution gauge $\approx_{\mathscr{D}'}$ *is defined by:*

$$f \approx_{\mathscr{D}'} g \iff \forall \psi \in \mathscr{D} \int^* [f(\xi) - g(\xi)] \, \psi^*(\xi) \, d\xi \sim 0.$$

Now let $\mathscr{C}^\infty(\Omega) \subset \mathscr{C}^0(\Omega)$ be the set of infinitely differentiable functions on $\Omega \subseteq \mathbb{R}^n$, and let f and g denote generic elements of $\mathscr{C}^\infty(\Omega)^*$ (which are internal functions $f, g : \Omega^* \to \mathbb{R}^*$).

(5) *Let $1 \le p < +\infty$ and $m \in \mathbb{N}$. The* Sobolev gauge $\approx_{W^{m,p}}$ *is defined by:*

$$f \approx_{W^{m,p}} g \iff \sum_{|\alpha| \le m} \left(\int^* |D^\alpha(f(\xi) - g(\xi))|^p \, d\xi \right) \sim 0,$$

where $\alpha = (\alpha_1, \ldots, \alpha_n) \in \mathbb{N}_0^n$ denote multi-indexes.[3]

Notice that examples (2), (3) and (5) are induced by a metric, and examples (1) and (4) are induced by families of pseudometrics. Indeed, the gauge of *pointwise convergence* is induced by the family of pseudometrics $\{d_x \mid x \in \Omega\}$ on $\mathscr{C}^0(\Omega)$, where:[4]

$$d_x(\varphi, \phi) = |\varphi(x) - \phi(x)|.$$

---

[3] Recall the *multi-index* notation: $D^\alpha f = \frac{\partial^{\alpha_1}}{\partial x_1^{\alpha_1}} \cdots \frac{\partial^{\alpha_n}}{\partial x_n^{\alpha_n}} f$, and $|\alpha| = \sum_{i=1}^n \alpha_i$.

[4] The pseudometrics d_x are indeed pseudonorms.

By definition,

$$f \approx_p g \iff \forall x \in \Omega \;\; d_x^*(f, g) \sim 0.$$

The *distribution* gauge is induced by the family of pseudometrics $\{d_\psi \mid \psi \in \mathscr{D}\}$ where:

$$d_\psi(\varphi, \phi) = \left| \int [\varphi(x) - \phi(x)] \psi(x) dx \right|.$$

2. Gauge Abelian spaces

An important class of gauge spaces that includes almost all relevant examples that will be considered here, is obtained by considering Abelian groups.

DEFINITION 8.7. *A gauge Abelian space (GAS for short) is a pair $(X, \mathcal{I})$ where X is an abelian group and $\mathcal{I}$ is a subgroup of the hyper-extension X^*, called the* group of infinitesimals.[5]

In this case, the gauge space determined by $(X, \mathcal{I})$ is the gauge space $(X, \approx_{\mathcal{I}})$ where $\approx_{\mathcal{I}}$ is the equivalence relation determined by $\mathcal{I}$, namely:

$$\xi \approx_{\mathcal{I}} \eta \iff \xi - \eta \in \mathcal{I}.$$

REMARK 8.8. *If X is an Abelian group, then a gauge space $(X, \approx)$ is determined by a GAS $(X, \mathcal{I})$ if and only if $\mathcal{I} = \{\xi - \eta \mid \xi \approx \eta\}$ is a subgroup of X^*.*

For instance, the canonical gauge on $\mathbb{R}$ is the one determined by the subgroup $\mathcal{I} = \mathrm{mon}(0) \subset \mathbb{R}^*$ of infinitesimal numbers. Also the gauges induced in vector spaces by a norm or by a family of seminorms are of this type. *E.g.*, consider the vector spaces in Example 8.6: all gauges presented there are determined by suitable subgroups (actually, vector subspaces) of the hyper-extension of the Abelian groups $\mathscr{C}^0(\Omega)$ or $\mathscr{C}^\infty(\Omega)$.

EXAMPLE 8.9.

- *The gauge $\approx_p$ of* pointwise convergence *is determined by the subgroup:*

$$\mathcal{I}_p = \{f \in \mathscr{C}^0(\Omega)^* \mid \forall x \in \Omega \;\; f(x^*) \sim 0\}.$$

---

[5] In general, the group $\mathcal{I}$ of infinitesimal will be a *non*-internal subset of X^*.

- *The gauge $\approx_u$ of* uniform convergence *is determined by the subgroup:*

$$\mathcal{I}_u = \{f \in \mathscr{C}^0(\Omega)^* \mid \forall \xi \in \Omega^* \ f(\xi) \sim 0\}.$$

- *The L^p-gauge $\approx_{L^p}$ where $1 \le p < +\infty$ is determined by the subgroup:*

$$\mathcal{I}_{L^p} = \left\{f \in \mathscr{C}^0(\Omega)^* \mid \int^* |f(\xi)|^p d\xi \sim 0\right\}.$$

- *The* distribution gauge $\approx_{\mathscr{D}'}$ *is determined by the subgroup:*

$$\mathcal{I}_{\mathscr{D}'} = \left\{f \in \mathscr{C}^\infty(\Omega)^* \mid \forall \psi \in \mathscr{D}(\Omega) \ \int^* f(\xi)\psi^*(\xi)d\xi \sim 0\right\}.$$

- *The* Sobolev gauge $\approx_{W^{m,p}}$ *is determined by the subgroup:*

$$\mathcal{I}_{W^{m,p}} = \left\{f \in \mathscr{C}^\infty(\Omega)^* \mid \sum_{|\alpha| \le m} \left(\int^* |D^\alpha f(\xi)|^p d\xi\right) \sim 0\right\}.$$

3. Topological notions for gauge spaces

In gauge spaces, several topological notions can be formalized in a natural manner.

DEFINITION 8.10. *Let $(X, \approx_X)$ and $(Y, \approx_Y)$ be two gauge spaces, and let $f : X \to Y$. We say that $f : (X, \approx_X) \to (Y, \approx_Y)$ is* gauge-continuous *at the point $x \in X$ if*

$$\xi \approx_X x^* \implies f^*(\xi) \approx_Y f^*(x^*).$$

We simply say that f is gauge-continuous *if it is gauge-continuous at all points $x \in X$.*

DEFINITION 8.11. *Let $(X, \approx_X)$ and $(Y, \approx_Y)$ be two gauge spaces, and let $f : X \to Y$. We say that $f : (X, \approx_X) \to (Y, \approx_Y)$ is* uniformly gauge-continuous *if*

$$\xi, \eta \in X^* \text{ and } \xi \approx_X \eta \implies f^*(\xi) \approx_Y f^*(\eta).$$

Other fundamental topological notions that can be defined in gauge spaces are the following.

DEFINITION 8.12. *Let $(X, \approx)$ be a gauge space.*

- *X is* gauge-Hausdorff *or $(\approx)$-Hausdorff if $x \ne y \Rightarrow x^* \not\approx y^*$.*

- $A \subseteq X$ *is* gauge-open *or* $(\approx)$*-open if* $x \in A \Rightarrow [x^*] \subseteq A^*$;[6]
- $C \subseteq X$ *is* gauge-closed *or* $(\approx)$*-closed if* $[x^*] \cap C^* \neq \emptyset \Rightarrow x \in C$;
- $K \subseteq X$ *is* gauge-compact *or* $(\approx)$*-compact if every point* $\xi \in K^*$ *is "near-standard", i.e. there exists* $x \in X$ *such that* $\xi \approx x^*$.
- $C \subseteq X$ *is* gauge-connected *or* $(\approx)$*-connected if every partition* $C = C_1 \cup C_2$ *has a connecting point* $x \in C$, *that is, there exist* $\xi_1 \in C_1^*$ *and* $\xi_2 \in C_2^*$ *such that*

$$\xi_1 \approx x^* \approx \xi_2.$$

When the gauge space is clear from the context, we will simply say open, closed, continuous, *etc.*, instead of gauge-open, gauge-closed, gauge-continuous, *etc.*, respectively.

As we will see in Section 5 of this chapter, in many important cases the above notions coincide with the usual topological ones, but not in general.

We end this section by generalizing the notion of *standard part* that was defined for (finite) hyperreal numbers.

DEFINITION 8.13. *Let* $(X, \approx)$ *be a gauge space. A point* $\xi \in X^*$ *is called* near-standard *if there is* $x \in X$ *such that* $\xi \approx x^*$.

DEFINITION 8.14. *Let* $(X, \approx)$ *be a Hausdorff gauge, and let*

$$X_{\mathrm{ns}} = \{\xi \in X^* \mid \xi \text{ is near-standard}\}.$$

The gauge-standard part *map*

$$\mathrm{St}_{\approx} : X_{\mathrm{ns}} \longrightarrow X$$

is defined by setting $\mathrm{St}_{\approx}(\xi)$ *as the unique point* $x \in X$ *such that* $\xi \approx x^*$.

4. Topology theorems in gauge spaces

The topological spaces are mathematical structures that allow formalizing fundamental concepts such as continuity, compactness, connectedness, *etc.* They also allow extending the validity of fundamental results of calculus, such as Bolzano's Theorem and Weierstrass' Theorem, to a very general framework. The gauge spaces are different mathematical structures which play a similar role. Indeed, the statements of the various theorems sound the same, although they might

---

[6] Recall that $[x^*] = \{\xi \in X^* \mid \xi \approx x^*\}$ denotes the equivalence classes of x^* in the gauge $\approx$.

have a very different meaning when interpreted in different contexts (this point will be discussed in the next Section 5). Here we will prove some of those theorems; as we will see, their proofs are quite simple and intuitive, and reflect the analogous proofs of calculus theorems as carried out in Alpha-Calculus.

THEOREM 8.15 (Weierstrass – Extreme value).
$f : (K, \approx_K) \to (\mathbb{R}, \sim)$ be a gauge-continuous function. If $(K, \approx_K)$ is gauge-compact, then f attains a least and a greatest value.

PROOF. Let us argue for the maximum (the minimum case is proved similarly). Let M be the least upper bound of the range of f. For each $n \in \mathbb{N}$ pick a point x_n such that $f(x_n) > M - 1/n$. Now pass to the Alpha-limit and consider the point $\xi = \lim_{n \uparrow \alpha} x_n \in X^*$; then $\lim_{n \uparrow \alpha} f(x_n) = f^*(\xi) > M - 1/\alpha$. Since K is gauge-compact, there exists $x \in X$ such that $\xi \approx_K x^*$. By gauge-continuity, $f^*(\xi) \sim f^*(x^*) = f(x)$.[7] The conclusion $f(x) = M$ is obtained by the following inequalities:

$$M \geq f(x) = \mathrm{st}\,(f^*(\xi)) \geq \mathrm{st}\,(M - 1/\alpha) = M.$$

$\square$

THEOREM 8.16 (Weierstrass – generalized).
If $(K, \approx)$ is gauge-compact and $f : (K, \approx_K) \to (Y, \approx_Y)$ is gauge-continuous, then the image set $f(K)$ is gauge-compact.

PROOF. Take $\eta \in [f(K)]^* = f^*(K^*)$; we need to prove that there exists $y \in f(K)$ such that $y^* \approx_Y \eta$. To do this, take $\xi \in K^*$ with $f^*(\xi) = \eta$. By compactness, we can pick $x \in K$ such that $\xi \approx_K x^*$. But then $y = f(x) = f^*(x^*) \approx_Y f^*(\xi) = \eta$, by the continuity of f. $\square$

THEOREM 8.17 (Cantor). Let $f : (K, \approx_K) \to (Y, \approx_Y)$ be gauge-continuous. If $(K, \approx_K)$ is gauge-compact then f is uniformly gauge-continuous.

PROOF. Let $\xi, \eta \in K^*$ with $\xi \approx_K \eta$. Since K is compact, there exists $x \in K$ such that $\xi \approx_K x^* \approx_K \eta$. By continuity of f at the point x, we have that $f^*(\xi) \approx_Y f(x) \approx_Y f^*(\eta)$, as desired. $\square$

---

[7] As usual, we assume that $r^* = r$ for all real numbers r.

THEOREM 8.18 (Bolzano – Intermediate Value).
Let $(C, \approx_C)$ be gauge-connected, and let $f : (C, \approx_C) \to (\mathbb{R}, \sim)$ be a gauge-continuous function such that $f(a) < f(b)$ for some $a, b \in C$. Then for every intermediate value $f(a) < y < f(b)$ there exists $x \in C$ such that $f(x) = y$.

PROOF. Set:

$$C_1 = \{c \in C \mid f(c) \le y\} \quad \text{and} \quad C_2 = \{c \in C \mid f(c) > y\}.$$

Clearly, $C = C_1 \cup C_2$ is a partition and hence, by the connectedness property, there exist $x \in C$, $\xi_1 \in C_1^*$, $\xi_2 \in C_2^*$, such that $\xi_1 \approx_C x^* \approx_C \xi_2$. Since f is continuous, we get

$$y^* \ge f^*(\xi_1) \sim f^*(x^*) \sim f^*(\xi_2) > y^*.$$

So, it must be $[f(x)]^* = f^*(x^*) = y^*$, and hence $f(x) = y$. □

THEOREM 8.19 (Bolzano – generalized).
Let $f : (C, \approx_C) \to (Y, \approx_Y)$ be gauge-continuous. If $(C, \approx_C)$ is gauge-connected then also the image set $f(C)$ is connected (with respect to the gauge $\approx_Y$).

PROOF. Let $f(C) = Y_1 \cup Y_2$ be a partition. Since C is connected, there exist $\xi_1 \in [f^{-1}(Y_1)]^* = (f^*)^{-1}(Y_1^*)$, $\xi_2 \in [f^{-1}(Y_2)]^* = (f^*)^{-1}(Y_2^*)$, and $x \in f^{-1}(C)$ such that $\xi_1 \approx_C x^* \approx_C \xi_2$. Now, let $\eta_1 = f^*(\xi_1) \in Y_1^*$, $\eta_2 = f^*(\xi_2) \in Y_2^*$, and $y = f(x) \in f(C)$. Since f is continuous, we have that $\eta_1 \approx_Y y^* \approx_Y \eta_2$, as desired. □

5. Gauge spaces *versus* topological spaces

In this section we will describe the relationships between gauge spaces and topological spaces. We will assume the reader to be familiar with the fundamental notions and facts in general topology (classic references are the books [**64, 42**]).

As we already pointed out, basically we can think of a topological space as a set where the notion of continuity makes sense. Using Alpha-Theory, we can see that a topological space is in fact a set where a suitable notion of "infinitely close" makes sense.

DEFINITION 8.20. *Let (X, τ) be a topological space. The τ-monad of a point $x \in X$ is defined as:*

$$\mu_\tau(x) = \bigcap \{A^* \mid A \text{ is a neighborhood of } x\}.$$

EXAMPLE 8.21. Let $d : X \times X \to \mathbb{R}$ be a distance, and let τ_d be the topology on X induced by d. The τ_d-monad of a point $x \in X$ is the set of points that are placed at infinitesimal distance from x^*:

$$\mu_{\tau_d}(x) = \bigcap_{n \in \mathbb{N}} \{y \in X \mid d(y,x) < 1/n\}^*$$

$$= \bigcap_{n \in \mathbb{N}} \{\xi \in X^* \mid d^*(\xi, x^*) < 1/n\} = \{\xi \in X^* \mid d^*(\xi, x^*) \sim 0\}.$$

Notice that the hyper-extension $d^* : X^* \times X^* \to \mathbb{R}^*$ fulfills all properties of a distance, except it takes hyperreal values.

The Euclidean topology on $\mathbb{R}$ is induced by the distance $d(x,y) = |x - y|$ and so, as a particular case of the above example, we obtain the following.

EXAMPLE 8.22. Let τ be the usual Euclidean topology on $\mathbb{R}$. Then the τ-monad of a point x coincides with the monad of x in the sense of Definition 1.17.

Notice that if (X, τ) is a first-countable topological space[8] where every open set is infinite, then every τ-monad is infinite. To see this, given a point $x \in X$, fix a countable family $\{B_n \mid n \in \mathbb{N}\}$ of open neighborhoods of x. For every finite $F \subset X^*$, the countable family of internal sets $\{B_n^* \setminus F \mid n \in \mathbb{N}\}$ has the *finite intersection property* and so, by *countable saturation*, its intersection

$$\bigcap_{n \in \mathbb{N}} (B_n^* \setminus F) = \mu_\tau(x) \setminus F \neq \emptyset.$$

The main properties of a topological space can be characterized in terms of its monads. (In fact, the equivalences below were the motivation for Definition 8.12.)

PROPOSITION 8.23. *Let (X, τ) be a first-countable topological space. Then the following hold:*

(1) *X is Hausdorff $\Leftrightarrow$ distinct points have disjoint monads, that is, $x \neq y \Rightarrow \mu_\tau(x) \cap \mu_\tau(y) = \emptyset$;*
(2) *$A \subseteq X$ is open $\Leftrightarrow \mu_\tau(x) \subseteq A^*$ for every $x \in A$;*
(3) *$C \subseteq X$ is closed $\Leftrightarrow \mu_\tau(x) \cap C^* \neq \emptyset$ implies that $x \in C$;*
(4) *$K \subseteq X$ is compact $\Leftrightarrow \bigcup_{x \in X} \mu_\tau(x) = X^*$;*

---

[8] A topological space is called *first-countable* if every point has a countable neighbourhood basis.

(5) $C \subseteq X$ is connected $\Leftrightarrow$ for every partition $C = C_1 \cup C_2$ there exists $x \in C$ such that $\mu_\tau(x) \cap C_1^* \neq \emptyset$ and $\mu_\tau(x) \cap C_2^* \neq \emptyset$.

PROOF. Proofs of the characterizations itemized above are easily found in the literature on nonstandard analysis (see, e.g., [70]). As an example, we see here a proof of (1), and leave the others to the reader.

Assume first that (X, τ) is Hausdorff. Given $x \neq y$, pick open sets $A, B \in \tau$ such that $x \in A$, $y \in B$ and $A \cap B = \emptyset$. Then, by *transfer*, also $A^* \cap B^* = \emptyset$, and so $\mu_\tau(x) \cap \mu_\tau(y) \subseteq A^* \cap B^* = \emptyset$.

Conversely, assume that (X, τ) is *not* Hausdorff, and pick $x \neq y$ that do not have disjoint neighborhoods. By the first-countability axiom, we can fix countable basis of open neighborhoods $\{A_n \mid n \in \mathbb{N}\}$ and $\{B_n \mid n \in \mathbb{N}\}$ of x and y, respectively. Notice that the family $\{A_n \cap B_m \mid n, m \in \mathbb{N}\}$ has the finite intersection property. So, by Proposition 6.7, we can conclude that

$$\emptyset \neq \bigcap_{n,m \in \mathbb{N}} (A_n \cap B_m)^* = \left(\bigcap_{n \in \mathbb{N}} A_n^*\right) \cap \left(\bigcap_{m \in \mathbb{N}} B_m^*\right) = \mu_\tau(x) \cap \mu_\tau(y),$$

against the hypothesis. $\qquad\square$

REMARK 8.24. *In general, the monads of a topology τ on X do not determine a gauge. Indeed, a gauge is an equivalence relation on X^*, while the monads of a topology on X – provided it is Hausdorff – are the equivalence classes of an equivalence relation that is only defined on the subspace of near-standard points $\bigcup_{x \in X} \mu_\tau(x) \subseteq X^*$.*

Gauge spaces are more general than topological spaces, because they always determine a topology in a natural way.

DEFINITION 8.25. *Let $(X, \approx)$ be a gauge space. The topology induced by $\approx$ is the topology $\tau_\approx$ whose open sets are the $(\approx)$-open sets. We call $(X, \tau_\approx)$ the topological space induced by the gauge space $(X, \approx)$.*

By definition, the $(\approx)$-open sets are the same as the $\tau_\approx$-open sets, and hence also the $(\approx)$-closed sets are the same as the $\tau_\approx$-closed sets. For first-countable topologies, the above identifications extend to the properties of Hausdorff, compact, and connected set, and to the notions of continuity and uniform continuity. These facts are direct consequences of the definitions and of the nonstandard characterizations itemized in Proposition 8.23, and we omit their proofs.

PROPOSITION 8.26. *Let $(X, \approx)$ be a gauge space. If the induced topological space $(X, \tau_\approx)$ is first-countable, then:*

(1) $(X, \approx)$ *is $(\approx)$-Hausdorff iff $(X, \tau_\approx)$ is Hausdorff;*

(2) $K \subseteq X$ *is $(\approx)$-compact iff K is compact in $(X, \tau_\approx)$;*

(3) $C \subseteq X$ *is $(\approx)$-connected iff C is connected in $(X, \tau_\approx)$;*

Moreover, if $(Y, \approx')$ is another gauge space whose induced topology is first-countable, then:

(5) $f : (X, \approx) \to (Y, \approx')$ *is gauge-continuous at a point x iff it is continuous at x with respect to the induced topologies;*

(6) $f : (X, \approx) \to (Y, \approx')$ *is gauge-uniformly continuous iff it is uniformly continuous with respect to the induced topologies.*

DEFINITION 8.27. *Let X be a set equipped with a topology τ and a gauge $\approx$. We say that τ and $\approx$ are consistent if the following two conditions hold:*

(1) $\tau = \tau_\approx$, *that is, A is τ-open if and only if A is $(\approx)$-open;*

(2) $\mu_\tau(x) = [x^*]_\approx$ *for every $x \in X$.*

REMARK 8.28. *It is well possible that a gauge $\approx$ and the induced topology $\tau_\approx$ are not consistent. Indeed, one always has the inclusion $[x^*]_\approx \subseteq \mu_{\tau_\approx}(x)$ for every $x \in X$, but it might happen that $[x^*]_\approx \neq \mu_{\tau_\approx}(x)$ for some point x.[9] Of course, in these cases the notions of topological continuity and gauge-continuity do not coincide.*

Under the hypothesis of first-countability, condition (1) in the previous definition is implied by condition (2).

PROPOSITION 8.29. *Let $(X, \approx)$ be a gauge space, and let τ be a first-countable topology on X. If $\mu_\tau(x) = [x^*]_\approx$ for every $x \in X$, then $\tau = \tau_\approx$.*

PROOF. If A is τ-open, then for every $x \in A$ there exists a neighborhood I of x such that $x \in I \subseteq A$. It follows that $[x^*]_\approx = \mu_\tau(x) \subseteq I^* \subseteq A^*$, and this shows that A is $(\approx)$-open.

Conversely, assume that B is not τ-open. Then there exists a point $x \in A$ such that $I \nsubseteq A$ for any neighborhood I of x. Now fix a countable base $\{I_n \mid n \in \mathbb{N}\}$ of neighborhoods of x. Notice that the countable family $\{I_n \setminus A \mid n \in \mathbb{N}\}$ has the FIP. Indeed, given $n_1, \ldots, n_k$,

---

[9] An example of this phenomenon is given by the *Epsilon-gauges* $\approx_\varepsilon$, that will be introduced in the next Section 6.

pick $I_m \subseteq I_{n_1} \cap \ldots \cap I_{n_k}$; then $\emptyset \neq I_m \setminus A \subseteq \bigcap_{j=1}^{k}(I_{n_j} \setminus A)$. But then, by Proposition 6.7,

$$\emptyset \neq \bigcap_{n \in \mathbb{N}}(I_n \setminus A)^* = \left(\bigcap_{n \in \mathbb{N}} I_n^*\right) \setminus A^* = \mu_\tau(x) \setminus A^* = [x^*]_\approx \setminus A^*.$$

This shows that $[x^*]_\approx \not\subseteq A^*$, and so A is not $(\approx)$-open.   □

There are many important cases where a gauge and the induced topology are consistent. A class of relevant examples is given by metric spaces.

THEOREM 8.30. *Let (X, d) be a metric space, and let τ_d and $\approx_d$ be the topology and the gauge induced by d, respectively. Then τ_d and $\approx_d$ are consistent.*

PROOF. By the definitions, for every $x \in X$ one has the equalities:

$$\mu_{\tau_d}(x) = \{\xi \in X^* \mid d^*(\xi, x^*) \sim 0\} = \{\xi \in X^* \mid \xi \approx_d x^*\} = [x^*]_{\approx_d}.$$

Since the topology determined by a metric is first-countable, the conclusion follows by the previous proposition.   □

For example, the usual Euclidean topology of $\mathbb{R}$ is consistent with the gauge $\sim$ of infinitesimal distance, because they are both induced by the distance $d(x, y) = |x - y|$.

Although gauges spaces and topological spaces originated from the same need, namely extending the main notions of calculus to a more general setting, they are separate concepts. Indeed, they are equivalent in the simplest cases, such as in metric spaces, but they differ in more sophisticated situations. This is an example of the fact that "Mathematics with infinitesimals" can evolve and develop towards different directions with respect to "traditional Mathematics".

REMARK 8.31. *If one could assume a stronger version of saturation, namely κ-saturation for a "sufficiently large" cardinal κ, the hypothesis of first-countability in Propositions 8.23, 8.26, and 8.29, would not be needed. Such a stronger saturation is satisfied by suitable models of nonstandard analysis, but cannot be added directly to the axioms of the Alpha-Theory. We remark that the problem could be overcome by considering a generalization of the Alpha-Theory grounded on a notion of Alpha-limit for κ-sequences $\langle x_i \mid i \in \kappa \rangle$.*

6. The Epsilon-gauges

In this section, we present gauges that are *not* consistent with any topology, but are closely connected to the relevant notions of *Lipschitz continuity* and *Hölder continuity*. In our opinion, these examples show that the notion of gauge is interesting and worth of further investigations.

DEFINITION 8.32. *Let $\varepsilon \sim 0$ be a fixed positive infinitesimal number. The* Epsilon-gauge *is the gauge $\approx_\varepsilon$ on $\mathbb{R}$ where, for every $\xi, \zeta \in \mathbb{R}^*$:*

$$\xi \approx_\varepsilon \zeta \iff \frac{\xi - \zeta}{\varepsilon} \text{ is finite.}$$

It is easily verified that $\approx_\varepsilon$ is actually an equivalence relation. Moreover, the gauge space $(\mathbb{R}, \approx_\varepsilon)$ is determined by the gauge Abelian space $(\mathbb{R}, \mathcal{I}_\varepsilon)$ with the subgroup $\mathcal{I}_\varepsilon = \{\eta \in \mathbb{R}^* \mid \eta/\varepsilon \text{ finite}\}$.

If not explicitly stated otherwise, in the following ε denotes a fixed positive infinitesimal number.

PROPOSITION 8.33.

(1) *For every $x \in \mathbb{R}$, there is a proper inclusion $[x]_\varepsilon \subsetneq \mathrm{mon}(x)$ between the $(\approx_\varepsilon)$-monad of x and the monad $\mathrm{mon}(x)$ of hyperreal numbers that are infinitely close to x;*

(2) *A set A of real numbers is open in the usual Euclidean topology of $\mathbb{R}$ if and only if it is $(\approx_\varepsilon)$-open.*

PROOF. (1) is trivial, because $\xi \in [x]_\varepsilon \Leftrightarrow (\xi - x)/\varepsilon$ is finite. In consequence, ξ is infinitely close to x, that is, $\xi \in \mu(x)$. Moreover, the inclusion is proper because, *e.g.*, $\zeta = x + \sqrt{\varepsilon} \in \mu(x)$ but $(\zeta - x)/\varepsilon = 1/\sqrt{\varepsilon}$ is infinite.

(2). If $A \subseteq \mathbb{R}$ is open and $x \in A$ then, by the nonstandard characterization, the whole monad $\mu(x) \subseteq A^*$, and hence also $[x]_\varepsilon \subseteq A^*$. This shows that A is $(\approx_\varepsilon)$-open. Conversely, let us assume that A is $(\approx_\varepsilon)$-open. Then, for every $x \in A$, the number ε witnesses that the following property holds: "There exists $\varepsilon \in (\mathbb{R}^*)^+$ such that the ball $B(x, \varepsilon) = \{\xi \in \mathbb{R}^* \mid |\xi - x| < \varepsilon\} \subseteq A^*$." By *backward transfer*, we get the existence of a number $\epsilon \in \mathbb{R}^+$ such that $\{y \in \mathbb{R} \mid |y - x| < \epsilon\} \subseteq A$, and this shows that A is open. $\qquad\square$

COROLLARY 8.34. *The gauges $\approx_\varepsilon$ are not consistent with any topology on $\mathbb{R}$.*

PROOF. By (2), the only possible topology on $\mathbb{R}$ that may be consistent with $\approx_\varepsilon$ is the Euclidean topology, but this is not the case because of the proper inclusion (1).   □

Since gauges $\approx_\varepsilon$ are not equivalent to any topology, it is worth investigating the meaning of the $(\approx_\varepsilon)$-continuity and $(\approx_\varepsilon)$-uniform continuity. As a preliminary observation, let us notice that $(\approx_\varepsilon)$-gauge continuity for all ε implies continuity. Precisely:

PROPOSITION 8.35. Let $D \subseteq \mathbb{R}$ be a neighborhood of x. If for every positive infinitesimal number ε the function $f : (D, \approx_\varepsilon) \to (\mathbb{R}, \sim)$ is gauge-continuous at x, then $f : D \to \mathbb{R}$ is continuous at x (with respect to the Euclidean topology).

PROOF. Given $\xi \sim x$, we want to prove that $f^*(\xi) \sim f(x)$. If $\xi = x$ the assertion is trivial. Otherwise $\varepsilon = |\xi - x|$ is a positive infinitesimal number. Since $\xi \approx_\varepsilon x$, by the hypothesis we get $f^*(\xi) \sim f(x)$.   □

Let us now state a straight consequence that will be used in the proof of the next theorem.

COROLLARY 8.36. If for every positive infinitesimal number ε the function $f : ([a, b], \approx_\varepsilon) \to (\mathbb{R}, \sim)$ is gauge-continuous at all points, then f is bounded.

PROOF. By the previous proposition, $f : [a, b] \to \mathbb{R}$ is continuous at all points, and the conclusion follows by Weierstrass' Theorem.   □

REMARK 8.37. Closed bounded intervals $[a, b]$ are not $(\approx_\varepsilon)$-compact. Indeed, for any $r \in \mathbb{R}$, the number $\xi_r = r + \sqrt{\varepsilon}$ is not $(\approx_\varepsilon)$-near-standard, that is, $\xi_r \not\approx_\varepsilon s$ for any $s \in \mathbb{R}$. In consequence, $(\approx_\varepsilon)$-continuity on an interval $[a, b]$ does not imply boundedness nor $(\approx_\varepsilon)$-uniform continuity.

Recall the following notion.

DEFINITION 8.38. A function $f : [a, b] \to \mathbb{R}$ is called Lipschitz continuous if there exists a constant $L > 0$ such that for every $x, y \in [a, b]$:

$$|f(x) - f(y)| \leq L \cdot |x - y|.$$

We remark that, in spite of the name Lipschitz continuity, there is no topology that provides such form of continuity; however, we have the following characterization.

THEOREM 8.39. *A function $f : [a, b] \to \mathbb{R}$ is Lipschitz continuous if and only if $f : ([a, b], \approx_\varepsilon) \to (\mathbb{R}, \approx_\varepsilon)$ is uniformly gauge-continuous for every positive infinitesimal number ε.*

PROOF. Assume first that f is Lipschitz-continuous with constant L. Fix an infinitesimal number $\varepsilon > 0$, and let $\xi \approx_\varepsilon \zeta$. Then there exists a finite number K such that $|\xi - \zeta|/\varepsilon \le K$. By *transfer* from the hypothesis, it follows that $f^*(\xi) \approx_\varepsilon f^*(\zeta)$. Indeed, the following inequalities hold:

$$\frac{|f^*(\xi) - f^*(\zeta)|}{\varepsilon} \le \frac{L \cdot |\xi - \zeta|}{\varepsilon} \le L \cdot K.$$

Conversely, let us assume by contradiction that f is *not* Lipschitz continuous. Then we can pick sequences $\langle x_n \mid n \in \mathbb{N} \rangle$ and $\langle y_n \mid n \in \mathbb{N} \rangle$ in $[a, b]$ such that for every n:

$$|f(x_n) - f(y_n)| > n \cdot |x_n - y_n|.$$

Let $\xi = \lim_{n \uparrow \alpha} x_n$ and $\eta = \lim_{n \uparrow \alpha} y_n$. By passing the above inequality to the Alpha-limits, we obtain that

$$|f^*(\xi) - f^*(\eta)| > \boldsymbol{\alpha} \cdot |\xi - \eta|.$$

If the number $\xi - \eta$ is *not* infinitesimal, then $f^*(\xi) - f^*(\eta)$ is infinite, and hence f is unbounded. This contradicts Corollary 8.36, because $f : ([a, b], \approx_\varepsilon) \to (\mathbb{R}, \approx_\varepsilon)$ uniformly gauge-continuous trivially implies that $f : ([a, b], \approx_\varepsilon) \to (\mathbb{R}, \sim)$ is gauge-continuous at all points. If the number $\varepsilon = |\xi - \eta| > 0$ is infinitesimal, we get the desired contradiction by noticing that $\xi \approx_\varepsilon \eta$ and $f^*(\xi) \not\approx_\varepsilon f^*(\eta)$, since $|f^*(\xi) - f^*(\eta)|/\varepsilon > \boldsymbol{\alpha}$ is infinite. $\square$

A related notion to Lipschitz continuity is the following one.

DEFINITION 8.40. *Let a be a positive real number. A function $f : [a, b] \to \mathbb{R}$ is called* Hölder continuous *of exponent a if there exists a constant $H > 0$ such that for every $x, y \in [a, b]$:*

$$|f(x) - f(y)| \le H \cdot |x - y|^a.$$

By using the same arguments as in the proof of Theorem 8.39, also the following is proved.

THEOREM 8.41. *A function $f : [a, b] \to \mathbb{R}$ is Hölder continuous of exponent $a > 0$ if and only if $f : ([a, b], \approx_\varepsilon) \to (\mathbb{R}, \approx_{\varepsilon^a})$ is uniformly gauge-continuous for every positive infinitesimal number ε.*

Again by using similar arguments, the following local properties are proved equivalent.

EXERCISE 8.42. *The following statements are equivalent:*

(1) *The function $f : ([a, b], \approx_\varepsilon) \to (\mathbb{R}, \approx_\varepsilon)$ is gauge-continuous at all points for all positive infinitesimal numbers ε ;*

(2) *For every $x \in [a, b]$ there exists a positive constant L_x such that for every $y \in [a, b]$:*

$$|f(y) - f(x)| \leq L_x \cdot |y - x|.$$

EXERCISE 8.43. *Let $f : [0, 1] \to \mathbb{R}$ be the function defined by setting:*

$$f(x) = \begin{cases} x^2 \cdot \sin\left(\frac{1}{x^2}\right) & \text{if } x \neq 0 \\ 0 & \text{if } x = 0. \end{cases}$$

Then:

(1) *f is differentiable at all points $x \in [0, 1]$;*

(2) *For every positive infinitesimal ε, $f : ([0, 1], \approx_\varepsilon) \to (\mathbb{R}, \approx_\varepsilon)$ is gauge-continuous at all points $x \in [0, 1]$;*

(3) *For every positive infinitesimal ε, $f : ([0, 1], \approx_\varepsilon) \to (\mathbb{R}, \approx_\varepsilon)$ is not uniformly gauge-continuous.*

CHAPTER 9

Gauge Quotients

It is a typical procedure in mathematics to construct new objects from old ones, aiming at finding solutions to problems. One primary example of this sort is the construction of the real numbers from the rationals, with the goal of filling the gaps found in Dedekind cuts. The completion of a metric space is an other important example of this procedure.

In this chapter we will present a method to construct new objects from gauge spaces, namely the *gauge quotients*. Such spaces are very general and include, *e.g.*, the completions of uniform spaces. We remark that those completions are directly obtained as quotient sets, without appealing to the (quite involved) theory of uniformities.

1. Definition of gauge quotients

Given a gauge space $(X, \approx)$, a natural way of constructing new spaces is to consider quotients of suitable subspaces of X^*.

DEFINITION 9.1. *Let* $(X, \approx)$ *be a gauge-space, and let* $\widehat{X} \subseteq X^*$ *with* $X \subset \widehat{X}$.[1] *The set*

$$\overline{X} = \widehat{X}/\approx = \left\{ [\xi]_\approx \mid \xi \in \widehat{X} \right\}$$

is called gauge quotient *of* $(X, \approx)$ *relative to the subspace* $\widehat{X} \subseteq X^*$.

We have already seen a particular case of gauge quotient in Theorem 7.1, where we presented a possible way of constructing the real numbers

---

[1] Here we are identifying each $x \in X$ with its hyper-image $x^* \in X^*$, so that $X \subseteq X^*$.

starting from the hyperrational numbers; precisely, we considered the gauge quotient of the gauge $(\mathbb{Q}, \sim)$ relative to the subspace $\mathbb{Q}^*_{\text{fin}}$ of the bounded hyperrational numbers. That example can be directly generalized to metric spaces.

EXAMPLE 9.2. *Given a metric space* (X, d), *let* $(X, \approx_d)$ *be the corresponding metric gauge. For a fixed point* $P_0 \in X$, *we set*

$$\widehat{X} \;=\; \{\xi \in X^* \mid d^*(\xi, P_0^*) \text{ is finite}\}.$$

Then the gauge quotient $\widehat{X}/\approx_d$, *endowed with the metric* $\delta([\xi], [\eta]) = st(d^*(\xi, \eta))$, *is known in the literature of nonstandard analysis as the* nonstandard hull *of the space* (X, d).

   The nonstandard hull is a complete metric space, but in general, it is larger than the completion of (X, d) *(see* [**52**] *§18.3).*

EXAMPLE 9.3. *The* Lebesgue *spaces* $L^p(\Omega)$, $1 \le p < \infty$, *can be regarded as gauge quotients of* $(\mathscr{C}^0(\overline{\Omega}), \approx_{L^p})$ *with respect to the* L^p-*gauge defined in Example 8.6, where:*

$$\widehat{X} = \left\{\lim_{n \uparrow \alpha} \varphi(n) \mid \{\varphi(n)\} \subset \mathscr{C}^0(\overline{\Omega}) \text{ is a Cauchy sequence in the } L^p\text{-norm}\right\}.$$

In fact, every $u \in L^p(\Omega)$ *is the limit of a sequence* $\{u_n\}$ *in* $\mathscr{C}^0(\overline{\Omega})$, *and so* u *can be identified with* $[\lim_{n \uparrow \alpha} u_n]_{\approx_{L^p}}$.

EXAMPLE 9.4. *Similarly as in the previous example, the* Sobolev *spaces* $W_0^{m,p}(\Omega)$ *can be regarded as gauge quotients of* $(\mathscr{D}(\Omega), \approx_{W^{m,p}})$ *with respect to the* $W^{m,p}$-*gauge defined in Example 8.6, where:*

$$\widehat{X} = \left\{\lim_{n \uparrow \alpha} \varphi(n) \mid \{\varphi(n)\} \subset \mathscr{D}(\Omega) \text{ is a Cauchy sequence in the } W^{m,p}\text{-norm}\right\}.$$

   It is possible to prove that all completions of first-countable uniform spaces can be obtained as suitable gauge quotients. However, as suggested by the example of nonstandard hulls, the notion of gauge quotient cannot be reduced to the notion of completion.

   Given a gauge quotient $\overline{X} = \widehat{X}/\approx$, there is a *natural embedding* $\jmath_X : X \to \overline{X}$ given by $\jmath_X(x) = [x^*]_\approx$.

DEFINITION 9.5. *Let* $(X, \approx)$ *and* $(Y, \approx_Y)$ *be gauge spaces and let* $\overline{X} = \widehat{X}/\approx_X$ *and* $\overline{Y} = \widehat{Y}/\approx_Y$ *be gauge quotients. We say that a function* $\overline{f} : \overline{X} \to \overline{Y}$ *is a* gauge extension *of a function* $f : X \to Y$

if $\overline{f}([x]) = [f(x)]$ for every $x \in X$, that is, if the following diagram commutes:

$$
\begin{array}{ccc}
\overline{X} & \xrightarrow{\ \ \overline{f}\ \ } & \overline{Y} \\[2mm]
\Big\uparrow{\scriptstyle \jmath_X} & & \Big\uparrow{\scriptstyle \jmath_Y} \\[2mm]
X & \xrightarrow{\ \ f\ \ } & Y
\end{array}
$$

With a little abuse of language, we called $\overline{f}$ an *extension* of f because this is actually the case if one identifies X and Y with their images under the natural embeddings $\jmath_X$ and $\jmath_Y$, respectively.

It is a natural question to ask whether a given function admits a gauge extension. The following result provides a sufficient condition.

PROPOSITION 9.6. *Let $(X, \approx_X)$ and $(Y, \approx_Y)$ be gauge spaces, let $\overline{X} = \widehat{X}/\approx_X$ and $\overline{Y} = \widehat{Y}/\approx_Y$ be gauge quotients, and let $f : X \to Y$. Assume that:*

(1) *f is uniformly gauge-continuous on $\widehat{X}$, that is, $\xi \approx_X \zeta \Rightarrow f^*(\xi) \approx_Y f^*(\zeta)$ for all $\xi, \zeta \in \widehat{X}$;*

(2) *for every $\xi \in \widehat{X}$ there exists $\zeta \in \widehat{Y}$ such that $f^*(\xi) \approx_Y \zeta$.*

Then f has a gauge extension $\overline{f} : \overline{X} \to \overline{Y}$.

PROOF. For $\xi \in \widehat{X}$, define $\overline{f}([\xi]_{\approx_X}) = [f^*(\xi)]_{\approx_Y}$. It is readily verified that the hypotheses are precisely what is needed for that definition to be well-posed and yield a function $\overline{f} : \overline{X} \to \overline{Y}$. Moreover, if $x \in X$, then $\overline{f}([x]) = [f^*(x)] = [f(x)]$. □

2. The differential Epsilon-ring

For a fixed positive infinitesimal number ε, let us consider the gauge on $\mathbb{R}$ determined by the additive subgroup of those numbers that are infinitesimal with respect to ε, that is,

$$
\mathcal{I}_{\varepsilon+} = \left\{ \xi \in \mathbb{R}^* \,\Big|\, \frac{\xi}{\varepsilon} \sim 0 \right\}.
$$

So, if $(\mathbb{R}, \approx_{\varepsilon+})$ is the gauge space determined by $(\mathbb{R}, \mathcal{I}_{\varepsilon+})$, then

$$
\xi \approx_{\varepsilon+} \zeta \iff (\xi - \zeta)/\varepsilon \sim 0.
$$

We remark that $\approx_{\varepsilon+}$ is different from the Epsilon-gauge $\approx_\varepsilon$ of Definition 8.32. Indeed, the subgroup $\mathcal{I}_{\varepsilon+}$ contains infinitesimals of order *higher* than ε, while $\mathcal{I}_\varepsilon \supset \mathcal{I}_{\varepsilon+}$ contains infinitesimals of order *higher or equal* to ε.

DEFINITION 9.7. *For a fixed positive infinitesimal number ε, the differential ε-ring $\mathbb{R}_\varepsilon$ is the gauge quotient of $(\mathbb{R}, \approx_{\varepsilon+})$ relative to the subspace $\mathbb{R}[\varepsilon] := \{a_0 + a_1\varepsilon + \ldots + a_n\varepsilon^n \mid n \geq 0, a_i \in \mathbb{R}\}$:*

$$\mathbb{R}_\varepsilon = \mathbb{R}[\varepsilon] / \approx_{\varepsilon+} .$$

Notice that

$$\mathcal{I}_{\varepsilon+} \cap \mathbb{R}[\varepsilon] = \{a_2\varepsilon^2 + \ldots + a_n\varepsilon^n \mid n \geq 2, a_i \in \mathbb{R}\}$$

is an ideal of the ring $\mathbb{R}[\varepsilon]$, and hence also the quotient $\mathbb{R}_\varepsilon$ is a ring. Actually, $\mathbb{R}_\varepsilon$ is an algebra over $\mathbb{R}$ of dimension 2, generated by the cosets $[1]$ and $[\varepsilon]$, and where multiplication is ruled by the nilpotency of $[\varepsilon]$ of order 2, namely by the relation $[\varepsilon]^2 = [\varepsilon] \cdot [\varepsilon] = [0]$.[2] Indeed,

$$a_0 + a_1\varepsilon + \ldots + a_n\varepsilon^n \approx_{\varepsilon+} a_0' + a_1'\varepsilon + \ldots + a_m'\varepsilon^m \Leftrightarrow a_0 = a_0' \ \& \ a_1 = a_1'.$$

In other words, $\mathbb{R}_\varepsilon$ is the set of all equivalence classes $[a + b\varepsilon]$ where $a, b \in \mathbb{R}$, and we can identify:

$$\mathbb{R}_\varepsilon \cong \{a + b\varepsilon \mid a, b \in \mathbb{R}, \varepsilon^2 = 0\}.$$

The following property justified the name "differential" ε-rings.

PROPOSITION 9.8. *Every function $f \in \mathscr{C}^1(\mathbb{R})$ has a gauge extension $\overline{f} : \mathbb{R}_\varepsilon \to \mathbb{R}_\varepsilon$ to the differential ε-ring, where:*

$$\overline{f}(a + b\varepsilon) = f(a) + f'(a)\,b\,\varepsilon.$$

PROOF. We will use Proposition 9.6. If $\xi \in \mathbb{R}[\varepsilon]$, then we can write $\xi = a + b\varepsilon + \xi'\varepsilon^2$ for suitable $a, b \in \mathbb{R}$ and a bounded $\xi' \in \mathbb{R}^*$. Since f is differentiable at all points, there exist $\sigma \sim 0$ such that

$$f^*(\xi) = f(a) + f'(a)(b + \xi'\varepsilon)\varepsilon + \sigma(b + \xi'\varepsilon)\varepsilon.$$

Then

$$\frac{f^*(\xi) - f(a) - f'(a)\,b\,\varepsilon}{\varepsilon} = f'(a)\xi'\varepsilon + \sigma(b + \xi'\varepsilon) \sim 0,$$

and hence $f^*(\xi) \approx_{\varepsilon+} f(a) + f'(a)\varepsilon \in \mathbb{R}[\varepsilon]$.

Now take $\xi, \zeta \in \mathbb{R}[\varepsilon]$ with $\xi \approx_{\varepsilon+} \zeta$. Similarly as above, we can write $\xi = a + b\varepsilon + \xi'\varepsilon^2$ and $\zeta = c + d\varepsilon + \zeta'\varepsilon^2$ for suitable $a, b, c, d \in \mathbb{R}$ and bounded $\xi', \zeta' \in \mathbb{R}^*$. Since $\xi \approx_{\varepsilon+} \zeta$, it must be $a = c$ and $b = d$.

---

[2] Rings of this type, named *Fermat reals*, have been considered, *e.g.*, in [50, 51].

By differentiability, $f^*(\zeta) = f(a) + f'(a)(b + \zeta'\varepsilon)\varepsilon + \tau(b + \zeta'\varepsilon)\varepsilon$ for a suitable $\tau \sim 0$, and we finally obtain $f^*(\xi) \approx_{\varepsilon+} f^*(\zeta)$ because

$$\frac{f^*(\xi) - f^*(\zeta)}{\varepsilon} = f'(a)(\xi' - \zeta')\varepsilon + \sigma(b + \xi'\varepsilon) - \tau(b + \zeta'\varepsilon) \sim 0.$$

$\square$

The above considerations can be extended to the k-dimensional gauge space $(\mathbb{R}^k, \approx_{\varepsilon+})$, where the gauge $\approx_{\varepsilon+}$ is defined coordinate-wise:

$$(x_1, \ldots, x_k) \approx_{\varepsilon+} (y_1, \ldots, y_k) \iff \frac{x_i - y_i}{\varepsilon} \sim 0 \text{ for every } i = 1, \ldots, k.$$

In this case, the corresponding k-dimensional differential ε-ring is defined as the gauge quotient

$$\mathbb{R}_\varepsilon^k = (\mathbb{R}[\varepsilon])^k / \approx_{\varepsilon+}.$$

Also in this case, we can identify:

$$\mathbb{R}_\varepsilon^k \cong \{\mathbf{x} + \varepsilon\,\mathbf{v} \mid \mathbf{x}, \mathbf{v} \in \mathbb{R}^k, \varepsilon^2 = 0\}.$$

Arguing as in the previous proposition, it is proved that every function $F : \mathbb{R}^k \to \mathbb{R}^h$ of class $\mathscr{C}^1(\mathbb{R}^k)$ has a gauge extension to a function $\overline{F} : \mathbb{R}_\varepsilon^k \to \mathbb{R}_\varepsilon^h$ where:

$$\overline{F}(\mathbf{x} + \varepsilon\,\mathbf{v}) := F(\mathbf{x}) + F'(\mathbf{x})[\varepsilon\,\mathbf{v}]$$

with $F'(\mathbf{x})$ the *Jacobian matrix* computed at $\mathbf{x}$.

REMARK 9.9. *The above ideas can be extended to manifolds, along the lines pursued in the work by P. Giordano and his coauthors (see, e.g., [50, 51]).*

3. Distributions as a gauge quotient

In this section we introduce the theory of distributions grounding on the notion of gauge quotient.

It is well-known that *nonstandard analysis* is a suitable framework for the study of distributions. This fact was realized quite early, *e.g.*, with the Dirac function represented as a function taking infinite values and whose support is an infinitesimal interval (see *e.g.* [83, §5.3]). Indeed, the definition of distributions as suitable limits of sequences (see, *e.g.* [76]) can be easily reformulated in the nonstandard language by means of infinitesimals and of standard parts (see, *e.g.*, [81]). Here we follow an approach that is more similar to that of [19] (the reader is referred to that paper for more details). A proof of the equivalence of our definition below with the classical definition of Schwartz [**87**] will be given in the next Section 4.

DEFINITION 9.10. *The space $\mathscr{D}'(\Omega)$ of distributions on the open set $\Omega \subseteq \mathbb{R}^N$ is the gauge quotient of $(\mathscr{C}^\infty(\Omega), \approx_{\mathscr{D}'})$ determined by the subspace*

$$\widehat{\mathscr{C}^\infty(\Omega)} = \left\{ f \in \mathscr{C}^\infty(\Omega)^* \Big| \forall \varphi \in \mathscr{D}, \int^* f(x)\varphi(x)dx \text{ is bounded}\right\}.$$

There is a canonical embedding

$$\Psi : \mathscr{C}^\infty(\Omega) \to \mathscr{D}'(\Omega)$$

given by $f \mapsto [f^*]_{\mathscr{D}'}$. Since the gauge space $(\mathscr{C}^\infty(\Omega), \approx_{\mathscr{D}'})$ is Hausdorff, the equivalence class $[f^*]_{\mathscr{D}'}$ contains only one hyper-image, namely f^*. So, sometimes we will identify $[f^*]_{\mathscr{D}'}$ with f.

EXAMPLE 9.11. *Probably, the most popular distribution that is not a function, is the* Dirac delta function $\delta \in \mathscr{D}'(\mathbb{R})$. *Historically, δ was originally defined as the function that is zero everywhere except in zero, and its integral equals 1. Clearly, a real function with those properties does not exist; however, in our theory the delta function can be defined as*

$$\delta(x) = \left[\frac{1}{\pi} \cdot \frac{\alpha}{1+\alpha^2 x^2}\right]_{\mathscr{D}'},$$

or by means of any other function in $\mathscr{C}^\infty(\Omega)^*$ in the same equivalence class, such as, for example,

$$\delta(x) = \left[\left(\frac{2\pi}{\alpha}\right)^{-\frac{n}{2}} \cdot e^{-\frac{1}{2}|x|^2\alpha}\right]_{\mathscr{D}'}.$$

Let us now see how one can define operations between our distributions, including multiplication.

DEFINITION 9.12. *For $[f]_{\mathscr{D}'}, [g]_{\mathscr{D}'} \in \mathscr{D}'(\Omega)$, $\lambda \in \mathbb{R}$, and $\varphi \in \mathscr{D}(\Omega)$, we set*

- Sum: $[f]_{\mathscr{D}'} + [g]_{\mathscr{D}'} = [f+g]_{\mathscr{D}'}$;
- Product by a scalar: $\lambda \cdot [f]_{\mathscr{D}'} = [\lambda \cdot f]_{\mathscr{D}'}$;
- Product by a function $\varphi \in \mathscr{D}(\Omega)$: $\varphi \cdot [f]_{\mathscr{D}'} = [\varphi^* \cdot f]_{\mathscr{D}'}$.

Moreover, if $[f(x)]_{\mathscr{D}'} \in \mathscr{D}'(\Omega_1)$ and $[g(y)]_{\mathscr{D}'} \in \mathscr{D}'(\Omega_2)$, we also set

- Tensor product: $[f]_{\mathscr{D}'} \otimes [g]_{\mathscr{D}'} = [f(x) \cdot g(y)]_{\mathscr{D}'} \in \mathscr{D}'(\Omega_1 \times \Omega_2)$.

It is easy to check that the above definitions are well-posed (we leave it to the reader as an exercise).

Let us now define derivative of distributions.

DEFINITION 9.13. *If* $[f(x_1, \ldots, x_N)]_{\mathscr{D}'} \in \mathscr{D}'(\Omega)$ *where* $\Omega \subseteq \mathbb{R}^N$, *for every* $j = 1, \ldots, N$ *we set*

$$\frac{\partial}{\partial x_j}[f]_{\mathscr{D}'} = \left[\frac{\partial f}{\partial x_j}\right]_{\mathscr{D}'}.$$

Let us check that the above definition is well-posed. Given $f \approx_{\mathscr{D}'} g$, we have to show that $\frac{\partial g}{\partial x_j} \approx_{\mathscr{D}'} \frac{\partial f}{\partial x_j}$. For every $\varphi \in \mathscr{D}(\Omega)$, one has that $\frac{\partial \varphi^*}{\partial x_j} \in \mathscr{D}(\Omega)^*$, and so

$$\int^* \left(\frac{\partial g(x)}{\partial x_j} - \frac{\partial f(x)}{\partial x_j}\right) \varphi^*(x)\, dx = -\int^* (g(x) - f(x)) \frac{\partial \varphi^*(x)}{\partial x_j}\, dx \sim 0.$$

4. Distributions as functionals

The modern theory of distribution has been developed by Schwartz. In his theory, a distribution is defined as an element of the topological dual of the space $\mathscr{D}(\Omega)$. We will show that our definition of distribution is actually equivalent to Schwartz's definition.

In this section, we will denote by $\mathscr{D}'_G(\Omega)$ the distributions as gauge quotient that we defined in the previous section (see Definition 9.10); and we will denote by $\mathscr{D}'_S(\Omega)$ the Schwartz distributions (see [**87**]), namely:

DEFINITION 9.14. *The space* $\mathscr{D}'_S(\Omega)$ *of Schwartz distributions on the open set* $\Omega \subseteq \mathbb{R}^N$ *is the topological dual of* $\mathscr{D}'(\Omega)$.

We have the following result:

THEOREM 9.15. *There is a linear isomorphism*

$$\Phi : \mathscr{D}'_G(\Omega) \to \mathscr{D}'_S(\Omega)$$

defined by the following formula:

$$\forall \varphi \in \mathscr{D}(\Omega), \ \left\langle \Phi\left([u]_{\mathscr{D}'_G}\right), \varphi \right\rangle = \text{st}\left(\int^* u\, \varphi^*\, dx\right).$$

Moreover, if $T = \Phi\left([u]_{\mathscr{D}'}\right)$, *we have that*

$$\partial_k T = \Phi\left([\partial_k u]_{\mathscr{D}'_G}\right).$$

PROOF. The map Φ is well-defined because $u \approx_{\mathscr{D}'_G} v \Rightarrow \Phi\left([u]_{\mathscr{D}'_G}\right) = \Phi\left([u]_{\mathscr{D}'}\right)$; moreover, it is linear, and its range is included in $\mathscr{D}'_S(\Omega)$, as one can easily verify.

It is also immediate to see that Φ is injective. Let us show that Φ is onto. To see this let $T \in \mathscr{D}'(\Omega)$; we want to find a function $u \in \mathscr{C}^\infty(\Omega)^*$ such that

$$\Phi\left([u]_{\mathscr{D}'_G}\right) = T. \tag{5}$$

Since $\mathscr{C}^\infty(\Omega)$ is dense in $\mathscr{D}'(\Omega)$ with respect to the weak topology, there is a sequence $\psi_n \to T$ weakly. We claim that

$$u_T := \lim_{n \uparrow \alpha} \psi_n$$

satisfies (5). In fact, for every $\varphi \in \mathscr{D}(\Omega)$,

$$\langle T, \varphi \rangle = \lim_{n \to \infty} \int \psi_n \varphi \, dx = st\left(\lim_{n \uparrow \alpha} \int \psi_n \varphi \, dx\right) = st\left(\int^* u_T \varphi^* \, dx\right).$$

Now let us prove the last statement. For every $\varphi \in \mathscr{D}(\Omega)$, we have that

$$\begin{aligned}
\langle \partial_k T, \varphi \rangle &= -\lim_{n \to \infty} \int \psi_n \partial_k \varphi \, dx = -st\left(\lim_{n \uparrow \alpha} \int \psi_n \partial_k \varphi \, dx\right) \\
&= st\left(\lim_{n \uparrow \alpha} \int \partial_k \psi_n \varphi \, dx\right) = st\left(\int^* \partial_k u_T \varphi^* dx\right) \\
&= \left\langle \Phi\left([\partial_k u_T]_{\mathscr{D}'_G}\right), \varphi \right\rangle.
\end{aligned}$$

$\square$

EXAMPLE 9.16. *The Dirac delta function as defined in Example 9.11 corresponds to Schwartz's definition:*

$$\begin{aligned}
\langle \delta, \varphi \rangle &= st\left(\int_{\mathbb{R}^*} \frac{1}{\pi} \cdot \frac{\alpha}{1 + \alpha^2 x^2} \cdot \varphi^*(x) \, dx\right) \\
&= st\left(\frac{1}{\pi} \int_{\mathbb{R}^*} \frac{1}{1 + y^2} \cdot \varphi^*\left(\frac{y}{\alpha}\right) dx\right) \\
&= st\left(\frac{1}{\pi} \int_{\mathbb{R}^*} \frac{1}{1 + y^2} \cdot \varphi^*(0) \, dy\right) \\
&\quad + st\left(\frac{1}{\pi} \int_{\mathbb{R}^*} \frac{1}{1 + y^2} \cdot \left[\varphi^*\left(\frac{y}{\alpha}\right) - \varphi^*(0)\right] dy\right) \\
&= st\left(\varphi^*(0) \cdot \frac{1}{\pi} \int_{\mathbb{R}^*} \frac{1}{1 + y^2} \, dy\right) = \varphi^*(0) = \varphi(0).
\end{aligned}$$

5. Grid functions and distributions

There are many ways to get distributions via a gauge quotient. In Section 3 we have seen that the space of distributions $\mathscr{D}'(\Omega)$ can be obtained as the following Gauge quotient:

$$\widehat{\mathscr{C}^\infty(\Omega)}/\approx_{\mathscr{D}'}.$$

It is not difficult to show that we can obtain $\mathscr{D}'(\Omega)$ as the following quotient:

$$\widehat{L^1_{loc}(\Omega)}/\approx_{\mathscr{D}'}$$

where

$$\widehat{L^1_{loc}(\Omega)} = \left\{ f \in L^1_{loc}(\Omega)^* \,\Big|\, \forall\varphi \in \mathscr{D}, \int^* f\varphi\,dx \text{ is bounded} \right\}.$$

Indeed, the map

$$\Psi : \widehat{L^1_{loc}(\Omega)}/\approx_{\mathscr{D}'} \longrightarrow \widehat{\mathscr{C}^\infty(\Omega)}/\approx_{\mathscr{D}'}$$

defined by

$$\Psi([f]) := [f]_{\mathscr{D}'} \cap \mathscr{C}^\infty(\Omega)^* = \{g \in \mathscr{C}^\infty(\Omega)^* \mid g \approx_{\mathscr{D}'} f\}$$

is a linear isomorphism, and we set $\partial_i([f]) := \partial_i([g])$ where $g \in \mathscr{C}^\infty(\Omega)^* \cap [f]$.

In this section, we will present an alternative way to define the distributions via the set of grid functions, which is still closely related to the notion of gauge quotient. This approach is useful in many situations; see the theory of stochastic differential equations in Chapter 10; see also, *e.g.*, [**32, 92, 27**] and references therein for other applications.

Let $\Omega \subset \mathbb{R}^N$. We define on the set of grid functions $\mathfrak{G}(\Omega)$ the following equivalence relation:

$$\xi \approx_{\mathfrak{G}} \zeta \iff \forall\varphi \in \mathscr{D}, \eta^N \cdot \sum_{x \in \mathbb{H}^N \cap \Omega} [\xi(x) - \zeta(x)]\,\varphi(x) \sim 0$$

and we put

$$\widehat{\mathfrak{G}(\Omega)} = \left\{ \xi \in \mathfrak{G}(\Omega) \,\Big|\, \forall\varphi \in \mathscr{D}, \eta^N \cdot \sum_{x \in \mathbb{H}^N \cap \Omega} \xi(x)\varphi(x) \text{ is bounded} \right\}.$$

DEFINITION 9.17. *The space* $\mathscr{D}'_{\mathfrak{G}}(\Omega)$ *of grid distributions on the open set* Ω *is defined as follows:*

$$\mathscr{D}'_{\mathfrak{G}}(\Omega) = \widehat{\mathfrak{G}(\Omega)}/\approx_{\mathfrak{G}}.$$

The equivalence class of u in $\mathscr{C}^\infty(\Omega)^*$ will be denoted by $[u]_{\mathfrak{G}}$.

We have the following result, that trivially follows from the definitions.

THEOREM 9.18. *There is a linear isomorphism*

$$\Phi : \mathscr{D}'_{\mathfrak{G}}(\Omega) \to \widehat{L^1_{loc}(\Omega)}/\approx_{\mathscr{D}'} \cong \mathscr{D}'(\Omega)$$

defined by the following formula:

$$\Phi\left([\xi]_{\mathfrak{G}}\right) = \left[\sum_{t\in\mathbb{H}^N\cap\Omega} \xi(t)\chi_t\right]_{\mathscr{D}'}$$

where

$$\chi_t(x) = \chi_{[t_1,t_1+\eta)}(x_1)\cdot\ldots\cdot\chi_{[t_N,t_N+\eta)}(x_N).$$

So, by identifying as usual $[\xi]_{\mathfrak{G}}$ with $\Phi([\xi]_{\mathfrak{G}})$, we have that

$$\langle[\xi]_{\mathfrak{G}},\varphi\rangle = \mathrm{st}\left(\eta\cdot\sum_{t\in\mathbb{H}^N\cap\Omega}\xi(t)\varphi(t)\right).$$

Given a grid function f, we recall that we use notation $D_{x_i}f$, $D_{x_i}^+f$, and $D_{x_i}^-f$ for the grid derivatives (see Definition 7.8), while we will use the notation $\partial_{x_i}T$ for the usual partial derivative of a distribution. We use symbols $\Delta_{\mathfrak{G}}$ and Δ for the grid Laplacian operator and the usual Laplacian operator, respectively.

PROPOSITION 9.19. *If* $f \in \mathfrak{G}(\Omega)$ *is a grid function and* $[f]_{\mathscr{D}'} \in \mathscr{D}'$, *then we have*

$$\partial_{x_i}[f]_{\mathscr{D}'} = [D_{x_i}f]_{\mathscr{D}'} = \left[D_{x_i}^+f\right]_{\mathscr{D}'} = \left[D_{x_i}^-f\right]_{\mathscr{D}'},$$

and

$$\Delta[f]_{\mathscr{D}'} = [\Delta_{\mathfrak{G}}f]_{\mathscr{D}'}.$$

PROOF. Let us prove that $\partial_{x_i}[f]_{\mathscr{D}'} = \left[D_{x_i}^+f\right]_{\mathscr{D}'}$. All the rest is a trivial consequence. With some abuse of notation, we use the same symbol for the grid function f and the step function

$$\sum_{t\in\mathbb{H}^N\cap\Omega} f(t)\chi_t,$$

and similarly for D^+f. Let $f_\infty \in \mathscr{C}^\infty(\Omega)$ be a function such that $f_\infty \approx_{\mathscr{D}'} f$; we have to prove that $\partial_{x_i}f_\infty \approx_{\mathscr{D}'} D_{x_i}^+f$, namely that for every $\varphi \in \mathscr{D}(\Omega)$:

$$\int \left(\partial_{x_i}f_\infty - D_{x_i}^+f\right)\varphi dx \sim 0.$$

We have that

$$(6) \quad \forall \varphi \in \mathscr{D}, \int \partial_{x_i} f_\infty \varphi \, dx = -\int f_\infty \partial_{x_i} \varphi \, dx \sim -\int f \partial_{x_i} \varphi \, dx.$$

On the other hand,

$$\int_\Omega D^+_{x_i} f \varphi \, dx = \frac{1}{\eta} \int_\Omega [f(\ldots, x_i + \eta, \ldots) - f(\ldots, x_i, \ldots) \varphi(\ldots, x_i, \ldots)] \, dx$$

$$= \frac{1}{\eta} \int f(\ldots, x_i, \ldots) \varphi(\ldots, x_i - \eta, \ldots) dx$$

$$- \frac{1}{\eta} \int f(\ldots, x_i, \ldots) \varphi(\ldots, x_i, \ldots) \, dx$$

$$= \frac{1}{\eta} \int f(\ldots, x_i, \ldots) [\varphi(\ldots, x_i - \eta, \ldots) - \varphi(\ldots, x_i, \ldots)] \, dx.$$

Now, by the Taylor's formula with Lagrange remainder, there exists $\theta \in (0, 1)_{\mathbb{R}^*}$ such that

$$\int D^+_{x_i} f \varphi \, dx = \frac{1}{\eta} \int f(\ldots, x_i, \ldots) \left[-\eta \partial_{x_i} \varphi(\ldots, x_i, \ldots) \right.$$

$$\left. + \frac{1}{2} \partial^2_{x_i} \varphi(\ldots, x_i - \theta\eta, \ldots) \eta^2 \right] dx$$

$$= -\int f \partial_{x_i} \varphi dx + \frac{1}{2} \eta \int f(.., x_i, ..) \partial^2_{x_i} \varphi(.., x_i - \theta\eta, ..) dx$$

$$= -\int f \partial_{x_i} \varphi dx - \frac{1}{2} \eta \int f(.., x_i + \theta\eta, ..) \partial^2_{x_i} \varphi(.., x_i, ..) dx$$

$$\sim -\int f \partial_{x_i} \varphi dx.$$

In the last step, we have used the fact that $\int f(.., x_i + \theta\eta, ..) \partial^2_{x_i} \varphi(.., x_i, ..)$ is bounded. So, using (6),

$$\int D^+_{x_i} f \varphi dx \sim -\int f \partial_{x_i} \varphi dx \sim \int \partial_{x_i} f_\infty \varphi dx.$$

$\square$

EXAMPLE 9.20. *In the framework of grid functions, the Dirac δ-function has a nice representative, namely*

$$\delta = [\alpha \delta_{0,t}]_{\mathscr{D}'}$$

where $\delta_{i,j}$ for $i, j \in \mathbb{H}$ is the Kronecker symbol:

$$\delta_{i,j} = \begin{cases} 1 & if \quad i = j \\ 0 & if \quad i \neq j . \end{cases}$$

Thus we are allowed to think of $\delta(t)$ as a function with value $\boldsymbol{\alpha}$ for $t = 0$, and with value 0 otherwise. Also its derivative takes a nice form, namely

$$\partial\delta = [\boldsymbol{\alpha} D\delta_{0,t}]_{\mathscr{D}'} = \left[\boldsymbol{\alpha}^2 \frac{\delta_{\eta,t} - \delta_{-\eta,t}}{2}\right]_{\mathscr{D}'}.$$

CHAPTER 10

Stochastic Differential Equations

In this chapter we will present an application of the Alpha-Theory to stochastic differential equations that was originally presented in [**16**].

1. Preliminary remarks on the white noise

Suppose that x is a physical quantity whose evolution is governed by a deterministic force which has small random fluctuations; such a phenomenon can be described by the following equation

$$(7) \qquad \dot{x} = f(x) + h(x)\xi(t)$$

where $\dot{x} = \frac{dx}{dt}$, and ξ is a "white noise". Intuitively, a *white noise* is the derivative of a *Brownian motion*, which is a continuous function which is not differentiable at any point.

There is no function ξ which has such a property; actually, the mathematical object which models ξ is a distribution. Thus equation (7) makes sense if it lives in the world of distributions. On the other hand the kind of problems which an applied mathematician asks are of the following type. Suppose that $x(0) = 0$ and that $\xi(t)$ is a random noise of which only the statistical properties are known. What is the probability distribution $P(t, x)$ of x at the time t? This question can be formalized by the theory of stochastic differential equations, and equation (7) takes the form

$$(8) \qquad dx = f(x)dt + h(x)dw.$$

Thus, the white noise dw is regarded as the "differential" of a *Wiener process* w. In this case, both $x(t)$ and $w(t)$ are not modelled by distributions, but rather by stochastic processes.

By the *Itô theory*, the above question can be solved rigorously: the probability distribution can be determined solving the *Fokker-Plank equation*:

$$(9) \qquad \frac{\partial P}{\partial t} = \frac{1}{2}\frac{\partial^2}{\partial x^2}\left(h(x)^2 P\right) + \frac{\partial}{\partial x}\left(f(x)P\right).$$

Equations (7) (or (8)) and (9) are very relevant in applications of Mathematics and the practitioners of mathematics such as engineers, physicists, economists, etc. make a large use of it. However the mathematics used in these equations is rather involved and many of them are not able to control it.

Usually people think of some intuitively simpler model. For example, $\xi(t)$ is considered as a force which acts at discrete instants of time t_i; it is supposed that the difference of two successive times $dt = t_{i+1}-t_i$ is infinitesimal and its strength is infinite; namely

$$(10) \qquad \xi(t) = \pm\frac{1}{\sqrt{dt}}.$$

The sign of this force is determined by a fair coin tossing. Clearly, equation (10) makes no sense and the gap between the rigorous mathematical description and the intuitive model is quite large.

The main purpose of this chapter is to show how this gap can be reduced by the use of Alpha-Theory. Using the notion of grid function, we are able to give a sense to (7) and (10) and, in this framework, we will deduce equation (9) rigorously. Our proof is relatively simple and very close to intuition. Our approach presents the following peculiarities:

(1) We will rewrite (7) as a "grid" differential equation (see Section 7.4):

$$(11) \qquad D_{\mathfrak{G}}^{+}x(t) = f(x) + \xi(t)$$

where $D_{\mathfrak{G}}^{+}$ denotes the grid derivative (see Definition 7.8). From this equation, it easy to recover both a distribution equation and a stochastic equation, and, at the same time, equation (11) has a very intuitive meaning.

(2) When equation (11) is considered from the stochastic point of view, the noise ξ is regarded as a *grid function* belonging to the space of all possible noises $\mathcal{R}$. If ξ is regarded as a random variable, the probability on the sample space $\mathcal{R}$ can be defined in a naive way, namely every noise has the same probability.[1]

---

[1] This is the basic idea of the *Loeb measure*, in the applications of NSA, but we do not need to use it. Actually, we do not need to introduce any kind of measure.

(3) By directly working with infinitesimal and skipping the step of making connections with the traditional approaches to problems, one gets simple new models of natural phenomena.

REMARK 10.1. *Many researchers working in nonstandard analysis have developed theories of nonstandard stochastic processes (most notably, see the work* [1] *by S. Albeverio and his coauthors, and see H.J. Keisler's work* [61] *and related bibliography). The peculiarities of the approach presented here, are described above.*

2. Stochastic grid equations

We begin this section with a definition of Brownian motion and white noise which is quite natural in the context of grid functions and which is close to our intuition.

DEFINITION 10.2. *A grid function $b(t)$ is called* Brownian notion *if it satisfies the following:*

- $|D_{\mathfrak{G}}^+ b(t)| = \sqrt{\alpha}$;
- $|b(t_2) - b(t_1)| \leq \frac{1 + |t_2 - t_1|}{|t_2 - t_1|^{1/2}}$.

Of course, this is not the only possible definition of "Brownian motion" for a single grid function. In any case, this definition is very simple and has the main characteristics which we should expect from a Brownian motion.

DEFINITION 10.3. *A grid function $\xi(t)$ is called* white noise *if it is the grid derivative of a Brownian motion; more precisely if it satisfies the following assumptions:*

- $|\xi(t)| = \sqrt{\alpha}$;
- $\eta \cdot \sum_{t \in [t_1, t_2]_{\mathbb{H}}} \xi(t) \leq \frac{1 + |t_2 - t_1|}{|t_2 - t_1|^{1/2}}$.

In our approach, a *stochastic differential equation* consists of a set of grid differential equations. Each differential equation has a *noise* term and gives a trajectory which can be considered as a realization of a process.

Let $\mathcal{R} \subset \mathfrak{G}[0,1]$ be an hyperfinite set of grid functions on $[0,1]_{\mathbb{H}} = [0,1]^* \cap \mathbb{H}$ and consider the class of Cauchy problems

(12)
$$\begin{cases} D_{\mathfrak{G}}^+ x(t) = f(t,x) + h(t,x)\xi, \\ x(0) = x_0, \\ \xi(t) \in \mathcal{R}, \end{cases}$$

where
$$f, h : [0, 1]_{\mathbb{H}} \times \mathbb{R}^* \to \mathbb{R}^*.$$

We want to study the statistical behavior of the set of solutions of the above Cauchy problems
$$\mathcal{S} = \{x_\xi(t) \mid \xi \in \mathcal{R}\}.$$

More precisely, we want to describe the behavior of the density grid function
$$\rho : [0, 1]_{\mathbb{H}} \times \mathbb{H}^* \to \mathbb{Q}^*$$

defined as follows

(13)
$$\rho(t, x) = \frac{|\{x_\xi \in \mathcal{S} \mid x \leq x_\xi(t) < x + \eta\}|}{\eta \cdot |\mathcal{R}|}$$

where $| \cdot |$ denotes the internal hyperfinite cardinality.

We are interested in the case in which $\mathcal{R}$ is the set of *white noises* according to Definition 10.3 or, more in general, in the case in which $\mathcal{R}$ is a normal stochastic class according to the definition below.

DEFINITION 10.4. *A set of grid functions is called a* normal stochastic class *if it satisfies the following properties:*

- (FIN) *the set* $\mathcal{R}$ *is hyperfinite;*
- (EXP) $\forall t \in \mathbb{H}$, $\sum_{\xi \in \mathcal{R}} \xi(t) = 0$;
- (VAR) $\forall \xi \in \mathcal{R}$, $\forall t \in \mathbb{H}$, $|\xi(t)|^2 = \alpha$.

Clearly, the family $\mathcal{R}$ of all white noises according to Definition 10.3 is a normal stochastic class. Another example of stochastic class is given by $\mathcal{R}_\alpha := \lim_\alpha \mathcal{R}_n$ where
$$\mathcal{R}_n := \{-\sqrt{n}, +\sqrt{n}\}^{[0,1]_{\mathbb{H}_n}}.$$

Hence $\mathcal{R}_\alpha$ is the set of all the grid functions that take values in $\{-\sqrt{\alpha}, +\sqrt{\alpha}\}$. So, $\mathcal{R}_\alpha$ is a hyperfinite set with internal cardinality $|\mathcal{R}_\alpha| = 2^{\alpha+1}$; any stochastic class is a subset of $\mathcal{R}$. Besides, it is possible to prove that the "big majority" of functions in $\mathcal{R}_\alpha$ are white noises according to Definition 10.3. In the next section, we will study the statistic properties of the solution of (12). The properties which we will analyze will be the same if $\mathcal{R}$ is the set of white noises or any other normal stochastic class.

We can give a probabilistic interpretation to problem (12). In this case a noise ξ is regarded as a *stochastic process* belonging to a probability space $\mathcal{R}$. The probability on $\mathcal{R}$ can be defined in a naive way, namely every noise has the same probability. So, we get the following *non-Archimedean probability*

$$P_{NA}(E) = \frac{|E|}{|\mathcal{R}|}; \ (E \subset \mathcal{R}, \ E \text{ internal})$$

which takes values in $\mathbb{Q}^*$. Using the probabilistic language, the properties (EXP) and (VAR) mean that $\xi(t)$ has expectation 0 and variance α. However, we do not need any probabilistic interpretation of equation (12) and from now on we will ignore it.

3. Itô's formula for grid functions

In this section we show a relevant example of the flexibility of the grid functions. It is a proposition which is, in some sense, a variant of the Itô's formula in a very simple frame. As in the standard approach, this formula will be the main tool in the study of grid stochastic equations (see Section 2).

THEOREM 10.5 (Itô's formula). *Let $\varphi \in \mathscr{C}_0^3(\mathbb{R}^2)$ and let $x(t)$ be a grid function such that*

$$(14) \qquad\qquad |D_t^+ x(t)| \le \zeta \alpha^{2/3},$$

where $\zeta \sim 0$. Then

$$D_t^+ \varphi(t, x(t)) \sim \varphi_t(t, x(t)) + \varphi_x(t, x(t)) D_t^+ x(t) + \frac{\eta}{2} \cdot \varphi_{xx}(t, x(t)) \cdot \left(D_t^+ x(t)\right)^2.$$

Here φ_t, φ_x and φ_{xx} denote the usual partial derivatives of φ.

PROOF. By definition of grid derivative we have that

$$
\begin{aligned}
D_t^+ \varphi(t, x(t)) \ =\ & \frac{\varphi(t + \eta, x(t + \eta)) - \varphi(t, x(t + \eta))}{\eta} \\
& + \frac{\varphi(t, x(t + \eta)) - \varphi(t, x(t))}{\eta} \\
\sim\ & \varphi_t(t, x(t + \eta)) + \frac{\varphi(t, x(t + \eta)) - \varphi(t, x(t))}{\eta} \\
\sim\ & \varphi_t(t, x(t)) + \frac{\varphi(t, x(t + \eta)) - \varphi(t, x(t))}{\eta}.
\end{aligned}
$$

Now,

$$\varphi(t, x(t + \eta)) = \varphi\left(t, x(t) + \eta D_t^+ x(t)\right),$$

Then, using the Taylor's formula, we have that

$$
\begin{aligned}
\varphi\left(t, x(t) + \eta D_t^+ x(t)\right) \ =\ & \varphi(t, x(t)) + \varphi_x(t, x(t)) \eta D_t^+ x(t) \\
& + \frac{1}{2} \varphi_{xx}(t, x(t)) \left(\eta D_t^+ x(t)\right)^2 \\
& + \frac{1}{3!} \varphi_{xxx}(t, x(t) + \theta_t) \left(\eta D_t^+ x(t)\right)^3, \\
& 0 < \theta_t < \eta D_t^+ x(t);
\end{aligned}
$$

and hence

$$\frac{\varphi(t, x(t + \boldsymbol{\eta})) - \varphi(t, x(t))}{\boldsymbol{\eta}} = \varphi_x(t, x(t)) D_t^+ x(t)$$

$$+ \frac{\boldsymbol{\eta}}{2} \varphi_{xx}(t, x(t)) \cdot \left(D_t^+ x(t) \right)^2$$

$$+ \frac{\boldsymbol{\eta}^2}{6} \varphi_{xxx}(t, x(t) + \theta_t) \cdot \left(D_t^+ x(t) \right)^3.$$

By assumption (14), the last term is infinitesimal and we get the desired result.   □

4. The Fokker-Plank equation

Now, we can state the main result of this chapter:

THEOREM 10.6. *Assume that $\mathcal{R}$ is a white noise and that $f(t, x)$ and $h(t, x)$ are continuous functions. Then the distribution $[\rho]$ relative to the grid density function ρ is a measure and satisfies the Fokker-Plank equation*

$$(15) \qquad \frac{d\,[\rho]}{dt} + \frac{d}{dx}\left(f(t, x)\,[\rho]\right) - \frac{1}{2}\frac{d^2}{dx^2}\left(h(t, x)^2\,[\rho]\right) = 0.$$

$$(16) \qquad\qquad\qquad [\rho]_{t=0} = \delta_{x_0}$$

where we have written $[\rho]$ instead of $[\rho]_{\mathscr{D}'}$ for short.

REMARK 10.7. *We recall that (15) and (16) "in the sense of distributions" mean that $[\rho]$ satisfies the following equation for every $\varphi \in \mathscr{D}\left([0, 1) \times \mathbb{R}\right)$:*

$$(17) \qquad \iint \left(\varphi_t + f\varphi_x + \varphi_{xx}h^2\right)\rho\,dx_\eta\,dt_\eta + \varphi(0, x_0) = 0.$$

Actually, we will prove Theorem 10.6 just proving the above equation.

REMARK 10.8. *If $f(t, x)$ and $h(t, x)$ are smooth functions, by standard results in PDE, we know that, for $t > 0$, the distribution $[\rho]$ coincides with a smooth function $u(t, x)$. Then, for any $t > 0$, ρ defines a smooth function u by the formula*

$$\forall \varphi \in \mathscr{D}\left((0, 1) \times \mathbb{R}\right), \quad \iint \rho\varphi\,dx_\eta dt_\eta = \iint u\varphi\,dx\,dt$$

and u satisfies the Fokker-Plank equation in $(0, 1) \times \mathbb{R}$ in the usual sense.

REMARK 10.9. *We will see in the proof of Theorem 10.6 that if the functions $f(t,x)$ and $h(t,x)$ are not continuous, but only bounded on compact sets, the equation (17) still holds. However, in this case equation (17) cannot be interpreted so easily. For example, if $f(t,x)$ and $h(t,x)$ are not measurable, there is no simple standard interpretation.*

If Γ is a hyperfinite set and $\Phi : \Gamma \to \mathbb{R}^*$ is an internal function, the *mean value* of Φ in Γ is defined as follows:

$$\mathbb{E}_{\xi \in \Gamma} [\Phi] = \frac{1}{|\Gamma|} \sum_{\xi \in \Gamma} \Phi(\xi).$$

In particular, the mean value of a grid function in the set $[x,y)_{\mathbb{H}}$ is the following:

$$\mathbb{E}_{[x,y)} [f] = \frac{\eta}{(y-x)} \sum_{t \in [x,y]_{\mathbb{H}}} f(t).$$

Let us now see a basic property of the density function.

LEMMA 10.10. *Let $g \in \mathscr{C}^0 ([0,1] \times \mathbb{R})$ and let $\langle x_\xi \mid \xi \in \mathcal{R} \rangle$ be an internal family of grid functions. Then*

$$\mathbb{E}_{\xi \in \mathcal{R}} [g(t, x_\xi(t))] \sim \mathbb{E}_{x \in \mathbb{H}} [g(t,x)\rho(t,x)].$$

where $\rho(t,x)$ is defined by (13).

PROOF. We have

$$\mathbb{E}_{\xi \in \mathcal{R}} [g(t, x_\xi(t))] = \frac{1}{|\mathcal{R}|} \sum_{\xi \in \mathcal{R}} g(t, x_\xi(t)) = \frac{1}{|\mathcal{R}|} \sum_{x \in \mathbb{H}} \left[\sum_{x \leq x_\xi(t) < x + \eta} g(t, x_\xi(t)) \right]$$

$$= \frac{1}{|\mathcal{R}|} \sum_{x \in \mathbb{H}} \left[\sum_{x \leq x_\xi(t) < x + \eta} (g(t,x) + \vartheta_\xi(t)) \right]$$

where $\vartheta_\xi(t) := g(t, x_\xi(t)) - g(t,x)$ is infinitesimal since $x \leq x_\xi(t) < x + \eta$. Hence,

$$\frac{1}{|\mathcal{R}|} \sum_{\xi \in \mathcal{R}} \vartheta_\xi(t) \sim 0$$

and so, using the definition of ρ,

$$\mathbb{E}_{\xi\in\mathcal{R}}\left[g(t,x_\xi(t))\right] \sim \frac{1}{|\mathcal{R}|}\sum_{x\in\mathbb{H}}\left[\sum_{x\le x_\xi(t)<x+\eta} g(t,x)\right]$$

$$= \frac{1}{|\mathcal{R}|}\sum_{x\in\mathbb{H}}[g(t,x)\cdot|\{x_\xi\in\mathcal{S}:x\le x_\xi(t)<x+\eta\}|]$$

$$= \frac{1}{|\mathcal{R}|}\sum_{x\in\mathbb{H}}[g(t,x)\rho(t,x)\cdot\eta\cdot|\mathcal{R}|]{=}\eta\sum_{x\in\mathbb{H}}[g(t,x)\rho(t,x)]$$

$$= \mathbb{E}_{x\in\mathbb{H}}\left[g(t,\rho)x(t,x)\right].$$

$\square$

LEMMA 10.11. *Let* $F : [0,1]_\mathbb{H}\times\mathbb{R}^* \to \mathbb{R}^*$ *be an internal bounded function. Then, for every* $t\in[0,1]_\mathbb{H}$

$$\mathbb{E}_{\xi\in\mathcal{R}}\left[F(t,x_\xi(t))\cdot\xi(t)\right] = 0.$$

PROOF. Given $t\in[0,1]_\mathbb{H}$, we set

$$\mathcal{R}[0,t) = \left\{\xi|_{[0,t)}\mid\xi\in\mathcal{R}\right\};$$

namely, $\mathcal{R}[0,t)$ is the set of the restrictions of the functions of $\mathcal{R}$ to $[0,t)_\mathbb{H}$. Moreover, for $\tau\in\mathcal{R}[0,s)$, we set

$$\mathcal{R}_\tau[s,1] = \{\xi\in\mathcal{R}\mid\xi(t)=\tau(t)\text{ for }t<s\}.$$

So we have the following decompositions for every $s\in[0,1)$:

$$\mathcal{R} = \bigcup_{\tau\in\mathcal{R}[0,s)}\mathcal{R}_\tau[s,1].$$

Then,

$$\mathbb{E}_{\xi\in\mathcal{R}}\left[F(t,x_\xi(t))\cdot\xi(t)\right] = \frac{1}{|\mathcal{R}|}\sum_{\tau\in\mathcal{R}[0,t)}\sum_{\xi\in\mathcal{R}_\tau[t,1]}F(t,x_\xi(t))\cdot\xi(t).$$

Since $x_\xi(t)$ does not depend on $\xi(s)$ for $s\ge t$, we have that

$$\mathbb{E}_{\xi\in\mathcal{R}}\left[F(t,x_\xi(t))\cdot\xi(t)\right] = \frac{1}{|\mathcal{R}|}\sum_{\tau\in\mathcal{R}[0,t)}\sum_{\xi\in\mathcal{R}_\tau[t,1]}F(t,x_\tau(t))\cdot\xi(t)$$

$$= \frac{1}{|\mathcal{R}|}\sum_{\tau\in\mathcal{R}[0,t)}\left(F(t,x_\tau(t))\sum_{\xi\in\mathcal{R}_\tau[t,1]}\xi(t)\right).$$

Now, it is sufficient to remark that one half of the functions in $\mathcal{R}_\tau\,[t,1]$ take the value $\sqrt{\alpha}$ in the point t and the other half take the value $-\sqrt{\alpha}$; then

$$\sum_{\xi\in\mathcal{R}_\tau[t,1]} \xi(t) = 0$$

and hence

$$\mathbb{E}_{\xi\in\mathcal{R}}\left[F(t,x_\xi(t))\cdot\xi(t)\right] = 0.$$

$\square$

PROOF OF THEOREM 10.6. Chosen an arbitrary function $\varphi \in \mathscr{C}^0([0,1]\times\mathbb{R})$ bounded in the second variable, by Theorem 7.11 we have that

$$\varphi(1,x_\xi(1)) - \varphi(0,x_0) = \eta \sum_{t\in[0,1-\eta]_\mathbb{H}} D_t^+\varphi(t,x_\xi(t)).$$

Now we assume that $\varphi \in \mathscr{D}([0,1)\times\mathbb{R}))$. We apply Itô's formula and we use the fact that $x_\xi(t)$ solves equation (12):

$$-\varphi(0,x_0) = \eta \sum_{t\in[0,1)_\mathbb{H}} \left[\varphi_t + \varphi_x \cdot D_t^+ x + \frac{\eta}{2}\varphi_{xx}\cdot\left(D_t^+ x\right)^2\right] + \eta\theta$$

$$= \eta \sum_{t\in[0,1)_\mathbb{H}} \left[\varphi_t + \varphi_x\cdot(f+h\xi) + \frac{\eta}{2}\varphi_{xx}\cdot(f+h\xi)^2\right] + \eta\theta$$

$$= \eta \sum_{t\in[0,1)_\mathbb{H}} \left[(\varphi_t+f\varphi_x)+(\varphi_x h+\eta\varphi_{xx}f)\,\xi+\frac{\eta}{2}\varphi_{xx}f+\frac{\eta}{2}\varphi_{xx}h^2\xi^2\right]$$
$$\quad + \eta\theta$$

$$= \eta \sum_{t\in[0,1)_\mathbb{H}} \left[\varphi_t+f\varphi_x+\frac{1}{2}\varphi_{xx}h^2\right]+\eta \sum_{t\in[0,1)_\mathbb{H}} [(\varphi_x h+\eta\varphi_{xx}f)\,\xi]$$

$$\quad + \eta \sum_{t\in[0,1)_\mathbb{H}} \left[\frac{\eta}{2}\varphi_{xx}f\right] + \eta\theta.$$

Now, for every $t \in [0,1]_\mathbb{H}$, we want to compute $\mathbb{E}_{\xi\in\mathcal{R}}$ of each term on the right hand side of the above equation considered as a function of t. By Lemma 10.10,

$$\mathbb{E}_{\xi\in\mathcal{R}}\left[\varphi_t+f\varphi_x+\frac{1}{2}\varphi_{xx}h^2\right] = \mathbb{E}_{x\in\mathbb{H}}\left[\left(\varphi_t+f\varphi_x+\frac{1}{2}\varphi_{xx}h^2\right)\rho\right].$$

Let us consider the second piece: $(\varphi_x h + \eta\varphi_{xx}f)\,\xi$. If we set

$$F(t,x_\xi(t)) = \varphi_x(t,x_\xi(t))h(t,x_\xi(t)) + \eta\varphi_{xx}(t,x_\xi(t))f(t,x_\xi(t)),$$

it turns out that $F(t, x_\xi(t))$ is bounded. In fact, if $x_\xi(t)$ is bounded, then $\varphi_x(t, x_\xi(t))$, $h(t, x_\xi(t))$, $\varphi_{xx}(t, x_\xi(t))$ and $f(t, x_\xi(t))$ are bounded since they are standard functions; if $x_\xi(t)$ is unbounded, $\varphi_x(t, x_\xi(t)) = 0$. Then we can apply Lemma 10.11 and get that

$$\mathbb{E}_\xi\left[(\varphi_x h + \eta \varphi_{xx} f)\, \xi\right] = 0.$$

Moreover,

$$\mathbb{E}_\xi\left[\frac{\eta}{2}\varphi_{xx} f\right] \sim 0$$

since $\frac{\eta}{2}\varphi_{xx} f \sim 0$. Then, by the previous equalities,

$$
\begin{aligned}
-\varphi(0, x_0) &= \mathbb{E}_{t\in[0,1]_\mathbb{H}, \xi\in\mathcal{R}}\left[-\varphi(0, x_0)\right] \\
&\sim \eta \sum_{t\in[0,1)_\mathbb{H}} \mathbb{E}_{\xi\in\mathcal{R}}\left[\varphi_t + f\varphi_x + \frac{1}{2}\varphi_{xx}h^2\right] + \mathbb{E}_{\xi\in\mathcal{R}}\left[(\varphi_x h + \delta\varphi_{xx} f)\,\xi\right] \\
&\quad + \eta \sum_{t\in[0,1)_\mathbb{H}} \mathbb{E}_{\xi\in\mathcal{R}}\left[\frac{\eta}{2}\varphi_{xx} f\right] \\
&\sim \eta^2 \sum_{t\in[0,1)_\mathbb{H}} \left(\sum_{x\in\mathbb{H}} (\varphi_t + f\varphi_x + \varphi_{xx}h^2)\,\rho\right) \\
&\sim \iint (\varphi_t + f\varphi_x + \varphi_{xx}h^2)\,\rho\, dx\, dt.
\end{aligned}
$$

Since the first and the last term of this equation are standard, we have the equality and we get equality (17).   □

5. Remarks and comments

1. Our approach to stochastic equations is more elementary than those found in [2], [61] and [78]. The reason lies in the fact that we have not made a nonstandard theory of *stochastic differential equations*, but rather, we have replaced them with *stochastic grid equations*, which are much simpler mathematical objects. In this context, Itô's integral is replaced by the grid integral, and Itô's formula, a key point of this theory, reduces to an exercise.

The reason why a much simpler object can represent the processes of diffusion at microscopic level is that we have taken the infinitesimals seriously, and we have used them to model an aspect of the "physical reality." This is a typical example of our philosophy about non-Archimedean mathematics: *the advantages of a theory which includes infinitesimals rely more on the possibility of building new models rather than in the proving techniques.*

In the case considered in this chapter, the construction of the model appears quite natural: the stochastic grid equations describe the diffusion processes at microscopic level, namely, at "infinitesimal" level. The Fokker-Plank equation describes the diffusion processes at the macroscopic level and uses standard differential equations (and the theory of distributions when the data are not regular).

2. A final remark concerns the *theory of probability*. Everything can be kept at a very elementary level, and our variant of the Itô formula makes sense also in a context where probability does not appear. In any case, probability can be introduced in a very elementary way.

We may think of the stochastic class $\mathcal{R}$ as a sample space. The events are the hyperfinite sets $E \subset \mathcal{P}(\mathcal{R})^*$ and the probability P of an event is given by

$$(18) \qquad\qquad P(E) = \frac{|E|}{|\mathcal{R}|}.$$

For example, E might be the event that at a time t_0, the moving particle lies in the interval $[a, b]$, namely

$$E = \{x_\xi \in \mathcal{S} : x_\xi(t_0) \in [a, b]_{\mathbb{H}}\}.$$

In this case we have that

$$P(E) \sim \int_a^b \rho(t, x) \, dx_\Delta.$$

Obviously, this is the most natural extension to infinite sample spaces of the classic definition of probability. There is no doubt that definition (18) is the most simple and intuitive definition of probability. The price one has to pay is that P takes its values in $\mathbb{Q}^*$ and not in $\mathbb{R}$; thus, a probability space $(\mathcal{R}, P)$ is an internal object and, if it is infinite, it is not standard. The problem arises if you want to connect the nonstandard world with the standard one. This operation can be done in a very elegant way via the Loeb integral (see [**71**] or also [**61**]). A different way to make easy the connection between this two worlds has been proposed by Nelson [**78**]. However, we think that, in most cases, it is not necessary to make this connection. Only at the very end you may consider the standard part of the numbers which you have obtained. So, if we accept a mathematical description in which the infinitesimal exist, *the probability of an event is given by a hyperrational number,* and many theorems become simpler. This scheme avoids also some facts in probability theory which are in contrast with the common sense: for example the fact that the union of impossible events gives a possible one (the probability of a non-denumerable set might be non-null even if the probability of each singleton is null). For a deeper discussion on these aspects see also [**17, 18**].

Part 4

Foundations

This Part 4 is focused on the foundations of the Alpha-Theory. In particular, we will investigate its relationship with *nonstandard analysis*, and its strength as an axiomatic theory. Most notably, we will prove that every statement of mathematics is proved by ZFC if and only if it is proved by the Alpha-Theory.

Moreover, we will clarify the position of *Cauchy Infinitesimal Principle* CIP, and prove that it is independent from the other axioms. This means that, on the one hand, CIP cannot be proved by the Alpha-Theory plus the usual principles of mathematics as given by Zermelo-Fraenkel axiomatics and, on the other hand, that CIP can be safely added to the axioms.

CHAPTER 11

Ultrafilters and Ultrapowers

As instrumental tools for our foundational research, we will use *ultrafilters* and *ultrapowers*. In this chapter we will review their basic properties.

1. Filters and ultrafilters

The notions of ultrafilter and ultrapower are intimately connected with hyper-images. We will use them throughout this whole foundational chapter as the basic tool to construct models, to prove characterization results, and to determine the strength of the considered theories.

Let us start with the basic definitions.

DEFINITION 11.1. *A filter $\mathcal{F}$ on a set I is a nonempty family of subsets of I that satisfies the following three properties:*

(1) $\emptyset \notin \mathcal{F}$;

(2) *If $X \in \mathcal{F}$ and $Y \supseteq X$, then also $Y \in \mathcal{F}$;*

(3) *If $X, Y \in \mathcal{F}$ then $X \cap Y \in \mathcal{F}$.*

Trivial examples of filters are given by the *principal filters*, that is the filters $\mathcal{F}_X$ generated by a fixed nonempty subset $X \subseteq I$.

$$\mathcal{F}_X = \{ Y \subseteq I \mid Y \subseteq X \}.$$

The fundamental example of a *non-principal filter* is the following.

DEFINITION 11.2. *The* Frechet filter $Fr(I)$ *over an infinite set I is the family of the cofinite subsets of I.*

$$Fr(I) = \{X \subseteq I \mid I \setminus X \text{ is finite}\}.$$

REMARK 11.3. *One may think of a filter as determining a notion of "largeness" that is stable under finite intersections. However, we remark that this intuition may be misleading. Given any infinite $X \subset I$, as "small" as it gets, it is easy to construct non-principal filters on I that contains X.*[1] *Clearly, with respect to such filters, X would be "large" while its complement $X^c = I \setminus X$ would be not. In our opinion, it seems more appropriate to think of the sets in a given filter as the "relevant" or "qualified" sets, that is, as those sets that "really count".*

DEFINITION 11.4. *We say that a family of sets $\mathcal{F}$ has the* finite intersection property (FIP) *if every finite intersection of elements in $\mathcal{F}$ is nonempty:*

$$\forall \, X_1, \ldots, X_n \in \mathcal{F} \quad X_1 \cap \ldots \cap X_n \neq \emptyset.$$

So, a filter $\mathcal{F}$ is a nonempty family of nonempty sets that has the FIP and is closed under supersets. In general:

DEFINITION 11.5. *If a family $\mathcal{G}$ has the FIP, then*

$$\langle \mathcal{G} \rangle = \{Y \mid \exists X_1, \ldots, X_n \in \mathcal{G} \text{ s.t. } Y \supseteq X_1 \cap \ldots \cap X_n\}$$

is the filter generated *by $\mathcal{G}$.*

We remark that the name "filter generated" is fully justified by the facts that $\langle \mathcal{G} \rangle$ is actually a filter, and moreover it is the smallest one that contains all elements of $\mathcal{G}$.

Given an element $i \in I$ and a subset $X \subseteq I$, trivially one has that either $i \in X$ or $i \notin X$. So, the principal filter $\mathcal{F}_{\{i\}}$ generated by a singleton satisfies the condition that for any $X \subseteq I$, either $X \in \mathcal{F}_{\{i\}}$ or its complement $X^c = I \setminus X \in \mathcal{F}_{\{i\}}$. This special property that filters may have will play a crucial role.

PROPOSITION 11.6. *Let $\mathcal{F}$ be a filter over a set I. The following properties are equivalent:*

---

[1] *E.g., if $X = \{x_n\}$ is countable, one can take the filter*

$$\mathcal{F} \; = \; \{Y \subseteq I \mid Y \supseteq \{x_n\}_{n \geq k} \text{ for some } k\}.$$

(1) $X \notin \mathcal{F} \Rightarrow X^c \in \mathcal{F}$;

(2) $X_1 \cup \ldots \cup X_n \in \mathcal{F} \Rightarrow \exists i \; X_i \in \mathcal{F}$;

(3) $\mathcal{F}$ *is a* maximal filter *on I with respect to inclusion.*

PROOF. $(1) \Rightarrow (2)$. Assume towards a contradiction that $X_i \notin \mathcal{F}$ for all $i = 1, \ldots, n$. Then, by the hypothesis (1), the complements $X_i^c \in \mathcal{F}$ for all i, and hence the intersection

$$X_1^c \cap \ldots \cap X_n^c = (X_1 \cup \ldots \cup X_n)^c \in \mathcal{F}.$$

This is a contradiction because we are assuming that $X_1 \cup \ldots \cup X_n \in \mathcal{F}$, and two disjoint sets cannot belong to the same filter.

$(2) \Rightarrow (3)$. If $\mathcal{F}$ was not maximal, then one could take a filter $\mathcal{F}'$ that properly includes $\mathcal{F}$. Now pick any $X \in \mathcal{F}' \setminus \mathcal{F}$; the complement $X^c \notin \mathcal{F}'$, and so also $X^c \notin \mathcal{F}$. We conclude that $I = X \cup X^c \in \mathcal{F}$ is the union of two sets neither of which belongs to $\mathcal{F}$, against the hypothesis.

$(3) \Rightarrow (1)$. By contradiction, let $X, X^c \notin \mathcal{F}$. We want to show that the following family:

$$\mathcal{G} = \{Y \cap X \mid Y \in \mathcal{F}\}$$

has the FIP. If $Y_1, \ldots, Y_n \in \mathcal{F}$, then $(Y_1 \cap X) \cap \ldots \cap (Y_n \cap X) = Y \cap X$ where $Y = Y_1 \cap \ldots \cap Y_n \in \mathcal{F}$. Clearly, $Y \cap X \neq \emptyset$, otherwise $Y \subseteq X^c$ would imply that $X^c \in \mathcal{F}$, against our hypothesis. Now, let us consider the filter $\mathcal{F}' = \langle \mathcal{G} \rangle$ generated by $\mathcal{G}$. Notice that $\mathcal{F}'$ properly includes $\mathcal{F}$. In fact, for every $Y \in \mathcal{F}, Y \supseteq Y \cap X \in \mathcal{G} \Rightarrow Y \in \mathcal{F}'$. Moreover, $X \notin \mathcal{F}$ but $X = I \cap X \in \mathcal{G} \subseteq \mathcal{F}'$. This contradicts the maximality of $\mathcal{F}$.   □

DEFINITION 11.7. *A filter that satisfies one (and hence all) of the three properties above, is called an* ultrafilter.

Notice that for every infinite set I, the Frechet filter $\mathrm{Fr}(I)$ is *not* an ultrafilter. To see this, take any infinite subset $X \subset I$ whose complement X^c is also infinite. Then $I = X \cup X^c \in \mathrm{Fr}(I)$, while both $X, X^c \notin \mathrm{Fr}(I)$.

PROPOSITION 11.8. *A principal filter $\mathcal{F}_X$ is an ultrafilter if and only if $X = \{i\}$ is a singleton.*

PROOF. If X contains at least two elements, then one can partition $X = X_1 \cup X_2$ into two disjoint nonempty set. By the property of ultrafilter, one would have that either $X_1 \in \mathcal{F}_X$, or $X_2 \in \mathcal{F}_X$. Both cases are contradictory, because X_1, X_2 are proper subsets of X, while

$\mathcal{F}_X$ only contains supersets of X. The reverse implication is trivial, because

$$X \notin \mathcal{F}_{\{i\}} \Leftrightarrow i \notin X \Leftrightarrow i \in X^c \Leftrightarrow X^c \in \mathcal{F}_{\{i\}}.$$

$\square$

The principal ultrafilters $\mathcal{F}_{\{i\}}$ generated by a singleton $\{i\}$ will be denoted by

$$\mathfrak{U}_i = \{\, Y \subseteq I \mid i \in Y \,\}.$$

In this case, we will simply say that $\mathfrak{U}_i$ is the *principal ultrafilter generated by i*.

PROPOSITION 11.9.

(1) *All ultrafilters on finite sets are principal;*

(2) *An ultrafilter $\mathcal{U}$ on an infinite set I is non-principal if and only if it contains no finite sets, and hence if and only if $\mathcal{U} \supseteq Fr(I)$.*

PROOF. (1). Let $\mathcal{U}$ be an ultrafilter on the finite set $I = \{i_1, \ldots, i_n\}$. Then the union $\{i_1\} \cup \ldots \cup \{i_n\} = I \in \mathcal{U}$ implies that exactly one $\{i_k\} \in \mathcal{U}$. It follows that $\mathcal{U} = \mathcal{U}_{i_k}$ is principal.

(2). Assume first that $Fr(I) \subseteq \mathcal{U}$. Then for every $i \in I$, one has $\{i\}^c \in \mathcal{U}$, hence $\{i\} \notin \mathcal{U}$, and $\mathcal{U}$ is non-principal. Conversely, assume by contradiction that $Fr(I) \nsubseteq \mathcal{U}$, and pick a cofinite set $A \notin \mathcal{U}$. Then its complement

$$A^c = \{a_1, \ldots, a_n\} = \{a_1\} \cup \ldots \cup \{a_n\} \in \mathcal{U},$$

and by the properties of ultrafilter, it follows that $\{a_i\} \in \mathcal{U}$ for some i. We conclude that $\mathcal{U} = \mathfrak{U}_{a_i}$ is principal, against the hypothesis. $\square$

In Alpha-Calculus, we already met with a non-principal ultrafilter.

EXAMPLE 11.10. *The family of* qualified sets

$$\mathcal{Q} = \{A \subseteq \mathbb{N} \mid \boldsymbol{\alpha} \in A^*\}$$

is a non-principal ultrafilter on $\mathbb{N}$ (see Theorem 2.22).

Let us now turn to the general existence problem, and show that every filter can be extended to an ultrafilter. The proof needs the axiom of choice, in its equivalent formulation known as *Zorn's lemma*.

THEOREM 11.11. *Every filter can be extended to an ultrafilter.*

PROOF. Let the filter $\mathcal{F}$ over the set I be given. The family

$$\mathbb{G} = \{\mathcal{G} \text{ filter on } I \mid \mathcal{G} \supseteq \mathcal{F}\}$$

is partially ordered by inclusion. In order to apply Zorn's lemma, we need to show that every chain $\langle \mathcal{G}_j \mid j \in J \rangle$ in $\mathbb{G}$ has an upper bound. Trivially, the union $\mathcal{G} = \bigcup_{j \in J} \mathcal{G}_j$ includes all $\mathcal{G}_j$ (and hence, it also includes $\mathcal{F}$). Moreover $\mathcal{G}$ is a filter, and so $\mathcal{G}$ is the desired upper bound. Indeed, let $X, Y \in \mathcal{G}$; then $X \in \mathcal{G}_{j_1}$ and $Y \in \mathcal{G}_{j_2}$ for suitable $j_1, j_2 \in J$. Since $\langle \mathcal{G}_j \mid j \in J \rangle$ is a chain, we can assume without loss of generality that $\mathcal{G}_{j_1} \subseteq \mathcal{G}_{j_2}$. So, $X, Y \in \mathcal{G}_{j_2} \Rightarrow X \cap Y \in \mathcal{G}_{j_2} \subseteq \mathcal{G}$. Similarly, it is shown that $\mathcal{G}$ is closed under supersets. We are finally ready to apply Zorn's lemma and get the existence of a maximal element $\mathcal{U} \in \mathbb{G}$. Clearly $\mathcal{U} \supseteq \mathcal{F}$ is the desired ultrafilter. $\square$

COROLLARY 11.12. *For every infinite set I, there exist non-principal ultrafilters on I.*

PROOF. Recall that an ultrafilter on I is non-principal if and only if it extends the Frechet filter $\mathrm{Fr}(I)$ (see Proposition 11.9). $\square$

2. Ultrafilters as measures and as ideals

An alternative way of presenting ultrafilters is by means of the notion of probability measure. Recall the following

DEFINITION 11.13. *A family $\mathcal{B}$ of subsets of I is an* algebra of sets *over I if it contains the whole set I, and is closed under complements, finite unions and finite intersections.*

A map $\mu : \mathcal{B} \to [0, 1]$ into the unit real interval is a finitely additive probability measure *on I if the following conditions hold:*

(1) $\mu(I) = 1$;
(2) $\mu(A \cup B) + \mu(A \cap B) = \mu(A) + \mu(B)$.[2]

We say that μ is non-principal *if $\mu(\{x\}) = 0$ for all points $x \in I$ (and hence, $\mu(F) = 0$ for all finite F).*

As straight consequences of the two properties above, one has the

- *Monotonicity property:* $A \subseteq B \Rightarrow \mu(A) \leq \mu(B)$.

---

[2] If not specified otherwise, usually a *measure* is assumed to satisfy the property of *countable additivity* (or *σ-additivity*): "*If $\{X_n\}_{n \in \mathbb{N}}$ is a family of pairwise disjoint sets, then $\mu(\bigcup_{n \in \mathbb{N}} X_n) = \sum_{n=1}^{\infty} \mu(X_n)$*".

An important example to our purposes is the following.

EXAMPLE 11.14. *The* Alpha-measure $\mu_\alpha : \mathcal{P}(\mathbb{N}) \to \{0,1\}$ *defined by*

$$\mu_\alpha(A) = \begin{cases} 1 & \text{if } \boldsymbol{\alpha} \in A^* \\ 0 & \text{if } \boldsymbol{\alpha} \notin A^* \end{cases}$$

is a non-principal finitely additive measure (see Theorem 2.19).

By considering those finitely additive measures that are defined on the algebra $\mathcal{B} = \mathcal{P}(I)$ of *all* subsets of I and only take value 0 or 1, one obtains an alternative equivalent definition of ultrafilter. The proof is straightforward from the definitions and it is left for the reader as an exercise.

EXERCISE 11.15. *Given an ultrafilter* $\mathcal{U}$ *on* I, *let*

$$\mu_\mathcal{U}(X) = \begin{cases} 1 & \text{if } X \in \mathcal{U}; \\ 0 & \text{if } X \notin \mathcal{U}. \end{cases}$$

Show that $\mu_\mathcal{U} : \mathcal{P}(I) \to \{0,1\}$ *is a finitely additive measure. Conversely, given a finitely additive measure* $\mu : \mathcal{P}(I) \to \{0,1\}$, *show that the family*

$$\mathcal{U}_\mu = \{X \subseteq I \mid \mu(X) = 1\}$$

is an ultrafilter on I. *Moreover, operations* $\mathcal{U} \mapsto \mu_\mathcal{U}$ *and* $\mu \mapsto \mathcal{U}_\mu$ *are one the inverse of the other, that is,* $\mathcal{U}_{\mu_\mathcal{U}} = \mathcal{U}$ *for all* $\mathcal{U}$, *and* $\mu_{\mathcal{U}_\mu} = \mu$ *for all* μ. *Finally, in this correspondence, the* non-principal *ultrafilters correspond to the* non-principal *measures.*

An example of the above correspondence is at hand.

EXAMPLE 11.16. *The non-principal ultrafilter that corresponds to the Alpha-measure* $\mu_\alpha : \mathcal{P}(\mathbb{N}) \to \{0,1\}$ *is the family* $\mathcal{Q}$ *of qualified sets. Indeed,* $A \in \mathcal{Q} \Leftrightarrow \mu_\alpha(A) = 1$.

A useful characterization of ultrafilters can also be given by using the familiar algebraic notion of a maximal ideal.

Let I be any nonempty set, and let

$$\mathfrak{Fun}(I, \mathbb{R}) = \{\varphi \mid \varphi : I \to \mathbb{R}\}$$

be the commutative ring of the real-valued I-*sequences* where sum and product are defined point wise.[3] Recall that a *principal* maximal ideal of $\mathfrak{Fun}(I, \mathbb{R})$ is a maximal ideal of the form

$$\mathfrak{m}_i = \{\varphi \in \mathfrak{Fun}(I, \mathbb{R}) \mid \varphi(i) = 0\}.$$

---

[3] That is, $(\varphi + \psi)(i) = \varphi(i) + \psi(i)$ and $(\varphi \cdot \psi)(i) = \varphi(i) \cdot \psi(i)$.

Denote by $Z(\varphi) = \{i \in I \mid \varphi(i) = 0\}$ the *zero-set* of a sequence $\varphi \in \mathfrak{Fun}(I, \mathbb{R})$.

The following characterization holds:

THEOREM 11.17. *If $\mathcal{U}$ is an ultrafilter on I, then*

$$\mathfrak{m}_{\mathcal{U}} = \{\varphi \mid Z(\varphi) \in \mathcal{U}\}$$

is a maximal ideal of $\mathfrak{Fun}(I, \mathbb{R})$. Conversely, if $\mathfrak{m}$ is maximal ideal of $\mathfrak{Fun}(I, \mathbb{R})$, then

$$\mathcal{U}_{\mathfrak{m}} = \{Z(\varphi) \mid \varphi \in \mathfrak{m}\}$$

is an ultrafilter on I. Moreover, operations $\mathcal{U} \mapsto \mathfrak{m}_{\mathcal{U}}$ and $\mathfrak{m} \mapsto \mathcal{U}_{\mathfrak{m}}$ are one the inverse of the other, i.e. $\mathcal{U}_{\mathfrak{m}_{\mathcal{U}}} = \mathcal{U}$ for every $\mathcal{U}$, and $\mathfrak{m}_{\mathcal{U}_{\mathfrak{m}}} = \mathfrak{m}$ for every $\mathfrak{m}$. Finally, in this correspondence, the principal ultrafilters $\mathcal{U}$ correspond to the principal maximal ideals.

PROOF. Let $\mathcal{U}$ be an ultrafilter on I. First of all, $Z(c_1) = \emptyset \notin \mathcal{U} \Rightarrow c_1 \notin \mathfrak{m}_{\mathcal{U}}$, so $\mathfrak{m}_{\mathcal{U}} \neq \mathfrak{Fun}(I, \mathbb{R})$ is proper.[4] Now notice that $Z(\varphi - \psi) \supseteq Z(\varphi) \cap Z(\psi)$. As a consequence, if $\varphi, \psi \in \mathfrak{m}_{\mathcal{U}}$, that is, if $Z(\varphi), Z(\psi) \in \mathcal{U}$, then also $Z(\varphi - \psi) \in \mathcal{U}$, and hence $\varphi - \psi \in \mathfrak{m}_{\mathcal{U}}$. Similarly, for every $\varphi \in \mathfrak{m}_{\mathcal{U}}$ and for every ϑ, one has that $\vartheta \cdot \varphi \in \mathfrak{m}_{\mathcal{U}}$, because $Z(\vartheta \cdot \varphi) \supseteq Z(\varphi)$. Finally, from the equality $Z(\varphi \cdot \psi) = Z(\varphi) \cup Z(\psi)$, we obtain that $\varphi, \psi \notin \mathfrak{m}_{\mathcal{U}} \Leftrightarrow Z(\varphi), Z(\psi) \notin \mathcal{U} \Leftrightarrow Z(\varphi) \cup Z(\psi) \notin \mathcal{U} \Leftrightarrow \varphi \cdot \psi \notin \mathfrak{m}_{\mathcal{U}}$. This completes the proof that $\mathfrak{m}_{\mathcal{U}}$ is a maximal ideal.

Conversely, let $\mathfrak{m}$ be a maximal ideal of $\mathfrak{Fun}(I, \mathbb{R})$. For every function $\varphi \in \mathfrak{Fun}(I, \mathbb{R})$, denote by:

$$\varphi'(n) = \begin{cases} 0 & \text{if } \varphi(n) = 0; \\ 1 & \text{otherwise.} \end{cases} \qquad \varphi''(n) = \begin{cases} 0 & \text{if } \varphi(n) = 0; \\ 1/\varphi(n) & \text{otherwise.} \end{cases}$$

Notice first that $\varphi \in \mathfrak{m} \Leftrightarrow \varphi' \in \mathfrak{m}$. In fact, by the property of ideal, $\varphi \in \mathfrak{m} \Rightarrow \varphi' = \varphi'' \cdot \varphi \in \mathfrak{m}$ and, conversely, $\varphi' \in \mathfrak{m} \Rightarrow \varphi = \varphi \cdot \varphi' \in \mathfrak{m}$.

The ideal $\mathfrak{m}$ is closed under sums, and so $\varphi, \psi \in \mathfrak{m} \Leftrightarrow \varphi', \psi' \in \mathfrak{m} \Rightarrow \varphi' + \psi' \in \mathfrak{m}$. As a consequence, $\mathcal{U}_{\mathfrak{m}}$ is closed under (finite) intersections, because $Z(\varphi) \cap Z(\psi) = Z(\varphi' + \psi') \in \mathcal{U}_{\mathfrak{m}}$. Now assume that $A \subseteq I$ is such that $A \notin \mathcal{U}_{\mathfrak{m}}$, and consider its characteristic function $\chi = \chi_A$. The neutral element $c_0 = \chi \cdot (c_1 - \chi) \in \mathfrak{m}$. Now, $c_1 - \chi \notin \mathfrak{m}$ because $Z(c_1 - \chi) = X \notin \mathcal{U}_{\mathfrak{m}}$. So, by the maximality of $\mathfrak{m}$, it must be $\chi \in \mathfrak{m}$, and hence $A^c = Z(\chi) \in \mathcal{U}_{\mathfrak{m}}$. Finally, the closure of $\mathcal{U}_{\mathfrak{m}}$ under supersets follows from the above properties. In fact, let $B \supseteq A \in \mathcal{U}_{\mathfrak{m}}$, and assume by contradiction that $B \notin \mathcal{U}_{\mathfrak{m}}$. Then $B^c \in \mathcal{U}_{\mathfrak{m}}$, and so also

---

[4] Recall that by c_r we denote the constant sequence with value $r \in \mathbb{R}$.

$Z(c_1) = \emptyset = B^c \cap A \in \mathcal{U}_\mathfrak{m}$. This is not possible, otherwise the invertible element $c_1 \in \mathfrak{m}$.

The identities $\mathcal{U}_{\mathfrak{m}_\mathcal{U}} = \mathcal{U}$ for every $\mathcal{U}$, and $\mathfrak{m}_{\mathcal{U}_\mathfrak{m}} = \mathfrak{m}$ for every $\mathfrak{m}$, directly follow from the definitions. Finally, it can be directly verified that for every $i \in I$, $\mathcal{U} = \mathcal{U}_i$ is the principal ultrafilter generated by i if and only if $\mathfrak{m}_\mathcal{U} = \mathfrak{m}_i$ is the principal maximal ideal on i. $\square$

Recall that the family of qualified sets in a model of Alpha-Calculus is a non-principal ultrafilter on $\mathbb{N}$ (see Theorem 2.22). As a corollary of the previous proposition, we now obtain that also a converse result holds. Precisely:

THEOREM 11.18. *For every non-principal ultrafilter $\mathcal{U}$ on $\mathbb{N}$, there exists an Alpha-morphism $J : \mathfrak{Fun}(\mathbb{N}, \mathbb{R}) \twoheadrightarrow \mathbb{R}^*$ such that the resulting family of qualified sets $\mathcal{Q} = \mathcal{U}$.*

PROOF. Let $\mathfrak{m} = \mathfrak{m}_\mathcal{U}$ be the non-principal maximal ideal that corresponds to $\mathcal{U}$, and let $J : \mathfrak{Fun}(\mathbb{N}, \mathbb{R}) \twoheadrightarrow \mathbb{R}^*$ be the canonical projection $\varphi \mapsto [\varphi]_\mathfrak{m}$ onto the residue field $\mathbb{R}^* = \mathfrak{Fun}(\mathbb{N}, \mathbb{R})/\mathfrak{m}$, where we identify cosets $[c_r]_\mathfrak{m} = r$ of constant sequences with the corresponding real numbers. Then, by the Characterization Theorem 2.55, such a J is an Alpha-morphism.

We now have to show that the resulting family $\mathcal{Q}$ of qualified sets coincides with $\mathcal{U}$. To this end, for every $A \subseteq \mathbb{N}$, consider its characteristic function χ_A. Then the following equivalences hold:

$$A \in \mathcal{Q} \iff \{n \mid \chi_A(n) = 1\} \in \mathcal{Q} \iff J(\chi_A) = J(c_1) \iff$$
$$\iff \chi_A - c_1 \in \ker(J) = \mathfrak{m} = \mathfrak{m}_\mathcal{U} \iff$$
$$\iff Z(\chi_A - c_1) \in \mathcal{U} \iff A \in \mathcal{U}.$$

$\square$

3. Ultrapowers, the basic examples

Ultrapowers are algebraic constructions based on ultrafilters. They play an important role in *model theory*, a branch of mathematical logic, because they provide a uniform and powerful method to extend structures in such a way that all "elementary" properties be preserved (see *e.g.* the survey by H.J. Keisler [**63**]). In particular, we will see that ultrapowers can be used to characterize the models of Alpha-Calculus and Alpha-Theory.

Given an ultrafilter $\mathcal{U}$ on the set I, consider the following relation $\equiv_{\mathcal{U}}$ on the class of functions with domain I:

$$f \equiv_{\mathcal{U}} g \iff \{i \in I \mid f(i) = g(i)\} \in \mathcal{U}.$$

By the properties of filter, it is readily verified that $\equiv_{\mathcal{U}}$ is an equivalence relation.

As customary in set theory, from this point on we will adopt the exponential notation and write A^I to indicate the set $\mathfrak{Fun}(I, A)$ of all functions $g : I \to A$.

DEFINITION 11.19. *The* ultrapower *of A modulo the ultrafilter $\mathcal{U}$ on the set I is the quotient set:*

$$A^I_{\mathcal{U}} = A^I/\equiv_{\mathcal{U}} = \{[f]_{\mathcal{U}} \mid f : I \to A\}$$

where $[f]_{\mathcal{U}} = \{g \in A^I \mid f \equiv_{\mathcal{U}} g\}$ denotes the $\equiv_{\mathcal{U}}$-equivalence class of f.

When the ultrafilter $\mathcal{U}$ is clear from the context, we will simply write $[f]$.

A set is canonically embedded into any of its ultrapowers.

DEFINITION 11.20. *Let $A^I_{\mathcal{U}}$ be an ultrapower of the set A. The corresponding* diagonal map *(or* canonical embedding)

$$d : A \longrightarrow A^I_{\mathcal{U}}$$

is the function where $d(a) = [c_a]$ is the $\mathcal{U}$-equivalence class of the constant function with value a.

Clearly d is 1-1, so one can identify each element $a \in A$ with its diagonal image $d(a)$. For simplicity, in the sequel we will always directly assume that the ultrapower $A^I_{\mathcal{U}}$ be a superset of A.

The fundamental property of ultrapowers $A^I_{\mathcal{U}}$ is that they preserve any "elementary property" that the ground set A may have.[5] Here we just consider the example given by ordered fields.

THEOREM 11.21. *Let $\mathbb{F} = \langle \mathbb{F}; 0; 1; +; \cdot; \leq \rangle$ be an ordered field, and let $\mathcal{U}$ be an ultrafilter on the set I. On the ultrapower $\mathbb{F}^I_{\mathcal{U}}$, define the operations*

$$[f] \oplus [g] = [f + g] ; \quad [f] \odot [g] = [f \cdot g]$$

---

[5] Here "elementary property" has the precise meaning of a property that is expressed as a *first order formula* in the *complete language* over the ground structure A, that is, in the language that contains one symbol for each constant, function and relation on A. (See, *e.g.*, [**29**, Ch.4].)

and the relation

$$[f] \preceq [g] \iff \{i \in I \mid f(i) \le g(i)\} \in \mathcal{U}.$$

Then $\mathbb{F}^I_{\mathcal{U}} = \langle \mathbb{F}^I_{\mathcal{U}} \, ; \, 0 \, ; \, 1 \, ; \, \oplus \, ; \, \odot \, ; \, \preceq \rangle$ *is an ordered superfield of* $\mathbb{F}$.

PROOF. Notice first that the above definitions of $\oplus, \odot$ are well-posed by the properties of filter. Indeed, if $f_1 \equiv_{\mathcal{U}} f_2$ and $g_1 \equiv_{\mathcal{U}} g_2$, that is, if $\Lambda = \{i \mid f_1(i) = f_2(i)\} \in \mathcal{U}$ and $\Gamma = \{i \mid g_1(i) = g_2(i)\} \in \mathcal{U}$, then clearly $f_1 + g_1 \equiv_{\mathcal{U}} f_2 + g_2$ because

$$\{i \mid f_1(i) + g_1(i) = f_2(i) + g_2(i)\} \supseteq \Lambda \cap \Gamma \in \mathcal{U}.$$

Similarly, $f_1 \cdot g_1 \equiv_{\mathcal{U}} f_2 \cdot g_2$. Finally, it is readily seen that $\oplus$, $\odot$ and $\preceq$ extend $+$, $\odot$ and $\le$, respectively.

The proof that $\mathbb{F}^I_{\mathcal{U}}$ satisfies the properties of a commutative ring, only require the hypothesis that $\mathcal{U}$ be a filter (we leave it as an easy exercise for the reader to check this). On the other hand, the existence of a multiplicative inverse for each $[f] \ne 0$, as well as the linearity of the relation $\preceq$, essentially need the property of ultrafilter: $A \notin \mathcal{U} \iff A^c \in \mathcal{U}$. Here are the details.

Assume that $[f] \ne 0$, that is $\Lambda = \{i \mid f(i) = 0\} \notin \mathcal{U}$. Pick any $g \in \mathbb{F}^I$ such that $g(i) = 1/f(i)$ for all $i \notin \Lambda$. Then

$$\Gamma = \{i \mid f(i) \cdot g(i) = 1\} \in \mathcal{U}$$

because Γ is a superset of the complement Λ^c, which in turn belongs to $\mathcal{U}$, by the property of ultrafilter.

Given $f, g \in \mathbb{F}^I$, the three sets

$$\Lambda_1 = \{i \mid f(i) < g(i)\}; \quad \Lambda_2 = \{i \mid f(i) = g(i)\}; \quad \Lambda_3 = \{i \mid f(i) > g(i)\}$$

form a partition of I. Then, by the properties of ultrafilter, exactly one of them belongs to $\mathcal{U}$, and so exactly one of the following relations holds, as desired: $[f] \prec [g]$, $[f] = [g]$, $[f] \succ [g]$. $\qquad \square$

4. Ultrapowers as models of Alpha-Calculus

Recall that, up to isomorphisms, the models of Alpha-Calculus can be identified with the residue fields $\mathbb{R}^* = \mathfrak{Fun}(\mathbb{N}, \mathbb{R})/\mathfrak{m}$, where Alpha-limit of a real sequence $\lim_{n \uparrow \alpha} \varphi(n) = [\varphi]_{\mathfrak{m}}$ is defined as its $\mathfrak{m}$-coset (see Theorem 2.55). Moreover, as we have seen in Section 2 of this chapter, giving a maximal ideal $\mathfrak{m}$ on the ring of real sequences $\mathfrak{Fun}(\mathbb{N}, \mathbb{R})$ is essentially the same thing as giving an ultrafilter $\mathcal{U}$ on $\mathbb{N}$ (see Theorem 11.17). By putting together these two results, one can characterize the models of Alpha-Calculus by means of ultrapowers $\mathbb{R}^{\mathbb{N}}_{\mathcal{U}}$ of the real numbers.

THEOREM 11.22. *Let $\mathcal{U}$ be a non-principal ultrafilter on $\mathbb{N}$. Then by letting $\mathbb{R}^* = \mathbb{R}^{\mathbb{N}}_{\mathcal{U}}$ and $\lim_{n\uparrow\alpha} \varphi(n) = [\varphi]_{\mathcal{U}}$ for every real sequence φ, one obtains a model of Alpha-Calculus where the family of qualified sets $\mathcal{Q} = \mathcal{U}$.*

Conversely, let a model of Alpha-Calculus be given, and let $\mathcal{Q}$ be the non-principal ultrafilter on $\mathbb{N}$ as given by the family of qualified sets. Then the map $\theta : \mathbb{R}^* \to \mathbb{R}^{\mathbb{N}}_{\mathcal{Q}}$ where $\theta : \lim_{n\uparrow\alpha} \varphi(n) \mapsto [\varphi]_{\mathcal{Q}}$ is an isomorphism of fields.

PROOF. By Proposition 11.21, the ultrapower $\mathbb{R}^{\mathbb{N}}_{\mathcal{U}}$ is an ordered field where the sum and product operations are defined by:

$$[\varphi]_{\mathcal{U}} + [\psi]_{\mathcal{U}} = [\varphi + \psi]_{\mathcal{U}} \text{ and } [\varphi]_{\mathcal{U}} \cdot [\psi]_{\mathcal{U}} = [\varphi \cdot \psi]_{\mathcal{U}}.$$

Recall that we agreed on identifying each diagonal image $d(r) = [c_r]$ with the corresponding real number r, and so $\mathbb{R}^{\mathbb{N}}_{\mathcal{U}}$ is a field that extends the real line. But then, since we let the Alpha-limit of a real sequence to be its equivalence class in the ultrapower, axioms (ACT1) and (ACT3) follow. Moreover, for every $k \in \mathbb{N}$, the set $\{n \in \mathbb{N} \mid \imath(n) = k\} = \{k\} \notin \mathcal{U}$ because $\mathcal{U}$ is non-principal, and so $\lim_{n\uparrow\alpha} \imath(n) = [\imath]_{\mathcal{U}} \neq [c_k]_{\mathcal{U}} = k$ and also (ACT2) is satisfied.

Conversely, by (ACT3), for all real sequences φ, ψ:

$$\theta\left(\lim_\alpha \varphi + \lim_\alpha \psi\right) = \theta\left(\lim_\alpha(\varphi + \psi)\right) = [\varphi + \psi]_{\mathcal{Q}} = [\varphi]_{\mathcal{Q}} + [\psi]_{\mathcal{Q}}$$
$$= \theta\left(\lim_\alpha \varphi\right) + \theta\left(\lim_\alpha \psi\right),$$

and similarly for products. This shows that θ is a field homomorphism. Notice that θ is trivially onto $\mathbb{R}^{\mathbb{N}}_{\mathcal{Q}}$ by definition. Finally, θ is 1-1 because $[\varphi]_{\mathcal{Q}} = [\psi]_{\mathcal{Q}} \Leftrightarrow \{n \in \mathbb{N} \mid \varphi(n) = \psi(n)\} \in \mathcal{Q}$, that is, $\varphi(n) = \psi(n)$ a.e. $\Leftrightarrow \lim_{n\uparrow\alpha} \varphi(n) = \lim_{n\uparrow\alpha} \psi(n)$. $\square$

As a general fact in Alpha-Theory, all hyper-extensions could be identified with ultrapowers modulo the ultrafilter of qualified sets. In fact, the following holds:

EXERCISE 11.23. *For any given nonempty set A, the function*

$$K_A : A^{\mathbb{N}}_{\mathcal{Q}} \longrightarrow A^*$$

mapping every $\varphi(\alpha) \in A^*$ to $[\varphi]_{\mathcal{Q}} \in A^{\mathbb{N}}/\mathcal{Q}$ is a bijection. Moreover, for all $\varphi(\alpha), \varphi(\alpha) \in A^*$, one has $\varphi(\alpha) \in \psi(\alpha) \Leftrightarrow \{n \mid \varphi(n) \in \psi(n)\} \in \mathcal{Q}$.

SOLUTION. The conclusion directly follows from the following equivalences, that hold for all sequences $\varphi, \psi : \mathbb{N} \to A$:

(1) $\varphi(\boldsymbol{\alpha}) = \psi(\boldsymbol{\alpha}) \Leftrightarrow \boldsymbol{\alpha} \in {}^*\{n \mid \varphi(n) = \psi(n)\} \Leftrightarrow$
$\{n \mid \varphi(n) = \psi(n)\} \in \mathcal{Q} \Leftrightarrow [\varphi]_{\mathcal{Q}} = [\psi]_{\mathcal{Q}}$;

(2) $\varphi(\boldsymbol{\alpha}) \in \psi(\boldsymbol{\alpha}) \Leftrightarrow \boldsymbol{\alpha} \in {}^*\{n \mid \varphi(n) \in \psi(n)\} \Leftrightarrow$
$\{n \mid \varphi(n) \in \psi(n)\} \in \mathcal{Q}$.

Indeed, by (1), the definition of K_A is well-posed, in that it does not depend on the representative chosen in the $\mathcal{Q}$-equivalence class; besides, K_A is 1-1. Finally, it directly follows from the definition that K_A is onto A^*. $\square$

CHAPTER 12

The Uniqueness Problem

The goal of this chapter is to investigate the problem of uniqueness of the hyperreal numbers and, more generally, of models of Alpha-Calculus.

1. Isomorphic models of Alpha-Calculus

Let us address the following question, which naturally arises:

Under which conditions are two models of Alpha-Calculus isomorphic?

First of all, we need to make such a notion of "isomorphism" precise. The intuition is clear: two models of the Alpha-Calculus are isomorphic when one is obtained from the other by simply "renaming" the corresponding hyperreal numbers, in such a way that the two notions of Alpha-limit coincide.

DEFINITION 12.1. *We say that the models of Alpha-Calculus as determined by two Alpha-morphisms*

$$J : \mathfrak{Fun}(\mathbb{N}, \mathbb{R}) \twoheadrightarrow \mathbb{R}^* \quad and \quad K : \mathfrak{Fun}(\mathbb{N}, \mathbb{R}) \twoheadrightarrow \mathbb{R}^\star$$

are isomorphic *when there exists a bijection* $\Psi : \mathbb{R}^* \to \mathbb{R}^\star$ *that makes the following diagram commute:*

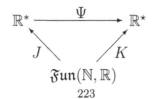

Remark that the above bijection Ψ is a *field isomorphism*. Indeed, given $\xi, \eta \in \mathbb{R}^*$, pick $\varphi, \psi \in \mathfrak{Fun}(\mathbb{N}, \mathbb{R})$ with $\xi = J(\varphi)$ and $\eta = J(\psi)$. Then:

$$\begin{aligned}
\Psi(\xi + \eta) &= \Psi(J(\varphi) + J(\psi)) = \Psi(J(\varphi + \psi)) = K(\varphi + \psi) \\
&= K(\varphi) + K(\psi) = \Psi(J(\varphi)) + \Psi(J(\psi)) = \Psi(\xi) + \theta(\eta),
\end{aligned}$$

and similarly for the product.

The following result formalizes the idea that a model of Alpha-Calculus is entirely described by specifying the family of its qualified sets.

THEOREM 12.2. *The models of Alpha-Calculus as determined by the Alpha-morphisms*

$$J : \mathfrak{Fun}(\mathbb{N}, \mathbb{R}) \twoheadrightarrow \mathbb{R}^* \quad and \quad K : \mathfrak{Fun}(\mathbb{N}, \mathbb{R}) \twoheadrightarrow \mathbb{R}^\star$$

are isomorphic if and only if the corresponding families of qualified sets $\mathcal{Q}_* = \mathcal{Q}_\star$ *coincide.*

PROOF. If $\mathcal{Q}_* = \mathcal{Q}_\star$, then $J(\varphi) = 0 \Leftrightarrow \{n \mid \varphi(n) = 0\} \in \mathcal{Q}_* \Leftrightarrow \{n \mid \varphi(n) = 0\} \in \mathcal{Q}_\star \Leftrightarrow K(\varphi) = 0$. Now let

$$\mathfrak{m} = \{\varphi \mid J(\varphi) = 0\} = \{\varphi \mid K(\varphi) = 0\}$$

be the common kernel of J and K. By the Characterization Theorem 2.55, there exist field isomorphisms θ_1, θ_2 that make the following diagram commute:

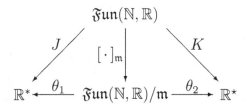

Clearly, $\Psi = \theta_2 \circ \theta_1^{-1} : \mathbb{R}^* \to \mathbb{R}^\star$ is a bijection such that $\Psi \circ J = K$, and so the two models of Alpha-Calculus are isomorphic.

Conversely, assume there exists a bijection $\Psi : \mathbb{R}^* \to \mathbb{R}^\star$ such that $\Psi \circ J = K$. For every $A \subseteq \mathbb{N}$, denote by χ_A its characteristic function. Since Ψ is a field isomorphism, and hence $\Psi(1) = 1$, we have the following chain of equivalences:

$$\begin{aligned}
A \in \mathcal{Q}_* &\Leftrightarrow \{n \mid \chi_A(n) = 1\} \in \mathcal{Q}_* \Leftrightarrow J(\chi_A) = J(c_1) = 1 \\
&\Leftrightarrow K(\chi_A) = \Psi(J(\chi_A)) = \Psi(1) = 1 = K(c_1) \\
&\Leftrightarrow \{n \mid \chi_A(n) = 1\} \in \mathcal{Q}_\star \Leftrightarrow A \in \mathcal{Q}_\star.
\end{aligned}$$

$\square$

2. Equivalent models of Alpha-Calculus

Recall that every non-principal ultrafilter $\mathcal{U}$ on $\mathbb{N}$ is the family of qualified sets in some model of Alpha-Calculus (see Theorem 11.18). In consequence, we have plenty of pairwise non-isomorphic models of Alpha-Calculus, one for each non-principal ultrafilter on $\mathbb{N}$.

We now weaken the notion of isomorphism, aiming to get closer to a uniqueness result. The starting observation is that two models of Alpha-Calculus are indistinguishable when, by suitably renaming the corresponding hyperreal numbers, one can make all hyper-extensions coincide. As it will be shown in the sequel, this may happen even if the families of qualified sets are different.

DEFINITION 12.3. *We say that the models of Alpha-Calculus as determined by two Alpha-morphisms*

$$J : \mathfrak{Fun}(\mathbb{N}, \mathbb{R}) \twoheadrightarrow \mathbb{R}^* \quad and \quad K : \mathfrak{Fun}(\mathbb{N}, \mathbb{R}) \twoheadrightarrow \mathbb{R}^\star$$

are equivalent *if exists a bijection* $\Psi : \mathbb{R}^* \to \mathbb{R}^\star$ *that is coherent with hyper-extensions, that is,*

- *For every $A \subseteq \mathbb{R}^k$ and for every $\xi_1, \ldots, \xi_k \in \mathbb{R}^*$:*

$$(\xi_1, \ldots, \xi_k) \in A^* \iff (\Psi(\xi_1), \ldots, \Psi(\xi_k)) \in A^\star;$$

- *For every $f : \mathbb{R}^k \to \mathbb{R}$ and for every $\xi_1, \ldots, \xi_k, \eta \in \mathbb{R}^*$:*

$$f^*(\xi_1, \ldots, \xi_k) = \eta \iff f^\star(\Psi(\xi_1), \ldots, \Psi(\xi_k)) = \Psi(\eta).$$

Similarly as with isomorphism, also the equivalence between models of Alpha-Calculus can be characterized in terms of qualified sets.

THEOREM 12.4. *The models of Alpha-Calculus as determined by the Alpha-morphisms*

$$J : \mathfrak{Fun}(\mathbb{N}, \mathbb{R}) \twoheadrightarrow \mathbb{R}^* \quad and \quad K : \mathfrak{Fun}(\mathbb{N}, \mathbb{R}) \twoheadrightarrow \mathbb{R}^\star$$

are equivalent if and only if there exists a permutation $\sigma : \mathbb{N} \to \mathbb{N}$ of the natural numbers such that $\Lambda \in \mathcal{Q}_* \iff \sigma(\Lambda) \in \mathcal{Q}_\star$.

PROOF. Assume first that there is a bijection $\sigma : \mathbb{N} \to \mathbb{N}$ such that $\Lambda \in \mathcal{Q}_* \iff \Lambda \in \mathcal{Q}_\star$. For every $\varphi \in \mathfrak{Fun}(\mathbb{N}, \mathbb{R})$, put

$$\Psi : J(\varphi) \longmapsto K(\varphi \circ \sigma^{-1}).$$

We claim that Ψ yields the equivalence we are looking for. For every φ, ψ, the following equivalences hold:

$$J(\varphi) = J(\psi) \iff \Lambda = \{n \mid \varphi(n) = \psi(n)\} \in \mathcal{Q}_*$$
$$\iff \sigma(\Lambda) = \{\sigma(n) \mid \varphi(n) = \psi(n)\}$$
$$= \{m \mid \varphi(\sigma^{-1}(m)) = \psi(\sigma^{-1}(m))\} \in \mathcal{Q}_\star$$
$$\iff K(\varphi \circ \sigma^{-1}) = K(\psi \circ \sigma^{-1}).$$

This shows that the definition of Ψ is well-posed, and that θ is 1-1. Moreover, for every φ, it is $\Psi(J(\varphi \circ \sigma)) = K(\varphi \circ \sigma \circ \sigma^{-1}) = K(\varphi)$, and so Ψ is onto $\mathbb{R}^\star$. As for the coherence with hyper-extensions, let $X \subseteq \mathbb{R}^k$, and let $\xi_i = J(\varphi_i) \in \mathbb{R}^*$ for $i = 1, \ldots, k$. Then

$$(\xi_1, \ldots, \xi_k) \in X^* \iff \Lambda = \{n \mid (\varphi_1(n), \ldots, \varphi_k(n)) \in X\} \in \mathcal{Q}_*$$
$$\iff \sigma(\Lambda) = \{m \mid (\varphi_1(\sigma^{-1}(m)), \ldots, \varphi_k(\sigma^{-1}(m))) \in X\} \in \mathcal{Q}_\star$$
$$\iff (K(\varphi_1 \circ \sigma^{-1}), \ldots, K(\varphi_k \circ \sigma^{-1})) = (\Psi(\xi_1), \ldots, \Psi(\xi_k)) \in X^\star.$$

The compatibility with functions $F : \mathbb{R}^k \to \mathbb{R}$ is proved in an entirely similar manner.

Conversely, given an equivalence $\Psi : \mathbb{R}^* \to \mathbb{R}^\star$, let $\tau \in \mathfrak{Fun}(\mathbb{N}, \mathbb{R})$ be such that $\Psi(J(\imath)) = K(\tau)$, where $\imath(n) = n$ is the identity sequence. Then for every $\Lambda \subseteq \mathbb{N}$:

$$\Lambda \in \mathcal{Q}_* \iff J(\imath) \in \Lambda^* \iff \theta(J(\imath)) = K(\tau) \in \Lambda^\star$$
$$\iff \{n \mid \tau(n) \in \Lambda\} = \tau^{-1}(\Lambda) \in \mathcal{Q}_\star.$$

Also $\Psi^{-1} : \mathbb{R}^\star \to \mathbb{R}^*$ is an equivalence, and so if the sequence σ is such that $J(\sigma) = \Psi^{-1}(K(\imath))$, then for all $\Gamma \subseteq \mathbb{N}$, we have the equivalence

$$\Gamma \in \mathcal{Q}_\star \iff \sigma^{-1}(\Gamma) \in \mathcal{Q}_*.$$

By combining the two equivalences that we just proved, one obtains that

$$\Lambda \in \mathcal{Q}_* \iff (\tau \circ \sigma)^{-1}(\Lambda) \in \mathcal{Q}_*.$$

We claim that the set $A = \{n \mid \tau(\sigma(n)) = n\} \in \mathcal{Q}_*$. If by contradiction $A \notin \mathcal{Q}_*$, then $J(\tau \circ \sigma) \neq J(\imath)$, and by Theorem 6.19 we could pick a set $B \subseteq \mathbb{N}$ such that $J(\imath) \in B^*$ but $J(\tau \circ \sigma) \notin B^*$. Now, $J(\imath) \in B^* \iff B \in \mathcal{Q}_*$, and so also $(\tau \circ \sigma)^{-1}(B) = \{n \mid \tau(\sigma(n)) \in B\} \in \mathcal{Q}_*$. But then we would conclude that $J(\tau \circ \sigma) \in B^*$, a contradiction.

Notice that the restriction $\sigma_{|A}$ is necessarily 1-1. Now split the infinite set $A = A_1 \cup A_2$ into two infinite disjoint subsets; exactly one of them is qualified, say $A_1 \in \mathcal{Q}_*$ and $A_2 \notin \mathcal{Q}_*$. Since both $\mathbb{N} \setminus A_1 \supseteq A_2$ and $\mathbb{N} \setminus \sigma(A_1) \supseteq \sigma(A_2)$ are infinite, we can pick a bijection $\upsilon : \mathbb{N} \to \mathbb{N}$ that extends the restriction $\sigma_{|A_1}$. Notice that $X \in \mathcal{Q}_* \iff X \cap A_1 \in \mathcal{Q}_*$.

Moreover, $\sigma^{-1}(v(X)) \cap A_1 = X \cap A_1$. To verify this, notice that if $x \in X \cap A_1$ then $\sigma(x) = v(x) \in v(X)$, and hence $x \in \sigma^{-1}(v(X))$. Conversely, if $x \in \sigma^{-1}(v(X)) \cap A_1$, then $\sigma(x) \in v(X)$. But $x \in A_1 \Rightarrow \sigma(x) = v(x) \in v(X)$ and hence, since v is a bijection, it must be $x \in X$. The theorem is finally proved, in consequence of the following equivalences:

$$v(X) \in \mathcal{Q}_\star \Leftrightarrow \sigma^{-1}(v(X)) \in \mathcal{Q}_\star$$
$$\Leftrightarrow \sigma^{-1}(v(X)) \cap A_1 = X \cap A_1 \in \mathcal{Q}_\star \Leftrightarrow X \in \mathcal{Q}_\star.$$

$\square$

We remark that also the notion of equivalence is really far from yielding a uniqueness result. To see this, let us first recall the following notion from ultrafilter theory.

DEFINITION 12.5. *Two ultrafilters* $\mathcal{U}, \mathcal{V}$ *on a set* I *are called* isomorphic *if there exists a bijection* $\sigma : I \to I$ *such that* $A \in \mathcal{U} \Leftrightarrow \sigma(A) \in \mathcal{V}$.

Thus the previous theorem says that two models of the Alpha-Calculus are equivalent if and only if the ultrafilters given by the respective families of qualified sets are isomorphic.

REMARK 12.6. *It is a well-known fact that there exist as many ultrafilters on an infinite set* I *as they can possibly be, namely* $2^{2^{|I|}}$*-many.*[1] *For every ultrafilter there are at most* $2^{|I|}$*-many ultrafilters that are isomorphic to it (indeed, at most one for each bijection* $\sigma : I \to I$*), and so there are* $2^{2^{|I|}}$*-many equivalence classes. In consequence, there exist* $2^{\mathfrak{c}}$*-many pairwise non-equivalent models of Alpha-Calculus. We also remark that the situation does not change by adding the* Qualified Set Axiom $(QSA)_Q$. *In fact, for every infinite* $Q \subseteq \mathbb{N}$*, there still exist* $2^{\mathfrak{c}}$*-many non-isomorphic ultrafilters on* $\mathbb{N}$ *that contain* Q.

3. Uniqueness up to countable equivalence

Although we do not have uniqueness in a strict sense, what we will prove next is the following property: "Two models of Alpha-theory cannot be distinguished by means of any countable language." Equivalently, it takes an uncountable language to detect that not all models of the Alpha-theory are equivalent.

---

[1] See, *e.g.*, [**30**, Corollary 7.4].

Let us further weaken our notion of isomorphism by requiring that only countably many hyper-extensions at a time can be considered.

DEFINITION 12.7. *We say that the models of Alpha-Calculus as determined by two Alpha-morphisms*

$$J : \mathfrak{Fun}(\mathbb{N}, \mathbb{R}) \twoheadrightarrow \mathbb{R}^* \quad and \quad K : \mathfrak{Fun}(\mathbb{N}, \mathbb{R}) \twoheadrightarrow \mathbb{R}^\star$$

are countably equivalent *if for every choice of countably many sets and countably many functions*

$$\{A_i \subseteq \mathbb{R}^{n_i} \mid i \in \mathbb{N}\}, \quad \{f_j : \mathbb{R}^{n_j} \to \mathbb{R} \mid j \in \mathbb{N}\}$$

there exists a bijection $\Psi : \mathbb{R}^* \to \mathbb{R}^\star$ *which is coherent with the corresponding hyper-extensions, that is,*

- *For every $i \in \mathbb{N}$ and for every $\xi_1, \ldots, \xi_{n_i} \in \mathbb{R}^*$:*

$$(\xi_1, \ldots, \xi_{n_i}) \in A_i^* \Leftrightarrow (\Psi(\xi_1), \ldots, \Psi(\xi_{n_i})) \in A_i^\star$$

- *For every $j \in \mathbb{N}$ and for every $\xi_1, \ldots, \xi_{n_j}, \eta \in \mathbb{R}^*$:*

$$f_j^*(\xi_1, \ldots, \xi_{n_j}) = \eta \Leftrightarrow f_j^\star(\Psi(\xi_1), \ldots, \Psi(\xi_{n_j})) = \Psi(\eta).$$

We can finally prove the desired uniqueness result.

THEOREM 12.8. *Under the* continuum hypothesis, *all models of Alpha-Calculus are* countably *equivalent.*

Notice that by taking as $A_1 = \{(x, y) \in \mathbb{R} \times \mathbb{R} \mid x < y\}$ the order relation on $\mathbb{R}$, and as $f_1, f_2 : \mathbb{R} \times \mathbb{R}$ the sum and product operations, respectively, one directly obtains the following uniqueness result for the hyperreals.

COROLLARY 12.9. *Under the* continuum hypothesis, *all hyperreal fields of Alpha-Calculus are isomorphic as ordered fields.*

In consequence, when the *continuum hypothesis* is assumed, one can refer to *the* hyperreal field, namely the unique hyperreal field up to isomorphism.

PROOF OF THEOREM 12.8. Recall that $|\mathbb{R}^*| = |\mathbb{R}^\star| = \mathfrak{c}$ (see Proposition 2.38) and so, by the *continuum hypothesis*, we can enumerate $\mathbb{R}^* = \{\xi_\alpha\}_{\alpha < \omega_1}$ and $\mathbb{R}^\star = \{\eta_\alpha\}_{\alpha < \omega_1}$. Since each function $f_j : \mathbb{R}^{n_j} \to \mathbb{R}$ is identified with its graph $\{(x, y) \in \mathbb{R}^{n_j} \times \mathbb{R} \mid f_j(x) = y\} \subseteq \mathbb{R}^{n_j+1}$, we can reduce to consider countable families $\mathcal{F}$ of sets of tuples $A \subseteq \mathbb{R}^n$.

For convenience, without loss of generality, we also assume the family $\mathcal{F}$ to contain all sets $\mathbb{R}^n$ and be closed under *definable sets*, that is, if

$\sigma(x_1, \ldots, x_n, y_1, \ldots, y_k)$ is an elementary formula and $A_1, \ldots, A_k \in \mathcal{F}$, then also

$$\{(x_1, \ldots, x_n) \in \mathbb{R}^n \mid \sigma(x_1, \ldots, x_n, A_1, \ldots, A_k)\} \in \mathcal{F}.$$

Recall that if $\mathcal{F}' = \mathrm{Def}(\mathcal{F})$ is the collection of all sets which are definable with constants in $\mathcal{F}$, then $\mathrm{Def}(\mathcal{F}') = \mathcal{F}'$ and $|\mathcal{F}'| = \max\{|\mathcal{F}|, \aleph_0\}$ (see Proposition 5.15 and Exercise 5.16).

We now proceed by transfinite induction and define two sequences

$$\langle \zeta_\alpha \mid \alpha < \omega_1 \rangle \quad \text{and} \quad \langle \vartheta_\alpha \mid \alpha < \omega_1 \rangle$$

of elements of $\mathbb{R}^*$ and $\mathbb{R}^\star$, respectively, in such a way that the following properties are satisfied:

(1) If $\alpha = \gamma \cdot 2$ is even then $\zeta_\alpha = \xi_\gamma$;
(2) If $\alpha = \gamma \cdot 2 + 1$ is odd then $\vartheta_\alpha = \eta_\gamma$;
(3) $\zeta_\alpha = \zeta_{\alpha'} \Leftrightarrow \vartheta_\alpha = \vartheta_{\alpha'}$ for every $\alpha, \alpha' < \omega_1$;
(4) $(\zeta_{\alpha_1}, \ldots, \zeta_{\alpha_n}) \in A^* \Rightarrow (\vartheta_{\alpha_1}, \ldots, \vartheta_{\alpha_n}) \in A^\star$ for every $A \in \mathcal{F}$ and for every $\alpha_1, \ldots, \alpha_n < \omega_1$.

We remark that the above conditions yield the conclusion. Indeed, by (1) and (2) it follows that $\{\zeta_\alpha \mid \alpha < \omega_1\} = \mathbb{R}^*$ and $\{\vartheta_\alpha \mid \alpha < \omega_1\} = \mathbb{R}^\star$ and so, by condition (3), the correspondence $\Psi : \zeta_\alpha \mapsto \vartheta_\alpha$ yields a bijection $\Psi : \mathbb{R}^* \to \mathbb{R}^\star$. Finally, condition (4) ensures that Ψ is an equivalence relative to the countable family $\mathcal{F}$, as desired. Indeed, if $A \subseteq \mathbb{R}^n$ belongs to $\mathcal{F}$ then also the complement

$$A^c = \{(x_1, \ldots, x_n) \in \mathbb{R}^n \mid (x_1, \ldots, x_n) \notin A\} \in \mathcal{F}$$

because we assumed $\mathcal{F}$ to be closed under definable sets. Then the reverse implications follow by noticing that $(\vartheta_{\alpha_1}, \ldots, \vartheta_{\alpha_n}) \in A^\star \Leftrightarrow (\vartheta_{\alpha_1}, \ldots, \vartheta_{\alpha_n}) \notin (A^c)^\star \Rightarrow (\zeta_{\alpha_1}, \ldots, \zeta_{\alpha_n}) \notin (A^c)^* \Leftrightarrow (\zeta_{\alpha_1}, \ldots, \zeta_{\alpha_n}) \in A^*$.

Given $\beta < \omega_1$, assume by inductive hypothesis that the partial sequences $\langle \zeta_\alpha \mid \alpha < \beta \rangle$ and $\langle \vartheta_\alpha \mid \alpha < \beta \rangle$ have been defined so that conditions (1), (2), (3) and (4) are satisfied. Let us first assume that $\beta = \gamma \cdot 2$ is even (if β is odd the proof is entirely similar; see below).

- Put $\zeta_\beta = \xi_\gamma$.

If we find an element $\vartheta_\beta \in \mathbb{R}^\star$ so that condition (4) is realized, then we are done (notice that condition (4) trivially implies (3)).

We need some notation. For $a = \{m_1 < \ldots < m_k\} \subseteq \{1, \ldots, n\}$, define the function $F_a^n : \mathbb{R} \times \mathbb{R}^k \to \mathbb{R}^n$ by setting

$$F_a^n : (y, x_1, \ldots, x_k) \longmapsto (z_1, \ldots, z_n)$$

where $z_{m_s} = x_s$ for $s = 1, \ldots, k$, and $z_m = y$ for $m \notin a$. This means that $x_1, \ldots, x_k$ are placed in order in the entries that correspond to indexes in a, and y is placed in the remaining entries. *E.g.*, if $a = \{2, 4, 5\}$, then $F_a^7(y, x_1, x_2, x_3) = (y, x_1, y, x_2, x_3, y, y)$.

We remark that every n-tuple $(\lambda_1, \ldots, \lambda_n)$ of elements in $\{\zeta_\alpha\}_{\alpha \leq \beta}$ can be written in the form

$$(\lambda_1, \ldots, \lambda_n) = (F_a^n)^*(\zeta_\beta, \zeta_{\alpha_1}, \ldots, \zeta_{\alpha_k})$$

for suitable $\alpha_1, \ldots, \alpha_k < \beta$, where $a = \{i \leq n \mid \lambda_i \neq \zeta_\beta\}$ and $k = |a|$. So, our goal is to find an element $\vartheta_\beta \in \mathbb{R}^*$ such that

- For every $a \subseteq \{1, \ldots, n\}$ with $|a| = k$, for every $\alpha_1, \ldots, \alpha_k < \beta$, and for every $A \subseteq \mathbb{R}^n$ that belongs to $\mathcal{F}$:

$$(F_a^n)^*(\zeta_\beta, \zeta_{\alpha_1}, \ldots, \zeta_{\alpha_k}) \in A^* \Rightarrow (F_a^n)^*(\vartheta_\beta, \vartheta_{\alpha_1}, \ldots, \vartheta_{\alpha_k}) \in A^*.$$

Now let $\{(A_i, a_i, \vec{\zeta}_i) \mid i \in I\}$ be the set of all triples such that $(F_{a_i}^{n_i})^*(\zeta_\beta, \vec{\zeta}_i) \in A_i^*$, where $A_i \subseteq \mathbb{R}^{n_i}$ belongs to $\mathcal{F}$, $a_i \subseteq \{1, \ldots, n_i\}$, $\vec{\zeta}_i = (\zeta_{\alpha_{i,1}}, \ldots, \zeta_{\alpha_{i,k(i)}})$ with $k(i) = |a_i|$ and $\alpha_{i,1}, \ldots, \alpha_{i,k(i)} < \beta$. For every $i \in I$, let

- $\Gamma_i = \{\eta \in \mathbb{R}^* \mid (F_{a_i}^{n_i})^*(\eta, \vec{\vartheta}_i) \in A_i^*\}$

where $\vec{\vartheta}_i = (\vartheta_{\alpha_{i,1}}, \ldots, \vartheta_{\alpha_{i,k(i)}})$ is the tuple of elements in $\{\vartheta_\alpha\}_{\alpha < \beta}$ that corresponds to $\vec{\zeta}_i$. We reach the conclusion by proving that there exists an element $\vartheta_\beta \in \bigcap_{i \in I} \Gamma_i$.

We remark that the family $\{\Gamma_i \mid i \in I\}$ is countable. Indeed, there are countably many possible sets $A_i \in \mathcal{F}$, and for every $A_i \subseteq \mathbb{R}^{n_i}$ there are finitely many $a_i \subseteq \{1, \ldots, n_i\}$. Moreover, the set $\{\vartheta_\alpha\}_{\alpha < \beta}$ is countable because $\beta < \omega_1$, and so also the possible tuples $\vec{\vartheta}_i$ are countably many. Now, observe that every set Γ_i is internal by the *internal definition principle* (see Theorem 5.17) and so, by *countable saturation* it suffices to show that $\{\Gamma_i \mid i \in I\}$ has the *finite intersection property*. The proof of this fact is not hard, but unfortunately it takes a heavy notation. To illustrate the idea, let us consider first an example.

With $\beta > 3$, assume that $(\zeta_\beta, \zeta_2, \zeta_1, \zeta_\beta) \in A_i^*$ and $(\zeta_2, \zeta_\beta) \in A_j^*$. We want to find an element $\eta \in \Gamma_i \cap \Gamma_j$, where

$$\Gamma_i = \{\eta \in \mathbb{R}^* \mid (\eta, \vartheta_2, \vartheta_1, \eta) \in A_i^*\} \text{ and } \Gamma_j = \{\eta \in \mathbb{R}^* \mid (\vartheta_2, \eta) \in A_j^*\}.$$

Notice that

$$(\zeta_1, \zeta_2) \in \{(x_1, x_2) \in \mathbb{R}^2 \mid \exists z \in \mathbb{R} \ (z, x_2, x_1, z) \in A_i \text{ and } (x_2, z) \in A_j\}^*,$$

where above set is defined by an elementary formula with parameters $\mathbb{R}, A_i, A_j \in \mathcal{F}$, and so it belongs itself to $\mathcal{F}$. Thus we can apply the inductive hypothesis and obtain that

$$(\vartheta_1, \vartheta_2) \in \{(x_1, x_2) \in \mathbb{R}^2 \mid \exists z \in \mathbb{R} \; (z, x_2, x_1, z) \in A_i \text{ and } (x_2, z) \in A_j\}^\star.$$

In consequence, there exists $\eta \in \mathbb{R}^\star$ such that $(\eta, \vartheta_2, \vartheta_1, \eta) \in A_i^\star$ and $(\vartheta_2, \eta) \in A_j^\star$, that is, $\eta \in \Gamma_i \cap \Gamma_j$, as desired.

Let us now consider the general case, and pick finitely many indexes $i_1, \ldots, i_N \in I$. Then

$$(F_{a_{i_s}}^{n_{i_s}})^*(\zeta_\beta, \vec{\zeta}_{i_s}) \in A_{i_s}^* \quad s = 1, \ldots, N.$$

Let $\{\gamma_1 < \ldots < \gamma_M\} \subseteq \beta$ be the finite set that contains all indexes of components of the tuples $\vec{\zeta}_{i_s}$, that is,

$$\{\zeta_{\gamma_1}, \ldots, \zeta_{\gamma_M}\} = \bigcup_{s=1}^{N} \left\{ \zeta_{\alpha_{i_s,1}}, \ldots, \zeta_{\alpha_{i_s,k(i_s)}} \right\}.$$

For every $\tau : \{1, \ldots, h\} \to \{1, \ldots, m\}$, denote by $G_\tau : \mathbb{R}^m \to \mathbb{R}^h$ the function that re-arrange components according to τ, that is

$$G_\tau : (x_1, \ldots, x_m) \mapsto (y_1, \ldots, y_h) \quad \text{where} \quad y_j = x_{\tau(j)} \quad \text{for } j = 1, \ldots, h.$$

We remark that all such functions G_τ are definable with set of constants $\{\mathbb{R}^n \mid n \in \mathbb{N}\}$, and hence they all belong to $\mathcal{F}$. For every $s = 1, \ldots, N$, let $\tau_s : \{1, \ldots, k(i_s)\} \to \{1, \ldots, M\}$ be such that $\zeta_{\alpha_{i_s,j}} = \zeta_{\gamma_{\tau_s(j)}}$ for all j, so that $G_{\tau_s}^*(\vec{\zeta}) = \vec{\zeta}_{i_s}$, where we denoted $\vec{\zeta} = (\zeta_{\gamma_1}, \ldots, \zeta_{\gamma_M})$. By the hypothesis we have that

$$\vec{\zeta} \in \left\{ (x_1, \ldots, x_M) \in \mathbb{R}^M \;\middle|\; \exists z \in \mathbb{R} \; \bigwedge_{s=1}^{N} (F_{a_{i_s}}^{n_{i_s}})(z, G_{\tau_s}(x_1, \ldots, x_M)) \in A_{i_s} \right\}^*.$$

The above set is definable with constants in $\mathcal{F}$, and so it belongs itself to $\mathcal{F}$. Thus we can apply the inductive hypothesis and obtain that

$$\vec{\vartheta} \in \left\{ (x_1, \ldots, x_M) \in \mathbb{R}^M \;\middle|\; \exists z \in \mathbb{R} \; \bigwedge_{s=1}^{N} (F_{a_{i_s}}^{n_{i_s}})(z, G_{\tau_s}(x_1, \ldots, x_M)) \in A_{i_s} \right\}^*,$$

where $\vec{\vartheta} = (\vartheta_{\gamma_1}, \ldots, \vartheta_{\gamma_M})$ is the tuple of elements of $\mathbb{R}^\star$ that correspond to $\vec{\zeta}$. In particular, there exists an element $\eta \in \mathbb{R}^\star$ such that $(F_{a_{i_s}}^{n_{i_s}})^*(\eta, G_{\tau_s}^*(\vec{\eta})) = (F_{a_{i_s}}^{n_{i_s}})^*(\eta, \vec{\vartheta}_{i_s}) \in A_{i_s}^\star$ for all $s = 1, \ldots, N$, that is, $\eta \in \bigcap_{i=1}^{N} \Gamma_{i_s}$, as desired. $\qquad\square$

REMARK 12.10. *The above Theorem 12.8 could be proved as a straight corollary of an important result in model theory, namely the fact that elementarily equivalent κ-saturated structures of equal cardinality κ are isomorphic, provided the language contains less than κ-many symbols.*[2] *In our case, all models of Alpha-Calculus are elementarily equivalent by transfer, they have cardinality $\mathfrak{c}$, and they are countably saturated (that is, $\aleph_1$-saturated). Since we are assuming the* continuum hypothesis $\mathfrak{c} = \aleph_1$, *and since the language we are considering is countable, Theorem 12.8 follows.*

In this chapter we focused on results that one *can prove* about uniqueness of models of Alpha-Theory; further on we will also see uniqueness properties that one *cannot disprove* (see 14.3).[3]

---

[2] See, *e.g.*, [**29**, Ch.5 §1].

[3] By saying that a property "cannot be disproved" we mean that its negation cannot be proved by the usual principles of mathematics, as formalized by the axioms of Zermelo-Fraenkel set theory ZFC. These matters will be discussed in Section 2 of this chapter.

CHAPTER 13

Alpha-Theory and Nonstandard Analysis

A basic fact in *model theory*, an important branch of mathematical logic, is that every infinite mathematical structure admits *nonstandard models*, that is, non-isomorphic structures that satisfy the same elementary (that is, first-order) properties of the starting structure. In other words, one can construct plenty of "almost isomorphic" structures that cannot be distinguished by only looking at their elementary properties.

At the end of the 50 years of the last century, by considering nonstandard models of the real numbers, Abraham Robinson was able to put the use of (actual) *infinitesimal numbers* on firm foundations, thus giving one possible solution to a century-old problem in mathematics. Developing that idea, and considering nonstandard models of arbitrary mathematical structures, he then presented a general theory called *nonstandard analysis* (see the book [**83**]).

We like to stress that the methods of nonstandard analysis do not produce *nonstandard* mathematics to be contrasted with *standard* mathematics. On the contrary, they provide powerful tools that are potentially applicable to many fields of mathematics, and whose strength is probably still far from being fully exploited. For this reason, many mathematicians (including the authors) now prefer to use the more general name *nonstandard methods*.

In this chapter we will briefly review the basic notions and terminology of nonstandard analysis, and show that Alpha-Theory is a more general theory, because it has a built-in model of nonstandard analysis.

1. Nonstandard analysis, a quick presentation

The theory of nonstandard analysis has been repeatedly extended and refined by several authors throughout the last 60 years, and the current introductions found in the literature are quite different from the original one.[1] The goal of this section is to give a quick introduction to nonstandard methods, according to the most widely used approach grounded on *superstructures*.[2] To this end, we will recall the fundamental notions and terminology, also aiming to make the reader ready to examine the extensive literature of nonstandard analysis. In regard to this, there are several good introductory monographs that are recommendable, including the classic book by K.D. Stroyan and W.A.J. Luxemburg [**90**], the books by H.J. Keisler [**60, 62**], or the more recent lecture notes by R. Goldblatt [**52**]. Updated reviews of the applications of the methods of nonstandard analysis across the whole spectrum of mathematics can be found in the volumes [**3, 34, 72**].

The basic ingredient of nonstandard analysis is the *star-map*, denoted by an asterisk $*$, that associates to every mathematical object A under study a "nonstandard extension" A^* in such a way that all "elementary properties" are preserved.

DEFINITION 13.1. *A* superstructure model *of nonstandard analysis is a triple* $\langle *; V(X); V(Y) \rangle$ *where:*

(1) $V(X)$ *and* $V(Y)$ *are the superstructures over the infinite sets of atoms* X *and* Y, *respectively;*

(2) $* : V(X) \to V(Y)$ *is a map where it is denoted* $*(A) = A^*$;

(3) *The following non-triviality condition holds: For every infinite set* $A \in V(X)$, *the inclusion* $\{a^* \mid a \in A\} \subsetneq A^*$ *is proper;*

(4) *The following* transfer principle *(also called* Leibniz principle*) holds: For every elementary formula* $\sigma(x_1, \ldots, x_k)$ *and for every* $a_1, \ldots, a_k \in V(X)$:

$$\sigma(a_1, \ldots, a_k) \iff \sigma(a_1^*, \ldots, a_k^*).$$

Without loss of generality, it is also usually assumed that $\mathbb{N} \subseteq X$, *that* $X^* = Y$, *and that* $x^* = x$ *for all* $x \in X$.

---

[1] Alpha-Theory itself could be used as an introduction to the methods of nonstandard analysis. In fact, as it will be shown in the next section §2 of this chapter, Alpha-Theory incorporates the full strength of nonstandard methods with countable saturation.

[2] The superstructure approach to nonstandard analysis was introduced by A. Robinson and E. Zakon in [**84**]. See also [**29**, §4.4] for a detailed exposition.

Below, we list the basic terminology as usually adopted in the literature of nonstandard analysis.

- $V(X)$ is called the *standard universe* ;
- $V(Y)$ is called the *nonstandard universe* ;
- The function $*$ is called the *nonstandard extension map*, or the *enlarging map*, or the *star-map* ;
- A^* is called the *nonstandard extension* of A, or the *hyper-extension* of A, or the *star* of A ;
- Elements in $V(X)$ are called *standard*, and elements in $V(Y)$ are called *nonstandard* ;
- Nonstandard elements are distinguished into three classes:
  - The sets that are nonstandard extensions A^* are called *standard* sets[3] ;
  - Elements $b \in A^*$ that belong to some nonstandard extension are called *internal*.
  - Elements $b \in V(Y)$ that are not internal, are called *external*.

Besides the star-map and the transfer principle, a third ingredient that is often used in nonstandard analysis is the saturation principle.

DEFINITION 13.2. *For an infinite cardinal κ, the κ-saturation principle is following property:*

- *Let $\mathcal{F} \subseteq A^*$ be a family of internal subsets of some nonstandard extension A^*, and suppose $|\mathcal{F}| < \kappa$. If $\mathcal{F}$ has the finite intersection property then $\bigcap_{F \in \mathcal{F}} F \neq \emptyset$.*

The $\aleph_0$-saturation property trivially holds in any model of nonstandard analysis, because a finite intersection of internal sets is internal. The $\aleph_1$-saturation property is usually called *countable saturation*, since it applies to countable families.

Roughly, the strategy of nonstandard analysis to prove (or disprove) a conjecture P about some mathematical structure S is the following.

---

[3] There is ambiguity here, as an element $A \in V(X)$ of the standard universe and its nonstandard extension $A^* \in V(Y)$ are both called *standard*. For this reason, some authors specify the name *internal-standard* for elements of the form A^*.

(1) Formalize the property P as an elementary (that is, first-order) formula σ, and consider a superstructure model of nonstandard analysis where the standard universe is the superstructure $V(S)$ (without loss of generality, we can assume that S be a set of atoms);

(2) Establish whether σ is true or not in the nonstandard universe $V(S^*)$, where additional tools such as infinitesimal numbers and saturation principle are available.

(3) Once the property P, as formalized by σ, has been proved (or disproved) in $V(S^*)$, by *backward transfer* one concludes that the property P about S is true (or false).

In some sense, the use of nonstandard models to prove "standard" theorems resembles the use of complex analysis in number theory to prove results about natural numbers. Even for those who are only interested in natural numbers and their properties complex analysis is important, at least as a tool to carry out proofs.

2. Alpha-Theory is more general than nonstandard analysis

In this section we will show that Alpha-Theory is more general than nonstandard analysis, because it has a built-in superstructure model $\langle *; V(\mathbb{R}); V(\mathbb{R}^*) \rangle$. To see this, we need the following fact about Alpha-limits and levels of the superstructures over $\mathbb{R}$ and $\mathbb{R}^*$.

LEMMA 13.3. *For every k, a sequence $\varphi(n) \in V_k(\mathbb{R})$ a.e. if and only if $\lim_{n \uparrow \alpha} \varphi(n) \in V_k(\mathbb{R}^*)$. In consequence, $V_k(\mathbb{R})^* \subseteq V_k(\mathbb{R}^*)$.*

PROOF. We proceed by induction. It directly follows from the *Field Axiom* that $\varphi(n) \in V_0(\mathbb{R}) = \mathbb{R}$ a.e. if and only if $\lim_{n \uparrow \alpha} \varphi(n) \in \mathbb{R}^* = V_0(\mathbb{R}^*)$, and so the basis $k = 0$ is proved. At the inductive step, we have that $\varphi(n) \in V_{k+1}(\mathbb{R}) = V_k(\mathbb{R}) \cup \mathcal{P}(V_k(\mathbb{R}))$ a.e. if and only if either $\varphi(n) \in V_k(\mathbb{R})$ a.e. or $\varphi(n) \subseteq V_k(\mathbb{R})$ a.e.. By the inductive hypothesis, the first case is equivalent to $\lim_{n \uparrow \alpha} \varphi(n) \in V_k(\mathbb{R}^*)$; and by also using the *Internal Set Axiom*, one shows that the latter case is equivalent to $\lim_{n \uparrow \alpha} \varphi(n) \subseteq V_k(\mathbb{R}^*)$. So, we conclude that $\varphi(n) \in V_{k+1}(\mathbb{R})$ a.e. if and only if either $\lim_{n \uparrow \alpha} \varphi(n) \in V_k(\mathbb{R}^*)$ or $\lim_{n \uparrow \alpha} \varphi(n) \subseteq V_k(\mathbb{R}^*)$, which in turn is equivalent to $\lim_{n \uparrow \alpha} \varphi(n) \in V_{k+1}(\mathbb{R}^*)$. $\square$

THEOREM 13.4. *Assume the axioms AT of Alpha-Theory, and let $* : A \mapsto A^*$ be the corresponding hyper-extension map. By restricting $*$ to the superstructure over the real numbers one obtains a superstructure* *model of nonstandard analysis $\langle V(\mathbb{R}), V(\mathbb{R}^*), * \rangle$ where the internal elements (in the sense of nonstandard analysis) are precisely the internal objects of $V(\mathbb{R}^*)^4$, and where $\aleph_1$-saturation holds.*

PROOF. For every $A \in V(\mathbb{R})$, pick k such that $A \in V_k(\mathbb{R})$. Then $A^* \in V_k(\mathbb{R})^* \subseteq V_k(\mathbb{R}^*) \subseteq V(\mathbb{R}^*)$, by the previous lemma. This shows that the hyper-extension map $* : V(\mathbb{R}) \to V(\mathbb{R}^*)$, and hence properties (1) and (2) of Definition 13.1 are fulfilled. Besides, property (3) is given by Proposition 4.24, and property (4) by Corollary 5.9.

Let us now show that the two notions of internal elements coincide. On the one hand, if b is internal in the sense of nonstandard analysis, then $b \in A^*$ for a suitable $A \in V(\mathbb{R})$. Pick k such that $A \subseteq V_k(\mathbb{R})$. Then $b \in A^* \subseteq V_k(\mathbb{R})^* \subseteq V_k(\mathbb{R}^*)$ is an element of the "nonstandard" superstructure $V(\mathbb{R}^*)$ which is an Alpha-limit, since it belongs to the hyper-image A^*. Conversely, let the Alpha-limit $\lim_{n \uparrow \alpha} \varphi(n)$ belong to $V(\mathbb{R}^*)$, and so $\lim_{n \uparrow \alpha} \varphi(n) \in V_k(\mathbb{R}^*)$ for a suitable k. Then, by the previous lemma, $\varphi(n) \in V_k(\mathbb{R})$ a.e., and hence $\lim_{n \uparrow \alpha} \varphi(n) \in V_k(\mathbb{R})^*$. Since $V_k(\mathbb{R}) \in V(\mathbb{R})$ belongs to the "standard" superstructure, this proves that $\lim_{n \uparrow \alpha} \varphi(n)$ is internal in the sense of nonstandard analysis.

Finally, $\aleph_1$-saturation is given by *countable saturation* (see Theorem 6.8). □

We close this section by itemizing several comments and remarks about a comparison between the Alpha-Theory and nonstandard analysis.

- As shown by Theorem 13.4, the Alpha-Theory is more general than nonstandard analysis with countable saturation.
- The usual approaches to nonstandard analysis are grounded on the existence of two distinct universes, namely the "standard" universe and the "nonstandard" universe.[5] In the Alpha-Theory there is no such distinction, as one works in a single universe, namely the universe of all mathematical objects; in a way, every object is "standard", including hyper-images A^*.
- Besides the usual mathematical objects, in the universe of the Alpha-Theory one finds new entities with respect to classic mathematics, namely the Alpha-limits. In this sense, the Alpha-Theory is more general than traditional mathematics.

---

[4] Recall that in the Alpha-Theory, *internal* objects were defined as Alpha-limits (see Definition 4.52).

[5] A construction of a superstructure model $\langle V(X), V(X), * \rangle$ where the standard and the nonstandard universe coincide was given in [**7**].

- In nonstandard analysis, the primary notion is that of a non-standard extension A^*, and internal objects are thereby defined as elements of nonstandard extensions. On the contrary, in Alpha-Theory the primary notion is that of Alpha-limit = internal object, and hyper-images A^* are defined as those special Alpha-limits that are obtained by taking constant sequences.

- One can consider hyper-extensions of any object under study, including the "non-classic" objects obtained as Alpha-limits. In consequence, one can iterate hyper-images and consider, e.g., the hyper-hypernatural numbers $\mathbb{N}^{**}$, or the hyper-image $[1, H]^*$ of a hyperfinite interval $[1, H] \subset \mathbb{N}^*$.

- The Alpha-Theory is grounded on the notion of Alpha-limit which is, in our opinion, a more intuitive concept with respect to that of a nonstandard extension. Indeed, the Alpha-limit conforms to the naïve idea of limit as a process that preserves "elementary properties".

- The Alpha-Theory consists in a small number of axioms that are all expressed in a simple familiar language. This allows us to unfold the theory in an elementary fashion, with no explicit need of tools from mathematical logic. Only later, one can reach a full development of the theory by proving the *transfer principle*, the logical property that formalizes the intuitive idea that "all elementary properties are preserved under Alpha-limits."

3. Remarks and comments

1. In *nonstandard analysis*, a typical use of κ-saturation (with uncountable κ) is done in topology. Assume that a Hausdorff topological space (X, τ) has *character* ν; this means that every point $x \in X$ has a base of neighborhoods $\mathcal{N}_x$ of cardinality at most ν. The family of internal sets $\{U^* \mid U \in \mathcal{N}_x\}$ has the finite intersection property and so, if we assume κ^+-saturation (κ^+ is the successor cardinal of κ), the following intersection is nonempty:

$$\mu(x) \;=\; \bigcap_{U \in \mathcal{N}_x} U^* \neq \emptyset.$$

In the literature, $\mu(x)$ is called the *monad* of x. By the Hausdorff property, it is proved that $\mu(x) \cap \mu(y) = \emptyset$ whenever $x \neq y$.

Monads are the basic ingredient of nonstandard methods in topology, grounding on the following characterizations:[6]

- $A \subseteq X$ is *open* if and only if $\mu(a) \subseteq A^*$ for every $x \in A$;
- $C \subseteq X$ is *closed* if and only if $\mu(x) \cap C^* = \emptyset$ for every $x \notin C$;
- $K \subseteq X$ is *compact* if and only if $K^* \subseteq \bigcup_{x \in K} \mu(x)$.

2. Sometimes in *nonstandard analysis*, the following intersection property is considered:

- κ-*enlarging property*: Let $\mathcal{F} \subseteq A$ be a family of subsets of some set A, and suppose $|\mathcal{F}| < \kappa$. If $\mathcal{F}$ has the finite intersection property then $\bigcap_{F \in \mathcal{F}} F^* \neq \emptyset$.

Clearly, the κ-*enlarging property* is implied by κ-*saturation* (just consider the family of internal sets $\mathcal{F}^\sigma = \{F^* \mid F \in \mathcal{F}\}$), but the converse implication does not hold. Indeed, for every uncountable κ, there exist models of nonstandard analysis that satisfies the κ-enlarging property, but where κ-saturation fails.

Notice that the κ^+-*enlarging property* suffices to prove that the *monads* $\mu(x)$ as considered in the previous item are nonempty.

3. The original presentation of *nonstandard analysis* by A. Robinson [83] made a heavy use of the logical formalism, and appeared not directly usable by many mathematicians without a good background in mathematical logic. Since the lecture notes [73] that W.A.J. Luxemburg wrote in the early sixties of the last century, several alternative approaches have been proposed in order to simplify matters. In a way, also our Alpha-Theory can be considered as one of those.

The first relevant contribution aimed to make the tools of nonstandard analysis available even at a freshman level is H.J. Keisler's college textbook [60]. Among the more recent works, there are the "gentle introduction" proposed by C.W. Henson [54], K.D. Stroyan's college textbook [91], and several alternative approaches which have been elaborated by the authors jointly with M. Forti (see the survey [13]).

4. Currently, most practitioners of *nonstandard analysis* adopt the *superstructure approach*, but a consistent group of mathematicians follow instead E. Nelson's *Internal Set Theory* IST [77].

---

[6] Proofs can be found, *e.g.*, in [70, Ch.III].

Roughly, the superstructure approach formalizes an *external* viewpoint, grounded on the existence of a star-map $*$ defined on the "standard" universe, which preserves all "elementary properties" of mathematical objects. Each "standard" set A is isomorphically embedded in its "nonstandard extension" A^*. Also the Alpha-Theory follows this external viewpoint, with the additional feature that *all* mathematical objects are indeed "standard", so that the star-map can be applied to *all* objects.

In the *internal* approach of IST, sets are identified with their "nonstandard extensions", and a predicate st is used in order to distinguish the "standard" elements. For instance, in IST the set of real numbers $\mathbb{R}$ already contains the infinitesimal numbers, the "standard" real numbers being those elements $r \in \mathbb{R}$ such that $\mathrm{st}(r)$. An axiom guarantees that the sub-universe of "standard" elements satisfy the same elementary properties as the whole universe. A shortcoming of IST is that "external" collections (*e.g.*, the set of infinitesimals) are not objects of the universe. As a consequence, there are foundational difficulties to formalize certain important constructions used in *nonstandard analysis*, the most relevant example being the *Loeb measure*.

A comparison between the external and internal viewpoints and their underlying philosophies, as well as a discussion on the common ground shared by the two approaches, can be found, *e.g.* in the papers [**56, 40**]. In particular, in [**40**] an explicit interpretation of IST in a *superstructure* is presented, thus supplying a bridge to connect followers of the two schools.

CHAPTER 14

Alpha-Theory as a Nonstandard Set Theory

In this chapter we formalize the Alpha-Theory as an axiomatic non-standard set theory, and investigate its foundational strength relative to the usual principles of mathematics, as axiomatized by Zermelo-Fraenkel set theory ZFC. We will also study the strength of *Cauchy Infinitesimal Principle* (CIP), and show that it is an independent property.

Throughout the chapter, we will assume the reader to be acquainted with the fundamental notions and results of set theory and model theory.[1] In particular, let us recall the following:

- *Completeness Theorem*: A first-order theory T *proves* a sentence σ if and only if $\mathfrak{M} \models \sigma$ (read: "σ holds in $\mathfrak{M}$") for every model $\mathfrak{M}$ of T. In this case we write T $\models \sigma$.

- A first-order theory T is *consistent* if there exists a model $\mathfrak{M} \models$ T, that is, if $\mathfrak{M} \models \sigma$ for every σ in T. A theory that is not consistent is called *contradictory*.

- Two first-order theories are *equiconsistent* if the consistency of one implies the consistency of the other (so, they are either both consistent or both contradictory).

- A mathematical property (formalized by the sentence) σ is called *consistent* relative to ZFC if its negation $\neg\sigma$ cannot be proved by ZFC; equivalently, if there exists a model of ZFC where σ holds.[2]

- A mathematical property σ is called *independent* if both σ and its negation $\neg\sigma$ are consistent relative to ZFC.

---

[1] Classic references are [**58, 65**] for set theory, and [**29**] for model theory.

[2] Here it is implicitly assumed that the theory ZFC be consistent.

1. The axioms of **AST**

As it was already pointed out in Remark 4.1, the axioms of the Alpha-Theory **AT** are really general in nature, as they rule properties of a notion of limit which is defined for *all* sequences of mathematical objects. In consequence, for a fully rigorous treatment, one needs to specify the underlying foundational theory of mathematics. The starting observation here is that the theory **ZFC** is not directly suitable to this purpose for the following two reasons:[3]

- The axioms of **AT** take into account the existence of *atoms*, that is, of primitive mathematical objects that are not set. On the contrary, **ZFC** is a "pure" set theory where every object is a set and atoms do not exist.

- The axioms of **AT** imply the existence of infinite $\in$-descending chains:

$$x_1 \ni x_2 \ni \ldots \ni x_n \ni x_{n+1} \ni \ldots .$$

(See Proposition 14.4.) On the contrary, the axiom of *Regularity* prevents the existence of such chains in **ZFC**.

We remark that the actual existence of atoms is not essential in formulating the axioms of the Alpha-Theory; indeed, all we need is the set of real numbers, and it is a well-known fact that the natural numbers $\mathbb{N}$ and the real numbers $\mathbb{R}$ can be constructed in **ZFC** as suitable sets. In consequence, a solution to the above problems is obtained by simply dropping the axiom of *Regularity* (that is never used in the construction of the real numbers), and by assuming

$$\mathsf{ZFC}_0 \ = \ \mathsf{ZFC} \setminus \{\text{Regularity}\}$$

as the underlying set theory of **AT**.

We are now ready to reformulate the Alpha-Theory as a general "pure" set theory.

DEFINITION 14.1. Alpha Set Theory *AST is the axiomatic theory whose axioms are the following:*

- **ZFC**$_0$, *the axioms of Zermelo-Fraenkel set theory with* Choice *but without* Regularity *;*
- (AST1) Existence Axiom. *Every sequence $\varphi(n)$ has a unique "Alpha-limit" $\lim_{n \uparrow \alpha} \varphi(n)$;*

---

[3] Throughout this section we will assume familiarity with *Zermelo-Fraenkel set theory* with *Choice* ZFC (a classic reference is T. Jech's book [**58**].)

- (AST2) **Alpha Number Axiom.** *The Alpha-limit of the identity sequence $\imath(n) = n$ is a "new" number denoted by $\boldsymbol{\alpha}$, that is $\lim_{n\uparrow\alpha} n = \boldsymbol{\alpha} \notin \mathbb{N}$;*
- (AST3) **Internal Set Axiom.** *If $c_{\emptyset}(n) = \emptyset$ is the constant sequence with value the empty set, then $\lim_{n\uparrow\alpha} c_{\emptyset}(n) = \emptyset$. If $\psi(n)$ is a sequence of nonempty sets, then*

$$\lim_{n\uparrow\alpha} \psi(n) = \left\{ \lim_{n\uparrow\alpha} \varphi(n) \,\Big|\, \varphi(n) \in \psi(n) \text{ for all } n \right\}.$$

- (AST4) **Composition Axiom.** *If two sequences take the same Alpha-limit $\lim_{n\uparrow\alpha} \varphi(n) = \lim_{n\uparrow\alpha} \psi(n)$, then $\lim_{n\uparrow\alpha} f(\varphi(n)) = \lim_{n\uparrow\alpha} f(\psi(n))$ for all functions f such that the compositions $f \circ \varphi$ and $f \circ \varphi'$ are defined.*
- (AST5) **Pair Axiom.** *If $\vartheta(n) = \{\varphi(n), \psi(n)\}$ for all n then $\lim_{n\uparrow\alpha} \vartheta(n) = \{\lim_{n\uparrow\alpha} \varphi(n), \lim_{n\uparrow\alpha} \psi(n)\}$.*

The *Pair Axiom* is a convenient replacement for the *Field Axiom*.[4] Indeed, the two properties are equivalent.

PROPOSITION 14.2. *Assume axioms (AST1), (AST2), (AST3) and (AST4). Then the following properties are equivalent:*

(1) *(AST5) Pair Axiom.*

(2) *Field Axiom: The set of all Alpha-limits of real sequences*

$$\mathbb{R}^* = \left\{ \lim_{n\uparrow\alpha} \varphi(n) \,\Big|\, \varphi : \mathbb{N} \to \mathbb{R} \right\}$$

is an ordered field, called the hyperreal field, *where:*

- $\lim_{n\uparrow\alpha} \varphi(n) + \lim_{n\uparrow\alpha} \psi(n) = \lim_{n\uparrow\alpha}(\varphi(n) + \psi(n))$
- $\lim_{n\uparrow\alpha} \varphi(n) \cdot \lim_{n\uparrow\alpha} \psi(n) = \lim_{n\uparrow\alpha}(\varphi(n) \cdot \psi(n))$

PROOF. Assume first that the *Field Axiom* holds. In particular, all properties proved by Alpha-Calculus also hold. Now let $\vartheta(n) = \{\varphi(n), \psi(n)\}$ for all n. That $\lim_{n\uparrow\alpha} \varphi(n), \lim_{n\uparrow\alpha} \psi(n) \in \lim_{n\uparrow\alpha} \vartheta(n)$ directly follows by the *Internal Set Axiom*. Conversely, again by the *Internal Set Axiom*, every $\xi \in \lim_{\alpha} \vartheta$ is of the form $\xi = \lim_{n\uparrow\alpha} \xi(n)$ where $\xi(n) \in \vartheta(n)$ for all n. This means that for all n one has that $\xi(n) = \varphi(n)$ or $\xi(n) = \psi(n)$. Then, by Proposition 2.5, it must be $\lim_{n\uparrow\alpha} \xi(n) = \lim_{n\uparrow\alpha} \varphi(n)$ or $\lim_{n\uparrow\alpha} \xi(n) = \lim_{n\uparrow\alpha} \psi(n)$, as desired.

---

[4] Here by "convenient" we mean "more suitable as an axiom of a set theory". In fact, the *Field Axiom* assumes the algebraic notion of an ordered field; on the contrary, the *Pair Axiom* has a really simple formulation in elementary set-theoretic terms.

Let us now turn to the converse implication and assume the *Pair Axiom*. For simplicity, in the following we will identify the neutral elements $0, 1 \in \mathbb{R}$ with the Alpha-limits in $\mathbb{R}^*$ of the corresponding constant sequences, namely with $\lim_{n \uparrow \alpha} c_0(n)$ and $\lim_{n \uparrow \alpha} c_1(n)$, respectively.[5]

With the only exception of the existence of inverses, all field properties of $\mathbb{R}^*$ directly follow from the definitions of the sum and product operations on $\mathbb{R}^*$ and from the corresponding properties of $\mathbb{R}$. *E.g.*, *associativity* is proved by the following equalities:

$$\lim_{n \uparrow \alpha} \varphi(n) \cdot \left(\lim_{n \uparrow \alpha} \psi(n) + \lim_{n \uparrow \alpha} \vartheta(n) \right) = \lim_{n \uparrow \alpha} \varphi(n) \cdot \lim_{n \uparrow \alpha} (\psi(n) + \vartheta(n))$$

$$= \lim_{n \uparrow \alpha} (\varphi(n) \cdot (\psi(n) + \vartheta(n))) = \lim_{n \uparrow \alpha} (\varphi(n) \cdot \psi(n) + \varphi(n) \cdot \vartheta(n))$$

$$= \lim_{n \uparrow \alpha} \varphi(n) \cdot \lim_{n \uparrow \alpha} \psi(n) + \lim_{n \uparrow \alpha} \varphi(n) \cdot \lim_{n \uparrow \alpha} \vartheta(n).$$

As for the existence of inverses, given a real sequence $\varphi(n)$ such that $\lim_{n \uparrow \alpha} \varphi(n) \neq 0$, define:

$$\psi(n) = \begin{cases} \frac{1}{\varphi(n)} & \text{if } \varphi(n) \neq 0 \\ 1 & \text{otherwise.} \end{cases}$$

Notice that $\varphi \cdot \psi = \chi_A$ where χ_A is the characteristic function of the set $A = \{ n \in \mathbb{N} \mid \varphi(n) \neq 0 \}$. If we prove that $\lim_{n \uparrow \alpha} \chi_A(n) = 1$ then we are done, because in this case

$$\lim_{n \uparrow \alpha} \varphi \cdot \lim_{n \uparrow \alpha} \psi(n) = \lim_{n \uparrow \alpha} (\varphi(n) \cdot \psi(n)) = 1 ,$$

and hence $\lim_{n \uparrow \alpha} \psi(n)$ is the inverse of $\lim_{n \uparrow \alpha} \varphi(n)$. By the *Pair Axiom*, the Alpha-limit of a characteristic function is either 0 or 1. Moreover, notice that $\chi_A \cdot \chi_{A^c} = c_0$ and $\chi_A + \chi_{A^c} = c_1$ are the constant functions with values 0 and 1, respectively. So, it follows that either $\lim_{n \uparrow \alpha} \chi_A(n) = 1$ and $\lim_{n \uparrow \alpha} \chi_{A^c} = 0$, or $\lim_{n \uparrow \alpha} \chi_A(n) = 0$ and $\lim_{n \uparrow \alpha} \chi_{A^c}(n) = 1$. Now, the latter case cannot hold as otherwise we would have

$$0 = \lim_{n \uparrow \alpha} c_0(n) = \lim_{n \uparrow \alpha} (\varphi(n) \cdot \chi_{A^c}(n)) = \lim_{n \uparrow \alpha} \varphi(n) \cdot \lim_{n \uparrow \alpha} \chi_{A^c}(n)$$

$$= \lim_{n \uparrow \alpha} \varphi(n) \cdot \lim_{n \uparrow \alpha} c_1(n) = \lim_{n \uparrow \alpha} (\varphi(n) \cdot c_1(n)) = \lim_{n \uparrow \alpha} \varphi(n),$$

against the hypothesis.

---

[5] In AST, one cannot assume that $r^* = r$ for all $r \in \mathbb{R}$. See Remark 14.5.

To show the property of *ordered* field we will use the same arguments as in the proof of Theorem 2.10. Precisely, we take as the positive elements the Alpha-limits of sequences of positive reals:

$$(\mathbb{R}^*)^+ = \left\{ \lim_{n\uparrow\alpha} \psi \mid \psi : \mathbb{N} \to \mathbb{R}^+ \right\}.$$

Then $(\mathbb{R}^*)^- = \{\lim_{n\uparrow\alpha} \vartheta(n) \mid \vartheta : \mathbb{N} \to \mathbb{R}^-\}$. We have to prove that $(\mathbb{R}^*)^-$, $\{0\}$, and $(\mathbb{R}^*)^+$ form a partition of $\mathbb{R}^*$. Given a real sequence $\varphi(n)$, define:

$$\varphi^+(n) = \begin{cases} \varphi(n) & \text{if } \varphi(n) > 0 \\ 1 & \text{otherwise} \end{cases} \qquad \varphi^-(n) = \begin{cases} \varphi(n) & \text{if } \varphi(n) < 0 \\ -1 & \text{otherwise.} \end{cases}$$

Trivially $\varphi(n) \in \{\varphi^-(n), 0, \varphi^+(n)\}$ for all n. Then, by the *Internal Set Axiom* and by the *Pair Axiom*, it follows that

$$\lim_{n\uparrow\alpha} \varphi(n) \in \left\{ \lim_{n\uparrow\alpha} \varphi^-(n), 0, \lim_{n\uparrow\alpha} \varphi^+(n) \right\} \subseteq (\mathbb{R}^*)^- \cup \{0\} \cup (\mathbb{R}^*)^+.$$

Finally, let f be the function defined on $\mathbb{R}$ and such that $f(0) = \emptyset$, $f(x) = \{\emptyset\}$ for $x \in \mathbb{R}^+$ and $f(x) = \{\{\emptyset\}\}$ for $x \in \mathbb{R}^-$. It is readily seen from the *Internal Set Axiom* that the Alpha-limits of the constant sequences

$$\lim_{n\uparrow\alpha} c_\emptyset(n) = \emptyset, \ \lim_{n\uparrow\alpha} c_{\{\emptyset\}}(n) = \{\emptyset\}, \ \lim_{n\uparrow\alpha} c_{\{\{\emptyset\}\}}(n) = \{\{\emptyset\}\}.$$

Thus, for every $\psi : \mathbb{N} \to \mathbb{R}^+$ and for every $\vartheta : \mathbb{N} \to \mathbb{R}^-$, the three Alpha-limits

$$\lim_\alpha (f \circ c_0) = \emptyset, \ \lim_\alpha (f \circ \psi) = \{\emptyset\}, \ \lim_\alpha (f \circ \vartheta) = \{\{\emptyset\}\}$$

are distinct and hence, by the *Composition Axiom*, also the three Alpha-limits $\lim_{n\uparrow\alpha} c_0 = 0$, $\lim_{n\uparrow\alpha} \psi(n)$, and $\lim_{n\uparrow\alpha} \vartheta(n)$ must be distinct. This shows that the three sets $(\mathbb{R}^*)^+$, $(\mathbb{R}^*)^-$ and $\{0\}$ are pairwise disjoint, and the proof is complete. □

Recall that in **ZFC**, the *Infinity Axiom* states the existence of an *inductive set*, that is, a set X that contains the empty set and that satisfies the inductive property: "$x \in X \Rightarrow x \cup \{x\} \in X$". The natural numbers, namely the *von Neumann natural numbers*, are then defined as those sets n that belong to all inductive sets. So, in our set-theoretic framework of **AST**,

- $0 = \emptyset$;
- $1 = 0 \cup \{0\} = \{\emptyset\}$;
- $2 = 1 \cup \{1\} = \{\emptyset, \{\emptyset\}\} = \{0, 1\}$; and so forth.

Inductively,

- $n + 1 = n \cup \{n\} = \{0, 1, \ldots, n\}$.

PROPOSITION 14.3 (AST). *If $k \in \mathbb{N}_0$ then $k^* = k$.*

PROOF. Proceed by induction on $k \in \mathbb{N}_0$. When $k = 0$, we have $0^* = \emptyset^* = \lim_{n\uparrow\alpha} c_\emptyset(n) = \emptyset$ by the *Internal Set Axiom* (AST3). At the inductive step, $(k+1)^* = (k \cup \{k\})^* = k^* \cup \{k\}^* = k^* \cup \{k^*\} = $ (by the inductive hypothesis) $= k \cup \{k\} = k + 1$. $\square$

As anticipated at the beginning of this section, the axioms of the Alpha-Theory imply the existence of $\in$-descending chains, and so the axiom of *Regularity* cannot be assumed in the underlying set theory.

PROPOSITION 14.4. *For every von Neumann natural number k, let*
$$\boldsymbol{\alpha} - k = \lim_{n\uparrow\alpha}(n - k),$$
where we agree that $n - k = \emptyset$ for $k \geq n$. Then one has the following infinite $\in$-descending chain:
$$\boldsymbol{\alpha} \ni \boldsymbol{\alpha} - 1 \ni \ldots \ni \boldsymbol{\alpha} - k \ni \boldsymbol{\alpha} - (k+1) \ni \ldots$$

PROOF. For every von Neumann natural number k, one has that $n - (k+1) \in n - k$ for all $n \geq k + 1$, and so by the *Internal Set Axiom*, $\boldsymbol{\alpha} - (k+1) = \lim_{n\uparrow\alpha}(n - (k+1)) \in \lim_{n\uparrow\alpha}(n - k) = \boldsymbol{\alpha} - k$. $\square$

Notice that the *Real Number Axiom* was not included in AST; indeed, while we just showed that $k^* = k$ for every von Neumann natural number k, the same property does not hold for real numbers.

REMARK 14.5. *The property $r^* = r$ for real numbers $r \in \mathbb{R}$ cannot be assumed in* AST. *Indeed, the definition of real number is usually given as a Dedekind cut of rational numbers, or as an equivalence class of Cauchy sequences of rational numbers. In both cases, a number $r \in \mathbb{R} \setminus \mathbb{Q}$ is defined as an infinite set, and therefore one necessarily has $r^* \neq r$ (see Proposition 4.24). In consequence, the set of hyperreal numbers*
$$\mathbb{R}^* = \left\{ \lim_{n\uparrow\alpha} \varphi(n) \, \middle| \, \varphi : \mathbb{N} \to \mathbb{R} \right\}$$
does not include $\mathbb{R}$ directly, but rather it includes its canonical image $\mathbb{R}^\sigma = \{r^* \mid r \in \mathbb{R}\}$ as a proper subfield (see Remark 4.20).

However, it is safe to say that in calculus one can directly identify $r^* = r$ for every real number r with no harm.

Before investigating the strength of the theory AST in the next section, some observations on the formalism are in order. Although expressed in a satisfactory manner for the "working mathematician", we remark that the axioms of AST as given above are not rigorous from a logical point of view. Indeed, one needs to formalize them as suitable first-order sentences.

To this end, one can consider a language where, in addition to the symbol $\in$ for the membership relation, there is a binary relation symbol J. Then a correct formalization of the *Existence axiom* (AST1) would state that J is a function defined on the class of all sequences, that is:

$$\forall \varphi \; \forall x \; \forall y \; [\text{``}\varphi \text{ is a sequence''} \;\rightarrow\; \exists! x \; J(\varphi, x)] \wedge$$
$$\forall \varphi \; \forall x \; [J(\varphi, x) \;\rightarrow\; \text{``}\varphi \text{ is a sequence''}]$$

Once the above is postulated, for each sequence φ one is allowed to write $\lim_{n \uparrow \alpha} \varphi(n)$ as a notation for the *unique* element x such that $J(\varphi, x)$. Then, reformulating the remaining axioms (AST2), (AST3), (AST4) and (AST5) as first-order sentences in the $\in$-J-language is done in a straightforward manner (we omit details).

As for the underlying axioms of set theory, namely ZFC_0, the *Replacement* and *Separation* schemata are postulated also for formulas where the symbol J appears. So, also (first-order) properties that mention the Alpha-limit are allowed when applying *Replacement* or *Separation*.

2. Alpha Set Theory *versus* ZFC

This section is focused on the consistency strength of the Alpha Set Theory AST, relative to ZFC. The goal is to make precise and then justify the following

- **Claim.** *A statement not involving Alpha-limits is proved by the Alpha Set Theory* AST *if and only if it is a theorem of "usual" mathematics.*

This means that the new methods incorporated in the Alpha-Theory are just additional tools that one can use with no harm, because they prove exactly the same results as the "usual" mathematics.

As a first observation, we notice that the superstructures $V(\mathfrak{R})$ that we constructed in §6.7 are not suitable as models of the full theory AST. Let us see why.

To begin with, Alpha-limits in $V(\mathfrak{R})$ only exist of "bounded" sequences that take values in some finite level $V_k(\mathfrak{R})$ (see Theorem 6.33).

Moreover, it is easily verified that superstructures do not satisfy the *Infinity axiom*, although it is fair to say that the existence of an infinite set of atoms replaces its role. A more relevant limitation from a set-theoretic point of view is that the *Replacement* schema fails in any superstructure $V(X)$ where X is infinite. For these reasons, we will have to look for more advanced set-theoretic constructions than superstructures.

Let us recall a few notions and facts from set theory.

DEFINITION 14.6. *A set A is* wellfounded *if either $A = \emptyset$ or there exists an "$\in$-minimal" element $a \in A$ such that $a \cap A = \emptyset$.*

Let us denote by wf(x) the elementary formula "x is wellfounded".

DEFINITION 14.7. *Let σ be an elementary formula. By σ^{wf} we mean the formula obtained from σ by replacing*

- *every existential quantifier $\exists x \ldots$ with $\exists x \, (\mathrm{wf}(x) \wedge \ldots)$, and*
- *every universal quantifier $\forall x \ldots$ with $\forall x \, (\mathrm{wf}(x) \rightarrow \ldots)$.*

The formula σ^{wf} is called the relativization *of σ to the class of wellfounded sets.*

The two properties below are well-known facts:

(1) Assume ZFC$_0$. Then the following are equivalent:
   (a) The axiom of *Regularity*;
   (b) Every set is wellfounded;
   (c) There are no $\in$-descending chains:
   $$x_1 \ni x_2 \ni \ldots \ni x_n \ni x_{n+1} \ni \ldots$$

(2) Assume ZFC$_0$. Then for every axiom σ of ZFC (including the axiom of *Regularity*) its relativization σ^{wf} is a theorem of ZFC$_0$.[6]

A convenient foundational framework for our purposes is given by *Zermelo-Fraenkel-Boffa* set theory ZFBC, the non-wellfounded variant of ZFC where the axiom of *Foundation* is replaced by Boffa's *Superuniversality* axiom, and where the axiom of *Choice* is strengthened to *Global Choice*.[7]

---

[6] This property is usually written "**WF** $\models$ ZFC" (read: "the class **WF** of wellfounded sets is a model of ZFC"). Recall that in ZFC, by a *class* one means a collection of sets that is defined by a formula; in this case **WF** $= \{x \mid \mathrm{wf}(x)\}$.

[7] The theory ZFBC was first used for the foundations of nonstandard methods by D. Ballard and K. Hrbàček in [**4**].

The axiom of *Superuniversality* extends the validity of *Mostowski's collapsing theorem* to all (possibly non-wellfounded) extensional structures in the following strong sense. If $\mathfrak{M} = (M, E)$ and $\mathfrak{M}' = (M', E')$ are extensional models (that is, structures in the language of set theory that satisfy the axiom of *Extensionality*), and $\mathfrak{M}$ is a transitive submodel of $\mathfrak{M}'$ (that is, $\{x \in M \mid xEa\} = \{x' \in M' \mid x'E'a\}$ for every $a \in M$), then every isomorphism $\tau : (M, E) \cong (T, \in)$ onto a transitive set T can be extended to an isomorphism $\tau' : (M', E') \cong (T', \in)$ onto a transitive superset $T' \supseteq T$.

The axiom of *Global Choice* takes an additional unary function symbol G to be formulated; it postulates that G is a global choice function, that is, $G(x) \in x$ for every $x \neq \emptyset$. In the theory ZFBC, the *Separation* and *Replacement* schemata are assumed also for formulas where the symbol G appears.

THEOREM 14.8.

(1) *Every countable model* $\mathfrak{M}$ *of* ZFC *is the wellfounded part*[8] *of a model* $\mathfrak{N}$ *of* ZFBC;

(2) ZFBC *proves the following: Given any non-principal ultrafilter* $\mathcal{U}$ *on* $\mathbb{N}$, *there is a definable correspondence that assigns an element* $\lim_{n \uparrow \alpha} \varphi(n)$ *to each sequence* φ *in such a way that all axioms of* AST *are satisfied, and the corresponding ultrafilter* $\mathcal{Q}$ *of qualified sets coincides with* $\mathcal{U}$.[9]

SKETCH OF PROOF. The proof consists in known arguments in non-standard set theory. Here we limit ourselves to give a sketch; more details can be found in [36, 10], and in §8.3 of the comprehensive monograph [59].

(1). A classic theorem by U. Felgner [47] states that every countable model of ZFC is the wellfounded part of a model of Zermelo-Fraenkel set theory plus *Global Choice*. By combining that result with Boffa's construction [25] of models of *Superuniversality*, one reaches the conclusion.

---

[8] We say that $\mathfrak{M}$ is the *wellfounded part* of $\mathfrak{N}$ to mean that $\mathfrak{M}$ is the submodel of $\mathfrak{N}$ whose universe is given by the $\mathfrak{N}$-wellfounded sets $\{x \in \mathfrak{N} \mid \mathfrak{N} \models \mathrm{wf}(x)\}$.

[9] By saying "there is a definable correspondence" we mean that there exists a formula $\sigma(x, y, z)$ such that for every non-principal ultrafilter $\mathcal{U}$, the formula with parameter $\sigma(x, y, \mathcal{U})$ defines a functional correspondence $J_{\mathcal{U}} : \varphi \mapsto \lim_{n \uparrow \alpha} \varphi(n)$ defined on all sequences, that is, $J_{\mathcal{U}}(x) = y \Leftrightarrow$ "x is a sequence" $\wedge \sigma(x, y, \mathcal{U})$. (Recall that by "sequence" we mean any function φ whose domain is the set of natural numbers.)

(2). Within the theory **ZFBC**, one can formalize the following construction. Let $\mathbf{V}^{\mathbb{N}}/\mathcal{U}$ be the ultrapower modulo $\mathcal{U}$ of the universal class $\mathbf{V}$ of all sets, and let

$$\pi : \mathbf{V}^{\mathbb{N}} \longrightarrow \mathbf{V}^{\mathbb{N}}/\mathcal{U}$$

be the canonical projection that maps every sequence φ to the corresponding $\mathcal{U}$-equivalence class $[\varphi]_{\mathcal{U}} \in \mathbf{V}^{\mathbb{N}}/\mathcal{U}$. By *Superuniversality* and *Global Choice*, define a *transitive collapse*

$$\tau : \mathbf{V}^{\mathbb{N}}/\mathcal{U} \xrightarrow{\;\cong\;} \mathbf{T}$$

of the extensional structure $\mathbf{V}^{\mathbb{N}}/\mathcal{U}$ onto a transitive class $\mathbf{T}$. Then define the Alpha-limit of any sequence φ by letting:

$$\lim_{n \uparrow \alpha} \varphi(n) \;=\; \tau(\pi(\varphi)).$$

By the properties of the ultrapower construction $\mathbf{V}^{\mathbb{N}}/\mathcal{U}$, it is verified that all axioms of **AST** are satisfied. Finally, the equality $\mathcal{Q} = \mathcal{U}$ is shown by the same arguments used in the proof of Theorem 11.22. $\square$

COROLLARY 14.9. *Let $\mathfrak{M}$ be a countable model of **ZFC** and assume that $\mathfrak{M} \models$ "$\mathcal{U}$ is a non-principal ultrafilter on $\mathbb{N}$". Then $\mathfrak{M}$ is the wellfounded part of a model $\mathfrak{N}_{\mathcal{U}}$ of **AST** such that $\mathfrak{N}_{\mathcal{U}} \models$ "$\mathcal{Q} = \mathcal{U}$".*

PROOF. By property (1) in the previous theorem, we can pick a model $\mathfrak{N}$ of **ZFBC** whose wellfounded part is $\mathfrak{M}$. By applying the definable construction in (2) within $\mathfrak{N}$, one obtains a model $\mathfrak{N}_{\mathcal{U}}$ of **AST** with same universe and same membership relation as in $\mathfrak{N}$, and such that $\mathfrak{N}_{\mathcal{U}} \models$ "$\mathcal{Q} = \mathcal{U}$". $\square$

We are finally ready to justify the informal claim that we made at the beginning of this section. In fact, the following theorem makes precise the intuition that the Alpha-Theory has the same logical strength as "ordinary" mathematics.

THEOREM 14.10. *The correspondence $\sigma \mapsto \sigma^{\mathrm{wf}}$ is a faithful interpretation of **ZFC** in **AST**, that is, a sentence σ in the language of set theory is a theorem of **ZFC** if and only if σ^{wf} is a theorem of **AST**:*[10]

$$\mathbf{ZFC} \vdash \sigma \iff \mathbf{AST} \vdash \sigma^{\mathrm{wf}}$$

*In particular, the two theories **ZFC** and **AST** are equiconsistent.*

---

[10] On the notion of *interpretability* of theories see, *e.g.*, Ch.V of W. Hodge's book [**55**].

PROOF. If σ^{wf} is not a theorem of AST, then we can pick a model $\mathfrak{N} \models$ AST where $\mathfrak{N} \models \neg\sigma^{\mathrm{wf}}$. Let $\mathfrak{M}$ be the wellfounded part of $\mathfrak{N}$. Then $\mathfrak{M} \models$ ZFC and $\mathfrak{M} \models \neg\sigma$, and hence σ is not a theorem of ZFC.

Conversely, if σ is not a theorem of ZFC, then there exists a model $\mathfrak{M} \models$ ZFC where $\mathfrak{M} \models \neg\sigma$. By applying the downward *Löwenheim-Skolem Theorem* if necessary, we can assume that $\mathfrak{M}$ is countable. Pick any $\mathcal{U}$ such that $\mathfrak{M} \models$ "$\mathcal{U}$ is a non-principal ultrafilter over $\mathbb{N}$", and consider the model $\mathfrak{N}_{\mathcal{U}} \models$ AST as given by previous corollary. Since $\mathfrak{M}$ is the wellfounded part of $\mathfrak{N}_{\mathcal{U}}$, we have

$$\mathfrak{M} \models \neg\sigma \iff \mathfrak{N}_{\mathcal{U}} \models \neg\sigma^{\mathrm{wf}},$$

and hence σ^{wf} is not a theorem of AST. (We remark that notation used here was not ambiguous because $\neg(\sigma^{\mathrm{wf}})$ is same as $(\neg\sigma)^{\mathrm{wf}}$.)   □

3. Cauchy infinitesimal principle and special ultrafilters

In this section we start investigating the foundations of *Cauchy Infinitesimal Principle* (CIP)

Along with Cauchy Infinitesimal Principle, we will also consider a weaker version of its where the monotonicity of the infinitesimal sequence is not assumed.

(CIP)$_{\mathrm{weak}}$ Cauchy Infinitesimal Principle - weak version
Every infinitesimal number is the Alpha-limit of some infinitesimal sequence.

(CIP) Cauchy Infinitesimal Principle
Every positive infinitesimal number is the Alpha-limit of some decreasing infinitesimal sequence.

We now relate the above principles to two special classes of ultrafilters on $\mathbb{N}$ that have been deeply studied by researchers in set theory, namely the "P-points" and the "selective" ultrafilters. Let us start by recalling the definitions.

DEFINITION 14.11. *Let $\mathcal{U}$ be a non-principal ultrafilter on $\mathbb{N}$.*

- *$\mathcal{U}$ is a P-point if for every disjoint union $\bigsqcup_{k\in\mathbb{N}} A_k \in \mathcal{U}$ where every $A_k \notin \mathcal{U}$ there exists a set $X \in \mathcal{U}$ such that every $X \cap A_k$ is finite.*
- *$\mathcal{U}$ is selective if for every disjoint union $\bigsqcup_{k\in\mathbb{N}} A_k \in \mathcal{U}$ where every $A_k \notin \mathcal{U}$ there exists a set $X \in \mathcal{U}$, namely the selector, such that every $X \cap A_k$ contains at most one point.*

Notice that, trivially, every selective ultrafilter is a P-point. The properties of P-point and selectiveness are stable under isomorphisms.

Let us put off for the moment the question of the existence of such ultrafilters, and see their connections with Cauchy's principles.

THEOREM 14.12. *Alpha-Calculus Theory* ACT *proves the following:*

(1) *(CIP)*$_{\text{weak}}$ *holds if and only if the ultrafilter* $\mathcal{Q}$ *of qualified sets is a P-point;*

(2) *(CIP) holds if and only if the ultrafilter* $\mathcal{Q}$ *of qualified sets is selective.*

PROOF. (1). Assume first that (CIP)$_{\text{weak}}$ holds, and let the disjoint union $A = \bigsqcup_{k \in \mathbb{N}} A_k \in \mathcal{Q}$ where every $A_k \notin \mathcal{Q}$. Define

$$\varphi(n) = \begin{cases} \frac{1}{k} & \text{if } n \in A_k \text{ for some } k\,; \\ 0 & \text{if } n \notin A. \end{cases}$$

For every $h > 1$,

$$\left\{ n \in \mathbb{N} \,\middle|\, \varphi(n) \geq \frac{1}{h} \right\} = A_1 \cup \ldots \cup A_h \notin \mathcal{Q},$$

and so $\lim_{n \uparrow \alpha} \varphi(n) < 1/h$. As h is arbitrary, we conclude that $\varepsilon = \lim_{n \uparrow \alpha} \varphi(n)$ is a positive infinitesimal. By the hypothesis, there exists a sequence $\psi(n)$ such that $\lim_{n \to \infty} \psi(n) = 0$ and $\lim_{n \uparrow \alpha} \psi(n) = \varepsilon$, and hence $X = \{n \in \mathbb{N} \mid \psi(n) = \varphi(n)\} \in \mathcal{Q}$. Now notice that $n \in X \cap A_k \Leftrightarrow \psi(n) = \varphi(n) = 1/k$. Since the sequence $\psi(n)$ converges to zero, the set $X \cap A_k$ must be finite.

Now assume that $\mathcal{Q}$ is a P-point, and let $\varepsilon = \lim_{n \uparrow \alpha} \varphi(n) > 0$ be a positive infinitesimal number. For each $k \in \mathbb{N}$, the set

$$A_k = \left\{ n \in \mathbb{N} \,\middle|\, \frac{1}{k+1} < \varphi(n) \leq \frac{1}{k} \right\} \notin \mathcal{Q}$$

because $\varepsilon \leq \frac{1}{k+1}$, but the disjoint union

$$A = \bigsqcup_{k \in \mathbb{N}} A_k = \{n \in \mathbb{N} \mid 0 < \varphi(n) \leq 1\} \in \mathcal{Q}$$

because $0 < \varepsilon \leq 1$. So, by the hypothesis, there exists $X \in \mathcal{Q}$ such that $X \cap A_k$ is finite for every k. Now define

$$\psi(n) = \begin{cases} \varphi(n) & \text{if } n \in X \cap A \\ 0 & \text{otherwise.} \end{cases}$$

Since φ and ψ agree on the qualified set $X \cap A$, they have the same Alpha-limit $\lim_{n \uparrow \alpha} \psi(n) = \lim_{n \uparrow \alpha} \varphi(n) = \varepsilon$. Moreover, for every $h > 1$, the set

$$\left\{ n \in \mathbb{N} \,\Big|\, |\psi(n)| > \frac{1}{h} \right\} = \bigsqcup_{k < h} X \cap A_k$$

is finite. Since h is arbitrary, it follows that $\lim_{n \to \infty} \psi(n) = 0$, as desired.

(2). The proof that (CIP) implies that $\mathcal{Q}$ is selective is entirely similar to the above proof that $(CIP)_{\text{weak}}$ implies that $\mathcal{Q}$ is a P-point. Indeed, in this case the infinitesimal sequence $\psi(n)$ can be taken to be decreasing, and so each set $X \cap A_k$ can contain at most one point.

Conversely, assume that $\mathcal{Q}$ is selective. We will use the following characterization (see condition (4) of Theorem 6.12):

- (CIP) $\Leftrightarrow$ for all $\nu \in \mathbb{N}_\infty$ there exists $\varphi : \mathbb{N} \to \mathbb{N}$ and a set $X \in \mathcal{Q}$ such that the restriction $\varphi_{|X}$ is 1-1 and $\lim_{n \uparrow \alpha} \varphi(n) = \nu$.

Given $\nu = \lim_{n \uparrow \alpha} \varphi(n) \in \mathbb{N}_\infty$, let $A_k = \varphi^{-1}(k) = \{n \mid \varphi(n) = k\}$. Notice that since $\nu \notin \mathbb{N}$, every $A_k \notin \mathcal{Q}$. Now, $\mathbb{N} = \bigsqcup_{k \in \mathbb{N}} A_k \in \mathcal{Q}$ and so, by the hypothesis, there exists $X \in \mathcal{Q}$ such that for every k the intersection $X \cap A_k$ contains at most one point. This means that $\varphi_{|X}$ is 1-1, as desired. □

We remark that by using the characterization (2) in the above theorem, equivalences of (CIP) as given in Theorem 6.12 can be reformulated as equivalences of the property of selectiveness.

EXERCISE 14.13. *Let $\mathcal{U}$ be a non-principal ultrafilter on $\mathbb{N}$. Prove that the following properties are equivalent:*

(1) *$\mathcal{U}$ is selective;*

(2) *For every $f : \mathbb{N} \to \mathbb{N}$ there exists $X \in \mathcal{U}$ such that the restriction $f_{|X}$ is either constant or 1-1;*

(3) *For every $f : \mathbb{N} \to \mathbb{N}$ there exists $X \in \mathcal{U}$ such that the restriction $f_{|X}$ is either constant or a bijection;*

(4) *For every $f : \mathbb{N} \to \mathbb{N}$ there exists $X \in \mathcal{U}$ such that the restriction $f_{|X}$ is either constant or increasing;*

(5) *For every $f : \mathbb{N} \to \mathbb{N}$ there exists $X \in \mathcal{U}$ such that the restriction $f_{|X}$ is non-decreasing.*

Thanks to the characterizations obtained in the previous theorem, one gets information about the strength of $(\mathsf{CIP})_{\text{weak}}$ and (CIP) by directly employing known results about P-points and selective ultrafilters, respectively. Since a treatment of ultrafilter theory is outside the scope of this book, here we will limit ourselves to state without proof a few known facts that are relevant to our purposes.[11]

THEOREM 14.14.

(1) *There exist non-principal ultrafilters on* $\mathbb{N}$ *that are not P-points;*

(2) *Under the* continuum hypothesis, *there exist selective ultrafilters, as well as P-points that are not selective. In consequence, in* ZFC *one cannot prove that selective ultrafilters do not exist;*

(3) *There exists a model of* ZFC *with no P-points. In consequence, in* ZFC *one cannot prove the existence of P-points;*

(4) *There exists a model of* ZFC *with exactly one P-point up to isomorphisms. Such a P-point is necessarily a selective ultrafilter.*

Examples that witness property (1) are not difficult to construct (see, *e.g.*, [**21**, Ch.1 §9]). A proof of (2) can be found in [**26**]. Finally, properties (3) and (4) are remarkable results due to S. Shelah (see [**88**, § VI.4] and [**88**, § XVIII.4], respectively).

REMARK 14.15. *By the above results (2) and (3), both the existence of P-points and of selective ultrafilters are* independent *of the axioms of* ZFC. *In other words, similarly as the* continuum hypothesis, *P-points and selective ultrafilters cannot be proved nor disproved to exist by the usual principles of mathematics.*

Notice that in consequence of (2) and (4), even by assuming the existence of P-points, ZFC *cannot prove the implication "P-point* $\Rightarrow$ *selective", nor its negation "P-point* $\nRightarrow$ *selective".*

4. The strength of Cauchy infinitesimal principle

In this section we show that *Cauchy infinitesimal principle* cannot be proved nor disproved by the axioms of the Alpha-Set-Theory. Thus (CIP) has an independent status with respect to AST, similar to that of the *continuum hypothesis* with respect to ZFC. In particular, this

---

[11] The literature on special ultrafilters is extensive: the interested reader can consult, *e.g.*, [**22**, Ch.9] or [**23**, §4], and references therein.

means that one can safely add (CIP) to the axioms of AST because it does not bring to contradictions.[12]

The "positive" results about the Cauchy principles are itemized in the following theorem, that is parallel to Theorem 14.10.

THEOREM 14.16. *The correspondence $\sigma \mapsto \sigma^{\mathrm{wf}}$ provides:*

(1) *A faithful interpretation of ZFC plus "there exists a P-point" in the theory AST plus (CIP)$_{\mathrm{weak}}$:*

$$ZFC + \text{"there exists a P-point"} \vdash \sigma \iff AST + (CIP)_{\mathrm{weak}} \vdash \sigma^{\mathrm{wf}};$$

(2) *A faithful interpretation of ZFC plus "there exists a selective ultrafilter" in the theory AST plus (CIP):*

$$ZFC + \text{"there exists a selective ultraf."} \vdash \sigma \iff AST + (CIP) \vdash \sigma^{\mathrm{wf}}.$$

In consequence, the three set theories ZFC, AST plus (CIP)$_{\mathrm{weak}}$, and AST plus (CIP), are equiconsistent.

PROOF. (1). The argument is entirely similar to the proof of Theorem 14.10. Let us see the details. Assume that σ^{wf} is not a theorem of AST plus (CIP)$_{\mathrm{weak}}$. Then we can pick a model $\mathfrak{N} \models AST$ such that $\mathfrak{N} \models (CIP)_{\mathrm{weak}}$ and $\mathfrak{N} \models \neg\sigma^{\mathrm{wf}}$. Notice that $\mathfrak{N} \models \text{"}\mathcal{Q}$ is a P-point", by Theorem 14.12 (1). If $\mathfrak{M}$ be the wellfounded part of $\mathfrak{N}$, then $\mathfrak{M}$ is a model of ZFC plus "there exists a P-point" (namely, the ultrafilter $\mathcal{Q}$), and moreover $\mathfrak{M} \models \neg\sigma$. This shows that σ is not a theorem of ZFC plus "there exists a P-point".

Conversely, assume that σ is not a theorem of ZFC plus "there exists a P-point", and pick a model $\mathfrak{M} \models ZFC$ such that $\mathfrak{M} \models \text{"there exists a}$ P-point" and $\mathfrak{M} \models \neg\sigma$. By applying the downward *Löwenheim-Skolem Theorem* if necessary, we can assume that $\mathfrak{M}$ is countable. Pick $\mathcal{U} \in \mathfrak{M}$ such that $\mathfrak{M} \models \text{"}\mathcal{U}$ is a P-point", and take the model $\mathfrak{N}_{\mathcal{U}}$ of AST as given by Corollary 14.9. Then $\mathfrak{N}_{\mathcal{U}}$ is a model of (CIP)$_{\mathrm{weak}}$ because $\mathfrak{N}_{\mathcal{U}} \models \text{"}\mathcal{Q} = \mathcal{U}\text{"}$; moreover, $\mathfrak{N}_{\mathcal{U}} \models \neg\sigma^{\mathrm{wf}}$. This shows that σ^{wf} is not a theorem of AST plus (CIP)$_{\mathrm{weak}}$.

Property (2) is proved exactly in the same way, by using the characterization given by Theorem 14.12 (2).   □

The next result clarifies the strength of Cauchy principles in the framework of Alpha-Theory.

---
[12] Unless ZFC is contradictory already.

THEOREM 14.17.

(1) *AST does not prove* $(CIP)_{weak}$;

(2) *AST plus* $(CIP)_{weak}$ *does not prove* (CIP).[13]

PROOF. (1). Assume by contradiction that $\mathsf{AST} \vdash (\mathsf{CIP})_{weak}$. Then, by Theorem 14.12 (1), AST proves the existence of a P-point, namely the ultrafilter $\mathcal{Q}$ of qualified sets. Finally, by Theorem 14.10, we can conclude that also ZFC proves the existence of a P-point, contradicting (3) of Theorem 14.14.

(2). Let $\mathfrak{M}$ be a countable model of ZFC where the *continuum hypothesis* holds. By Theorem 14.14 (2), we can pick $\mathcal{U} \in \mathfrak{M}$ such that $\mathfrak{M} \models$ "$\mathcal{U}$ is a P-point which is not selective". By Corollary 14.9, $\mathfrak{M}$ is the wellfounded part of some model $\mathfrak{N}_{\mathcal{U}} \models \mathsf{AST}$ where $\mathfrak{N}_{\mathcal{U}} \models$ "$\mathcal{Q} = \mathcal{U}$". But then, by Theorem 14.12, $(\mathsf{CIP})_{weak}$ holds in $\mathfrak{N}_{\mathcal{U}}$ but (CIP) fails, and the assertion is proved. $\square$

The existence of models where Cauchy infinitesimal principles hold is *independent* of ZFC.

THEOREM 14.18.

(1) *ZFC neither proves the existence nor the non-existence of models of ACT plus* $(CIP)_{weak}$;

(2) *ZFC neither proves the existence nor the non-existence of models of ACT plus* (CIP).

PROOF. If by contradiction one could prove in ZFC that there exists a model of ACT plus $(\mathsf{CIP})_{weak}$, then one could also prove the existence of a P-point, namely the ultrafilter of qualified sets, but this contradicts Theorem 14.14 (3).

In the other direction, let us assume by contradiction that the following is a theorem of ZFC: "(CIP) fails in every model of ACT." Given any non-principal ultrafilter $\mathcal{U}$, construct the corresponding model of ACT by considering the ultrapower of $\mathbb{R}^* = \mathbb{R}^{\mathbb{N}}_{\mathcal{U}}$ (see Theorem 11.22). Since (CIP) fails in that model, the ultrafilter of qualified sets $\mathcal{Q} = \mathcal{U}$ is *not* selective. But then we would have a proof in ZFC of the non-existence of selective ultrafilters on $\mathbb{N}$, against Theorem 14.14 (2).

Finally, notice that since $\mathcal{U}$ selective implies $\mathcal{U}$ P-point, the above arguments completely prove (1) and (2). $\square$

---

[13] In items (2) and (3), we are implicitly assuming the consistency of ZFC, and hence of AST plus $(\mathsf{CIP})_{weak}$ and of AST plus (CIP). In fact, recall that a contradictory theory proves everything!

We close this section with a consequence of (4) in Theorem 14.14. It is a consistency result about the existence of a "canonical" model of the Alpha-Calculus theory.

THEOREM 14.19. *The following property is consistent with* ZFC: *Up to equivalences, there exists a unique model of* ACT *plus* (CIP).

5. The strength of a Hausdorff S-topology

In this section we focus on the following property:

(Haus) Hausdorff Principle
The S-topology is Hausdorff on every hyper-image.

Recall that the S-topology is Hausdorff on every hyper-image if and only if it is Hausdorff on $\mathbb{N}^*$ (see Proposition 6.17). We also recall that, as proved in Theorem 6.20, (CIP) $\Rightarrow$ (Haus).

Similarly as (CIP)$_{\text{weak}}$ and (CIP), also property (Haus) corresponds to a special class of ultrafilters.

THEOREM 14.20. *Alpha-Calculus Theory* ACT *proves the following equivalence* Haus *holds if and only if the ultrafilter* $\mathcal{Q}$ *of qualified sets satisfies:*[14]

$(\star)$ *For every* $f, g : \mathbb{N} \to \mathbb{N}$,

$$f(\mathcal{Q}) = g(\mathcal{Q}) \implies \{n \mid f(n) = g(n)\} \in \mathcal{Q}.$$

PROOF. Let us start by observing that for every $A \subseteq \mathbb{N}$ and for every $\varphi : \mathbb{N} \to \mathbb{N}$, one has $A \in \varphi(\mathcal{Q}) \Leftrightarrow \varphi^{-1}(A) \in \mathcal{Q} \Leftrightarrow \varphi(n) \in A$ a.e. $\Leftrightarrow \lim_{n \uparrow \alpha} \varphi(n) \in A^*$.

Now assume that (Haus) holds, and let $f, g : \mathbb{N} \to \mathbb{N}$ be such that $\{n \mid f(n) = g(n)\} \notin \mathcal{Q}$. Then $f(n) \neq g(n)$ a.e., and so $\nu = \lim_{n \uparrow \alpha} f(n)$ and $\mu = \lim_{n \uparrow \alpha} g(n)$ are two distinct hypernatural numbers. By the hypothesis, there exists $A \subseteq \mathbb{N}$ such that $\nu \in A^*$ and $\mu \notin A^*$, and hence $f(\mathcal{Q}) \neq g(\mathcal{Q})$, because $A \in f(\mathcal{Q})$ and $A \notin g(\mathcal{Q})$ by the above equivalences.

Conversely, assume that $\mathcal{Q}$ satisfies $(\star)$. If $\nu = \lim_{n \uparrow \alpha} f(n)$ and $\mu = \lim_{n \uparrow \alpha} g(n)$ are two distinct hypernatural numbers, then $f(n) \neq g(n)$ a.e., and hence $\{n \mid f(n) = g(n)\} \notin \mathcal{Q}$. By the hypothesis, we have $f(\mathcal{Q}) \neq g(\mathcal{Q})$, and so there exists $A \subseteq \mathbb{N}$ such that $A \in f(\mathcal{Q})$

---

[14] Recall from ultrafilter theory that if $f : I \to J$ and $\mathcal{U}$ is an ultrafilter on I, then $f(\mathcal{U}) = \{A \subseteq J \mid f^{-1}(A) \in \mathcal{U}\}$ is an ultrafilter on J, named the *image ultrafilter* of $\mathcal{U}$ under the function f.

and $A \notin g(\mathcal{Q})$. The last two conditions are equivalent to $\nu \in A^*$ and $\mu \notin A^*$, respectively. $\square$

In consequence of the above equivalences, ultrafilters on $\mathbb{N}$ that satisfy condition $(\star)$ are called *Hausdorff* in [**37**].

The following interpretability result is parallel to Theorems 14.10 and 14.16, and it is proved in the same fashion.

THEOREM 14.21. *The correspondence $\sigma \mapsto \sigma^{\mathrm{wf}}$ is a faithful interpretation of ZFC plus "there exists a Hausdorff ultrafilter" in the theory AST plus (Haus):*

$$\mathsf{ZFC} + \text{"there exists a Hausdorff ultraf."} \vdash \sigma \iff \mathsf{AST} + (\mathsf{Haus}) \vdash \sigma^{\mathrm{wf}}.$$

We will need the following known facts:

THEOREM 14.22.

(1) *Every selective ultrafilter is a Hausdorff ultrafilter;*

(2) *Under the* continuum hypothesis, *there exist Hausdorff ultrafilters that are not P-points, as well as P-points that are not Hausdorff ultrafilters.*

Proofs of both (1) and (2) can be found in [**35**] (see also [**37**]).

In consequence of the above properties of Hausdorff ultrafilters, one obtains the following result. (The proof consists of the same arguments as used for Theorem 14.17.)

THEOREM 14.23.   *Assume AST. Then*

(1) *(CIP) implies (Haus)* ;[15]

(2) *AST plus (CIP)$_{\mathrm{weak}}$ does not prove (Haus);*

(3) *AST plus (Haus) does not prove (CIP)$_{\mathrm{weak}}$.*

As a final remark, it is worth mentioning that the existence of Hausdorff ultrafilters, and hence of models where S-topology is Hausdorff, is still open to this day (see [**37, 5**]).

- *Open problem*: Can one prove is **ZFC** that there exist Hausdorff ultrafilters?

---

[15] This property was already proved in Theorem 6.20.

6. Remarks and comments

1. According to the usual foundational framework of mathematics, namely Zermelo-Fraenkel set theory **ZFC**, every mathematical object is a set. The philosophical plausibility of such an assumption is disputable, but the foundational success of pure set theory is due to the fact that virtually all objects of mathematics can actually be coded as sets.

2. Following I. Lakatos [**66**], J. Cleave proposed in [**28**] an interpretation of Cauchy's conception of infinitesimals in the context of nonstandard analysis. Precisely, by considering models of the hyperreal line as given by ultrapowers $\mathbb{R}^{\mathbb{N}}_{\mathcal{U}}$, he proposed to interpret Cauchy's infinitesimals as the equivalence classes of infinitesimal sequences.

In their paper [**31**], N. Cutland, C. Kessler, E. Kopp and D. Ross discussed the mathematical content of Cleave's interpretation. Most notably, they pointed out that the assumption of an ultrapower $\mathbb{R}^{\mathbb{N}}_{\mathcal{U}}$ where *every* infinitesimal is the equivalence class of some infinitesimal sequence, requires the ultrafilter $\mathcal{U}$ to be a *P-point*, and hence it is independent of the axioms of **ZFC**.

As we have seen, the Alpha-Theory provides an appropriate framework for accommodating this idea of infinitesimals, that we formalized as (the weak version of) *Cauchy infinitesimal principle* (see §6.3 and §14.3).

Part 5

Numerosity Theory

The idea of counting the *"number of elements"* of a set is formalized by the Cantorian theory of cardinal and ordinal numbers. The theory of "numerosity" – as elaborated in the last years by the authors of this book jointly with M. Forti – can be seen as a complement to those theories.

It is well-known that when dealing with infinite cardinal and ordinal numbers one has the paradoxical phenomenon that proper subsets may have "equal size" as the whole set. The main motivation of a theory of numerosity is to explore the possibility of a sound counting system that extends finite cardinality to infinite sets in such a way that the ancient principle *"the whole is greater than the part"* is preserved.

In this Part 5 we will fully develop a theory of numerosity, and point out its close relationships with the Alpha-Theory. Here, we will only consider countable sets for the following three reasons. First of all, the countable case can be fully developed within the Alpha-Theory, while numerosities of uncountable sets would require a stronger theory where also Alpha-limits of sequences of uncountable length could be considered. The second reason is that the countable case is already general enough to provide all the fundamental aspects of our idea of a "numerosity". Finally, the third reason is incidental: the research for a general theory of uncountable numerosities is still fluid and under development. While several results have been already obtained (see [**12, 14, 38, 24, 15**]), a few fundamental aspects are still to be clarified at the moment when this part is being written.

CHAPTER 15

Counting Systems

1. The idea of counting

In order to count the elements of sets in a "universe" $\mathfrak{U}$, one needs an ordered set of numbers $(\mathfrak{N}, <)$ and a function $\mathfrak{n}$ that associates to every $A \in \mathfrak{U}$ the "number of its elements" $\mathfrak{n}(A) \in \mathfrak{N}$. More precisely, we can say that the operation of counting consists of a triple $(\mathfrak{U}, \mathfrak{N}, \mathfrak{n})$ where:

- $\mathfrak{U}$ is the "universe" of sets to be counted;
- $\mathfrak{N}$ is the ordered collection of "numbers" that are used to count;
- $\mathfrak{n}$ is a notion of "size" that assigns to every $A \in \mathfrak{U}$ its "numerosity" $\mathfrak{n}(A) \in \mathfrak{N}$.

The naive idea of a counting system is grounded on the intuition as originated from natural numbers and finite sets. In particular, our intuition of natural numbers comes with the idea of an order (a set has a bigger size than another) and with the operations of *sum* and *product* as corresponding to the set-operations of *disjoint union* and *Cartesian product*, respectively.

In the following discussion, we will adopt the following (semi-formal) definition.

DEFINITION 15.1. *A counting system is a triple* $(\mathfrak{U}, \mathfrak{N}, \mathfrak{n})$ *where:*

- $\mathfrak{U}$ *is a* universe, *that is, a nonempty family of sets such that:*

> – *If $B \subseteq A$ and $A \in \mathfrak{U}$ then $B \in \mathfrak{U}$;*
> – *If $A, B \in \mathfrak{U}$ then the disjoint union $A \sqcup B \in \mathfrak{U}$;*
> – *If $A, B \in \mathfrak{U}$ then the product $A \times B \in \mathfrak{U}$;*

- *$\mathfrak{N}$ is a linearly ordered set of numbers;*

- *$\mathfrak{n} : \mathfrak{U} \to \mathfrak{N}$ is a surjective map, called* numerosity function, *that assigns to every set $A \in \mathfrak{U}$ its "numerosity" $\mathfrak{n}(A) \in \mathfrak{N}$;*

and that satisfies the following properties:

(CS1) *Finite Set Principle: If A is finite, then $\mathfrak{n}(A) = |A|$ is the number of elements of A;*

(CS2) *Monotonicity Principle: If $A \subseteq B$ then $\mathfrak{n}(A) \le \mathfrak{n}(B)$;*

(CS3) *Union Principle: If $\mathfrak{n}(A) = \mathfrak{n}(A')$ and $\mathfrak{n}(B) = \mathfrak{n}(B')$ then the disjoint unions $\mathfrak{n}(A \sqcup B) = \mathfrak{n}(A' \sqcup B')$;*

(CS4) *Product Principle: If $\mathfrak{n}(A) = \mathfrak{n}(A')$ and $\mathfrak{n}(B) = \mathfrak{n}(B')$ then $\mathfrak{n}(A \times B) = \mathfrak{n}(A' \times B')$;*

(CS5) *Unit Principle: $\mathfrak{n}(\{P\} \times A) = \mathfrak{n}(A \times \{P\}) = \mathfrak{n}(A)$ for every singleton $\{P\}$.*

Caveat: The notions of "disjoint union" and of "product" mentioned in the above definition must be understood in a broad sense. For instance, although in a strict sense ordinals are neither closed under disjoint unions nor under Cartesian products, nevertheless they can be taken as a basic example of a counting system (see the next section).

Notice that by the *Finite Set Principle* and the *Monotonicity Principle*, the non-negative integers $\mathbb{N}_0$ are an initial segment of the ordered set of numerosities $(\mathfrak{N}, \le)$.

The coherence properties given by the *Union Principle* and the *Product Principle* make it possible to define a sum and a product operation on numerosities.

DEFINITION 15.2. *For any $A, B \in \mathfrak{U}$, set:*

- $\mathfrak{n}(A) + \mathfrak{n}(B) = \mathfrak{n}(A \sqcup B)$;
- $\mathfrak{n}(A) \cdot \mathfrak{n}(B) = \mathfrak{n}(A \times B)$.

A fundamental property suggested by our intuition of counting, as originated by finite sets, is the idea that sets have the same size if and only if they can be put in a 1-1 correspondence. This principle, attributed to Hume, was adopted by Cantor as his fundamental definition of *equipotency*.[1]

- **Hume's Principle (HP).** $\mathfrak{n}(A) = \mathfrak{n}(B)$ *if and only if there exists a bijection between A and B.*

A significant principle suggested by our intuition is the following strengthening of the *Monotonicity Principle*, which can be seen as a reformulation in a set-theoretic context of the ancient Euclidean principle that *"the whole is greater than the part"*.[2]

- **Euclid's Principle (EP).** *If A is a proper subset of B then $\mathfrak{n}(A) < \mathfrak{n}(B)$.*

A related principle where one postulates that the ordering on numerosities originates from the subset relation is the following:

- **Zermelo's Principle (ZP).** $\mathfrak{n}(A) \leq \mathfrak{n}(B)$ *if and only if $\mathfrak{n}(A) = \mathfrak{n}(A')$ for some subset $A' \subseteq B$.*

Recall that the comparability between well-ordered sets was established by Cantor, but it was Zermelo who proved that every set can be well-ordered (by using the axiom of choice). In consequence of Zermelo's result, cardinalities are arranged in a linear order, and for any two sets, it is always the case that one has the same cardinality as a subset of the other. Notice that **(ZP)** holds for Cantorian cardinalities, while **(EP)** fails badly.

The main goal of *numerosity theory* is to investigate the possibility of a counting system for infinite sets where *Euclid's Principle*, and possibly also *Zermelo's Principle*, are maintained. Another goal is extending the nice algebraic properties of natural numbers to the new numbers.

DEFINITION 15.3. *A numerosity system is a counting system where* Euclid's Principle *(EP) holds.*

DEFINITION 15.4. *A numerosity system is called* Zermelian *when* Zermelo's Principle *(ZP) holds.*

---

[1] The usual quotation one refers to is found in Part III of Book I of Hume's *Treatise of Human Nature*: "*When two numbers are so combined, as that the one has always a unit answering to every unit of the other, we pronounce them equal.*"

[2] That principle is explicitly formulated as Common Notion 5 in Euclid's *Elements*: "*The whole is greater than the part*" (translation by T.L. Heath [**46**]).

PROPOSITION 15.5. *A counting system* $(\mathfrak{U}, \mathfrak{N}, \mathfrak{n})$ *is Zermelian if and only if the following is satisfied:*

- **Trichotomic Property.** *For any* $A, B \in \mathfrak{U}$, *exactly one of the following holds:*
  - (1) $\mathfrak{n}(A) = \mathfrak{n}(B)$;
  - (2) $\mathfrak{n}(A) = \mathfrak{n}(A')$ *for some proper subset* $A' \subset B$;
  - (3) $\mathfrak{n}(B) = \mathfrak{n}(B')$ *for some proper subset* $B' \subset A$.

PROOF. One direction is easy; indeed the *Tricothomic property* directly implies both (ZP) and (EP).

Conversely, given $A, B \in \mathfrak{U}$, by combining (EP) with (ZP) it is readily seen that when condition (1) fails, then one of conditions (2) and (3) must hold.

Let us now show that the three cases (1), (2), (3) are pairwise incompatible. Assume by contradiction that $\mathfrak{n}(A) = \mathfrak{n}(B)$ and that $\mathfrak{n}(A) = \mathfrak{n}(A')$ for some proper subset $A' \subset B$; then $\mathfrak{n}(B) = \mathfrak{n}(A')$, against (EP). Similarly, it is not possible that the first and the third cases above hold simultaneously. Finally, if $\mathfrak{n}(A) = \mathfrak{n}(A')$ and $\mathfrak{n}(B) = \mathfrak{n}(B')$ for proper subsets $A' \subset B$ and $B' \subset A$; then by (EP) we would have that $\mathfrak{n}(B) = \mathfrak{n}(B') < \mathfrak{n}(A) = \mathfrak{n}(A') < \mathfrak{n}(B)$, a contradiction. $\square$

EXAMPLE 15.6. *If we take as universe* $\mathfrak{U} = \mathfrak{Fin}$ *the class of finite sets, as* $\mathfrak{N} = \mathbb{N}_0$ *the set of non-negative integers with the usual ordering, and as* $\mathfrak{n} = |\cdot|$ *the cardinality function, we obtain the* natural numbers counting system $(\mathfrak{Fin}, \mathbb{N}_0, |\cdot|)$. *Such a system is a* Zermelian numerosity *that also satisfies* Hume's Principle *(HP).*

It is a well-known and counter-intuitive phenomenon that *Euclid's Principle* and *Hume's Principle* cannot go together when considering infinite sets. Indeed, every infinite set can be put in 1-1 correspondence with a proper subset of its. With respect to the intuitive evidence of (EP), it seems appropriate mentioning the following quotation:

> *The possibility that whole and part may have the same number of terms is, it must be confessed, shocking to common sense.*
>
> (B. Russell, *Principles of Mathematics*, 1903, p. 358)

Historically, (EP) and (HP) revealed incompatible for infinite collections long before the celebrated *Galileo's paradox* that there should be simultaneously "equally many" and "much less" perfect squares than natural numbers.

> *But if I inquire how many roots there are, it cannot be denied that there are as many as the numbers because every number is the root of some square. This being granted, we must say that there are as many squares as there are numbers because they are just as numerous as their roots, and all the numbers are roots. Yet at the outset we said that there are many more numbers than squares, since the larger portion of them are not squares.*
>
> (Galileo, *Discorsi e Dimostrazioni Matematiche Intorno a Due Nuove Scienze*, 1638, translation by Crew and de Salvio)

The inconsistency of *Hume's* and *Euclide's Principles* together for infinite sets has been a relevant issue in the history of mathematics: let us only mention that it led Leibniz, an inventor of infinitesimal analysis who used actual infinitesimal numbers, to assert the impossibility of an infinite whole.

> *It appears to me that we must say either that the infinite is not truly one whole, or else that if the infinite is a whole, and yet is not greater than its part, then it is something absurd. Indeed I demonstrated many years ago that the number of the multitude of all numbers implies a contradiction if taken as a unitary whole.*
>
> (Leibniz, *Letter to Bernoulli*, 1698)

Much later, in the second half of the nineteenth century, Cantor had the great merit of realizing that, by dropping one of these assumptions, it is actually possible to construct consistent theories. In fact, by giving up *Euclid's principle*, he developed the then revolutionary theories of *cardinal numbers* and of *ordinal numbers*.

Let us now quickly review cardinal and ordinal numbers as counting systems.

2. Cardinals and ordinals as counting systems

With his theory of *transfinite cardinalities*, Cantor showed that constructing sound counting systems for the universe of all sets is actually possible, provided one abandons *Euclid's Principle* (EP).

EXAMPLE 15.7. *The Cantorian counting system of* cardinal numbers *is the system*

$$(\mathbf{V}, \mathbf{Card}, |\cdot|)$$

where $\mathfrak{U} = \mathbf{V}$ *is the universal class of all sets,* $\mathfrak{N} = \mathbf{Card}$ *is the class of cardinal numbers, and* $\mathfrak{n}(A) = |A|$ *is the Cantorian cardinality of* A.

Trivially, the Cantorian counting system satisfies *Hume's Principle*, as given by the very definition of the relation of *equipotency*.

Cantor also exploited another intuition that can be informally expressed as follows: When counting an infinite set, the result may depend on the "process" that is used for counting. In his theory of *ordinal numbers*, Cantor formalized the idea of counting the size of a set by putting its elements in a row. It is worth remarking that such a process applies only to *well-ordered sets*. With respect to this, it is interesting the fact that Cantor himself initially took it for granted that all sets were well-orderable. It took the work of Zermelo to clarify this matter and show that such an assumption needs the *axiom of choice*.

EXAMPLE 15.8. *The Cantorian counting system of* ordinal numbers *is the system*

$$(\mathbf{WO}, \mathbf{Ord}, \mathrm{ot})$$

where the universe $\mathfrak{U} = \mathbf{WO}$ *is the class of* well-ordered sets, *the numbers* $\mathfrak{N} = \mathbf{Ord}$ *are the ordinal numbers, and* $\mathfrak{n} = \mathrm{ot}$ *is the order-type function that associates to each well-ordered set the unique ordinal number isomorphic to it.*

Recall that the class of well-ordered sets is closed under disjoint unions and Cartesian products, provided the following rules are assumed:

- The well-ordering on the *disjoint union* $A \sqcup B$ is the order obtained by putting all elements of A before the elements of B;
- The well-ordering on the *Cartesian product* $A \times B$ is obtained by arranging the ordered pairs in an *anti-lexicographic* manner, that is, by putting $(a, b) < (a', b')$ if either $b < b'$, or $b = b'$ and $a < a'$.

The above operations of disjoint union and Cartesian product correspond to the sum and product operations of ordinals, respectively. It is worth recalling that both such operations are *not* commutative (see Example 15.9 below).

Recall that if $A \subset B$ is a *proper initial segment* of B then $\mathrm{ot}(A) < \mathrm{ot}(B)$.[3] However, ordinal numbers do not satisfy *Euclid's Principle*

---

[3] A subset $A \subseteq B$ is an initial segment of the ordered set $(B, <)$ if A is *downward closed*, that is, if $x \in B$ is such that $x < a \in A$ then $x \in A$.

(EP), since one easily finds examples of isomorphic well-ordered sets where one is properly included in the other (*e.g.* $\mathbb{N} \subset \mathbb{N}_0$).

Notice that sets with the same order types are necessarily in bijection, and so the following weaker version of *Hume's Principle* holds:

- **Hume's Half Principle (HHP).** If $\mathfrak{n}(A) = \mathfrak{n}(B)$ *then there exists a bijection between A and B.*

Let us remark that the full *Hume's Principle* (HP) fails for the counting system of ordinal numbers because there is plenty of well-ordered sets with same cardinality but different order types.[4] As a consequence, the two Cantorian counting systems of cardinal and ordinal numbers give different results when applied to infinite sets.[5]

The theories of ordinal and cardinal numbers were a great breakthrough in the history of mathematics, because they finally made it possible to deal with infinite quantities in a rigorous manner. However, Cantor's theories cannot be used to define infinitesimal numbers and develop infinitesimal analysis. The reason for this unfeasibility is the fact that the sum and product operations for cardinal and ordinal numbers do not have nice algebraic properties, and they indeed behave in an awkward way. Recall that whenever $\mathfrak{a}$ and $\mathfrak{b}$ are infinite cardinals, we have

$$\mathfrak{a} + \mathfrak{b} = \mathfrak{a} \cdot \mathfrak{b} = \max\{\mathfrak{a}, \mathfrak{b}\}.$$

The above equalities, which trivialize cardinal arithmetic, are consequences of the failure of *Euclid's Principle* (EP): one may add something to an infinite set without making it "bigger"!

About ordinal numbers, things are even worse because sum and product are not even commutative operations.

EXAMPLE 15.9. *By adding an element to an infinite well-ordered set, one gets a larger order-type if and only if the element is added on the right side. E.g., by adding a new element on the left side of $\mathbb{N}$ one gets the same order-type of $\mathbb{N}$, namely ω; while adding a new element on the right side of $\mathbb{N}$ one gets a larger order-type, namely $\omega + 1$.*

$$\mathrm{ot}(\{\star\} \sqcup \mathbb{N}) = 1 + \omega = \omega < \omega + 1 = \mathrm{ot}(\mathbb{N} \sqcup \{\star\}).$$

---

[4] Precisely, every infinite set of cardinality κ admits κ^+-many pairwise non-isomorphic well-orderings.

[5] In contemporary set theory, as given by Zermelo-Fraenkel axiomatics ZFC, the first infinite cardinal is the same object as the first infinite ordinal, namely the set of von Neumann natural numbers. However, even in this case, different notations are used to avoid confusion: $|\mathbb{N}| = \aleph_0$ and $\mathrm{ot}(\mathbb{N}) = \omega$.

Similar "absorption on the left side" phenomena also occur with prod-
ucts. For instance,

$$\mathrm{ot}(\{1,2\} \times \mathbb{N}) = 2 \cdot \omega = \omega < \omega + \omega = \omega \cdot 2 = \mathrm{ot}(\mathbb{N} \times \{1,2\}).$$

REMARK 15.10. *There is a simple way of defining a sum and a*
product operation on ordinal numbers that are associative, commuta-
tive, and satisfy distributivity. Such operations, the so-called Hessenberg *natural operations (see, e.g., [*89*, Ch.XIV §28]), are related to*
*sums and products of numerosities (see [*12, 15*]).*

3. Three different ways of counting

In everyday life, there are several possible equivalent ways of counting the number of elements of finite sets. However, when these ways of
counting are formalized and extended to infinite sets, they may give different counting systems. Basically, there are three different approaches.

The first way of counting consists in associating to each element of
a set an element of another one. In this way one gets a 1-1 correspondence, and then one claims that the two sets have the same number of
elements. This intuition corresponds to the equipotency relation and
to the Cantorian theory of cardinal numbers.

In the second way of counting, one arranges the elements of a given
set in a row, and then compares such a row with the sequence of natural
numbers. This intuition leads to the notion of order type and to the
theory of ordinal numbers.

However, there exists a third way of counting which consists in
arranging the elements of a given sets into smaller groups to be counted
separately. Let us see an example. Assume we are given a big bunch of
randomly chosen playing cards. Probably, the better strategy to count
the cards is to divide them into smaller decks. For example, one can
put all the aces in a deck, all the twos in another deck, and so forth.
This operation can be simplified by using numbered boxes: one puts
the aces in box number 1, the twos in box number 2, and so forth.
The total number of cards in the bunch is finally obtained by adding
up the number of cards found in each box. This process of counting
corresponds to the numerosity of *labelled sets*, that will be presented
in the next chapter.

The three ways of counting discussed above imply more and more
complex logical operations.

- The first way corresponds to the concept of number of a two years old kid, who associate numbers to sets of fingers of his hands; *e.g.*, the number 3 corresponds to the set
{index finger, middle finger, ring finger}.[6]

- The second way of counting corresponds to the concept of number of a four years old child: she/he has already memorized the sequence of the first natural numbers and she/he is able to count objects by arranging them in a row.

- The third way of counting is much more sophisticated and requires several operations, such as collecting similar objects together, and comparing different groups. This is the way of counting of a grown child.

Clearly, the third way of counting is only possible if the given set has a "structure" that allows us to collect "similar objects". That kind of structure is formalized by the notion of *labelled set* that will be considered in the next chapter.

4. The equisize relation

We close this chapter by presenting and discussing an alternative equivalent presentation of Zermelian systems. In analogy to the theory of cardinalities, which is is grounded on the equivalence relation of *equipotency*, here we take the equivalence relation of "equisize" as our primitive concept, and then postulate natural principles that such a notion should satisfy. The resulting framework will be shown to be equivalent to the one introduced in §15.1.

One nice point in favor of the equisize approach is the fact that the proposed five axioms seem to exactly match, if appropriately interpreted, the traditional principles for a theory of magnitudes as explicitly formulated in the classic Euclid's *Elements* (see below).

DEFINITION 15.11. *An equivalence relation* $\cong$ *over a universe* $\mathfrak{U}$ *is called* equisize *if the following properties are satisfied:*

(E1) *For every A and B, exactly one of the following holds:*

(1) $A \cong B$;

(2) $A \cong A'$ *for some proper subset $A' \subset B$;*

(3) $B \cong B'$ *for some proper subset $B' \subset A$;*

(E2) *If $A \cong A'$ and $B \cong B'$ then the disjoint unions $A \sqcup B \cong A' \sqcup B'$;*

---

[6] In some parts of Europe, thumb is used in place of the ring finger.

(E3) *If $A \subseteq B$ and $A' \subseteq B'$ then*

$$A \cong A' \text{ and } B \cong B' \;\Rightarrow\; B \setminus A \cong B' \setminus A';$$

(E4) $\{P\} \times A \cong A \times \{P\} \cong A$ *for all singletons* $\{P\}$;

(E5) *If $A \cong A'$ and $B \cong B'$, then*

$$A \times B \;\cong\; A' \times B'.$$

Let us now briefly discuss the above five axioms and their relationships with classical principles for a theory of magnitude. Let us recall the first five *Common Notions* from Euclid's *Elements* (see [**46**]):

- Common Notion 1. *"Things which are equal to the same thing are also equal to one another".*

- Common Notion 2. *"If equals be added to equals, the wholes are equal".*

- Common Notion 3. *"If equals be subtracted from equals, the remainders are equal".*

- Common Notion 4. *"Things applying [exactly] onto one another are equal to one another".*

- Common Notion 5. *"The whole is greater than the part".*

The first common notion can be formalized as the implication:

$$A \cong C \text{ and } B \cong C \;\Rightarrow\; A \cong B.$$

Notice that, if one takes reflexivity for granted, this property precisely corresponds to our initial assumption that equisize be an equivalence relation.

Axioms (E2) and (E3) can be seen as direct reformulations in a set-theoretic framework of the second and third common notion, respectively. We remark these two axioms trivially hold for equipotency between finite sets. However, while (E2) also holds in general, axiom (E3) fails badly for equipotency between infinite sets.

The *tricotomic property* (E1) combines two natural ideas. The first one is that, given two sets, one is equinumerous to some subset of the other. As a consequence, there is a natural ordering of numerosities that satisfies the implicit assumption of the classical theory that (homogeneous) magnitudes are always comparable. The second idea is that a set cannot be equinumerous to a proper subset, and this corresponds to the fifth common notion.

Axiom (E4) could be viewed as an instance of the fourth common notion, that says that "superpositions" preserve the magnitude. In fact, in a set-theoretic framework, it seems natural to assume that any transformation $A \mapsto \{P\} \times A$ or $A \mapsto A \times \{P\}$ preserves the numerosity of sets, in a similarly fashion as in geometry the "rigid motions" preserve the magnitude of figures.[7]

Axiom (E5) formalizes the idea that a natural definition of a product of numerosities of sets can be given by means of Cartesian products. Such a product admits the size of every singleton as an identity by (E4).

REMARK 15.12. *While addition, subtraction, and order of magnitudes are explicit in ancient mathematics, multiplication is lacking. Notice that the geometric idea of a product necessarily yields objects of higher dimension; e.g., the "product" of two lines produces a rectangle. It is worth stressing the fact that measuring* non-homogeneous *magnitudes by means of the same numbers, namely the real numbers $\mathbb{R}$, is an entirely modern viewpoint, alien to the ancient geometric thinking.*

We conclude this discussion on the given axiomatization by remarking that the *difference property* (E3) is actually redundant, in that it follows from axioms (E1) and (E2), as shown below. We decided to keep it as part of the definition of equisize because it exactly matches one of Euclid's common notions.

PROPOSITION 15.13. *Axiom (E3) is a consequence of axioms (E1) and (E2).*

PROOF. Assume $A \subseteq B$ and $A' \subseteq B'$ where $A \approx A'$ and $B \approx B'$. Notice that it cannot be $B \setminus A \approx C$ for any proper subset $C \subset B' \setminus A'$, otherwise we would have

$$B' \approx B = (B \setminus A) \cup A \approx \text{(by (E2))} \approx C \cup A' \subset B',$$

but a set B' cannot be equinumerous to a proper subset, by (E1). Similarly, it is shown that $B' \setminus A' \not\approx C'$ for any proper subset $C' \subset B \setminus A$. By applying the tricotomic property, the only possibility that is left is $B \setminus A \approx B' \setminus A$, as desired.   □

Let us now show that the proposed axiomatization for a notion of *equisize* is actually equivalent to the one given for a Zermelian system (see Definition 15.4).

---

[7] It should be stressed here that some set-theoretic limitation is needed to avoid the possibility that $\{P\} \times A$ be a proper subset of A. For instance, if we could take the set $A = \{a_n \mid n \in \mathbb{N}\}$ where inductively $a_{n+1} = (P, a_n)$, then $\{P\} \times A$ would be properly included in A.

THEOREM 15.14. *Given the equisize relation $\cong$ on the universe $\mathfrak{U}$, define:*

- $\mathfrak{N}_\cong = \mathfrak{U}/\!\cong$ *as the quotient of $\mathfrak{U}$ modulo $\cong$;*
- $\mathfrak{n}_\cong : \mathfrak{U} \to \mathfrak{N}_\cong$ *as the canonical projection $A \mapsto [A]_\cong$;*
- $\mathfrak{n}_\cong(A) < \mathfrak{n}_\cong(B) \iff A \cong A'$ *for some proper subset $A' \subset B$.*

Then $(\mathfrak{N}_\cong, <)$ is a linear ordering that includes (a copy of) the natural numbers $\mathbb{N}_0$ as an initial segment, and $(\mathfrak{U}, \mathfrak{N}_\cong, \mathfrak{n}_\cong)$ is a Zermelian numerosity system.

Conversely, given a Zermelian numerosity system $(\mathfrak{U}, \mathfrak{N}, \mathfrak{n})$, define:

- $A \cong_\mathfrak{n} B \iff \mathfrak{n}(A) = \mathfrak{n}(B)$.

Then $\cong_\mathfrak{n}$ is an equisize relation.

PROOF. Let $\cong$ be an equisize relation. We begin by proving the following property, that will be used in the sequel of the proof.

$(\star)$ If $A \cong B$, then for any $X \subset A$ there exists $Y \subset B$ such that $X \cong Y$.

Trivially X has equal size to a proper subset of A and so, by the trichotomy (E1), $X \not\cong A$, and hence $X \not\cong B$. Besides, $B \cong B' \subset X$ cannot happen for any B', because otherwise we would have $A \cong B'$ where $B' \subset A$, a contradiction. Again by the trichotomic property (E1) applied to the pair X, B, we conclude that $X \cong Y$ for some proper subset $Y \subset B$, as desired.

Now define the following relation on $\mathfrak{U}$:

$$A \prec B \iff A \cong A' \text{ for some proper subset } A' \subset B.$$

As a first step, let us verify that the given definition of $\prec$ is compatible with the equivalence classes modulo $\cong$. Assume that $A_0 \prec B_0$, $A_1 \cong A_0$ and $B_1 \cong B_0$. By definition, $A_0 \cong A_0'$ for some proper subset $A_0' \subset B_0 \cong B_1$. By the property $(\star)$ above, there exists a proper subset $A_1' \subset B_1$ such that $A_0' \cong A_1'$. We conclude that $A_1 \prec B_1$ because $A_1 \cong A_0 \cong A_0' \cong A_1' \subset B_1$.

Let us now show that $\prec$ yields a linear order $<$ on the equivalence classes $\mathfrak{N}_\cong = \mathfrak{U}/\!\cong$. The *irreflexive* property $A \not\prec A$ holds because we already noticed that a set cannot be equinumerous to a proper subset, by (E1). As for *transitivity*, assume that $A \prec B$ and $B \prec C$. Pick proper subsets $A' \subset B$ and $B' \subset C$ such that $A \cong A'$ and $B \cong B'$. By $(\star)$, there exists $A'' \subset B'$ such that $A' \cong A''$. Then $A \prec C$ because $A \cong A''$ where $A'' \subset C$. Finally, the *linearity* of $<$ is just a reformulation of the trichotomic property (E1).

For every $n \in \mathbb{N}_0$, fix a finite set $F_n \in \mathfrak{U}$ of cardinality n, and identify n with the equivalence class of F_n; in this way, we obtain a copy of the natural numbers included in $\mathfrak{N}_\approx$. To prove (CS1), we have to show that for every finite set G of cardinality n one has that $G \approx F_n$.

Let us proceed by induction. If $n = 0$ then the thesis is trivial, because it must be $G = F_0 = \emptyset$. Now let $n = 1$, and assume by contradiction that there exist two singletons such that $\{x\} \not\approx \{y\}$. Since a proper set of a singleton is necessarily empty, by (E1) it would follow that either $\{x\} \approx \emptyset$ or $\{y\} \approx \emptyset$. On the other hand, $\emptyset$ is a proper subset of any singleton and so, again by (E1), we would obtain that $\emptyset \not\approx \{x\}$ and $\emptyset \not\approx \{y\}$, a contradiction. If $|G| = n > 1$, pick an element $x \in G$ and an element $y \in F_n$, and let $G' = G \setminus \{x\}$ and $F' = F_n \setminus \{y\}$. By the inductive hypothesis, $|G'| = |F'| = n - 1 \Rightarrow G' \approx F_{n-1} \approx F'$. Since $\{x\} \approx \{y\}$ we can apply (E2) and get $G = G' \cup \{x\} \approx F' \cup \{y\} = F_n$, as desired. Finally, observe that any infinite set contains subsets of arbitrary finite cardinalities; therefore $\mathfrak{n}_\approx(F) < \mathfrak{n}_\approx(X)$ whenever F is finite and X is infinite, and $\mathbb{N}_0$ is an initial segment of $\mathfrak{N}_\approx$.

It is now easily seen that $(\mathfrak{U}, \mathfrak{N}_\approx, \mathfrak{n}_\approx)$ is a Zermelian numerosity system. Indeed, the *Monotonicity Principle* (CS2) and *Zermelo's Principle* (ZP) directly follow by the definition of the order relation $<$ and by axiom (E1). Moreover, the *Union Principle* (CS3), the *Product Principle* (CS4), and *Unit Principle* (CS5), precisely correspond to axioms (E2), (E5) and (E4), respectively.

Let us now turn to the converse implication, and assume that we are given a Zermelian numerosity system $(\mathfrak{U}, \mathfrak{N}, \mathfrak{n})$. We already noticed in Proposition 15.5 that for any $A, B \in \mathfrak{U}$, exactly one of the following three cases holds:

(1) $\mathfrak{n}(A) = \mathfrak{n}(B)$, that is, $A \approx_\mathfrak{n} B$;

(2) $\mathfrak{n}(A) < \mathfrak{n}(B)$, and hence $A \approx_\mathfrak{n} A'$ for some proper subset $A' \subset B$, by (ZP);

(3) $\mathfrak{n}(B) < \mathfrak{n}(A)$, and hence $B \approx_\mathfrak{n} B'$ for some proper subset $B' \subset A$, by (ZP).

This proves (E1). Moreover, axioms (E2), (E4) and (E5) correspond to properties (CS3), (CS5), and (CS4) of a counting system, respectively. Finally, by Proposition 15.13, also axiom (E3) holds, as a consequence of (E1) and (E2). $\square$

In our opinion, while the principles expressed by axioms (E2), (E3), (E4) and (E5) should always be assumed for a notion of equisize between sets, the *trichotomic property* (E1) is disputable. It is worth recalling that the principle of comparability of Cantorian cardinalities is a highly non-trivial result, and in fact it took a few decades before Zermelo gave it satisfying axiomatic grounds by using the full strength of the *axiom of choice*.[8]

---

[8] Precisely, by using the axiom of choice, Zermelo proved that every set can be well-ordered (the comparability between well-ordered sets had been already established by Cantor). In the axiomatic framework of Zermelo-Fraenkel set theory ZF, the axiom of choice is in fact equivalent to the property that, given two sets, one is equipotent to a subset of the other.

CHAPTER 16

Alpha-Theory and Numerosity

1. Labelled sets

In the previous chapter we reviewed cardinal and ordinal numbers as possible ways to extend the notion of finite cardinality to infinite sets. In particular, we recalled that one needs an order structure to be able to associate an ordinal number to a given set, and that the ordinal number of a set depends on the particular way its elements are arranged in a well-ordered manner (see Example 15.9).

Now, the question that arises naturally is the following one: Is there a way to count the elements of infinite sets in such a way that one obtains a *numerosity counting system*? In other words, is there a counting system that extends the finite cardinality to infinite sets by preserving the principle that *"the whole is larger than the part"*? Following the idea about a "third way of counting" as outlined in §15.3, we are asking whether there is some "reasonable" structure one can add to infinite sets, so that its elements can be splitted into finite sets to be counted separately.

For example, if one wants to count the inhabitants of the world, then he/she may divide them into nations, count the inhabitants of each nation, and finally add them up. If we want to apply this idea to infinite sets, it is necessary to have a criterium to arrange elements into smaller groups. To this end, every element should come with a "label" that distinguishes it from other elements. (In the previous example, such a label is the nationality). So, we are lead to the following

DEFINITION 16.1. *A labelled set* $\mathbf{A} = (A, \ell)$ *is a pair where A is a set and*

$$\ell : A \to \mathbb{N}_0$$

is a finite-to-one function, that is, all pre-images $\ell^{-1}(n)$ are finite.

The function ℓ is called the labelling function *of $\mathbf{A}$, and the number $\ell(x)$ is called the* label *of x. The set A is called the* domain *of $\mathbf{A}$.*

REMARK 16.2. *Since the labelling function is finite-to-one, the domain of a labelled set has at most countable cardinality.*

We will use boldface letters $\mathbf{A}, \mathbf{B}, \mathbf{C}, \ldots$ to denote labelled sets, and the corresponding capital letters $A, B, C, \ldots$ to denote their domains.

EXAMPLE 16.3. *Every set $A \subseteq \mathbb{N}_0$ comes with the* canonical label $\ell(n) = n$. *More generally, every tuple $(a_1, \ldots, a_k) \in \mathbb{N}_0^k$ of non-negative integers comes with a* canonical label, *namely:*

$$\ell(a_1, \ldots, a_k) \;=\; \max\{a_1, \ldots, a_k\}.$$

EXAMPLE 16.4. *Also tuples $(a_1, \ldots, a_k) \in \mathbb{Z}^k$ have canonical labels obtained by taking absolute values:*

$$\ell(a_1, \ldots, a_k) \;=\; \max\left\{|a_1|, \ldots, |a_k|\right\}.$$

DEFINITION 16.5. $\mathbf{A} = (A, \ell_A)$ *is a* labelled subset *of $\mathbf{B} = (B, \ell_B)$ if $A \subseteq B$ and $\ell_B(a) = \ell_A(a)$ for every $a \in A$. In this case we write $\mathbf{A} \subseteq \mathbf{B}$. We denote $\mathbf{A} \subset \mathbf{B}$ if $\mathbf{A} \subseteq \mathbf{B}$ and $A \neq B$.*

DEFINITION 16.6. *Two labelled set $\mathbf{A} = (A, \ell_A)$ and $\mathbf{B} = (B, \ell_B)$ are* isomorphic *if there exists a bijection $\Psi : A \to B$ that preserves the labels, that is, such that $\ell_B(\Psi(a)) = \ell_A(a)$ for every $a \in A$.*

DEFINITION 16.7. *The* disjoint union *of two labelled sets $\mathbf{A} = (A, \ell_A)$ and $\mathbf{B} = (B, \ell_B)$ where $A \cap B = \emptyset$ is the labelled set $\mathbf{A} \sqcup \mathbf{B} = (A \sqcup B, \ell_{A \sqcup B})$ where the labelling function is defined by:*

$$\ell_{A \sqcup B}(x) \;=\; \begin{cases} \ell_A(x) & \text{if } x \in A; \\ \ell_B(x) & \text{if } x \in B. \end{cases}$$

REMARK 16.8. *The above definition also applies to arbitrary (not necessarily disjoint) unions $\mathbf{A} \cup \mathbf{B}$ provided the labels are coherent on the intersection, that is, when $\ell_A(x) = \ell_B(x)$ for all $x \in A \cap B$.*

It is easily seen that for all $\mathbf{A}$ and $\mathbf{B}$ there exist isomorphic $\mathbf{A}' \cong \mathbf{A}$ and $\mathbf{B}' \cong \mathbf{B}$ where $A' \cap B' = \emptyset$.

There is a natural labelling for ordered pairs and, in consequence, one has a natural notion of product of labelled sets.

DEFINITION 16.9. *The* product *of two labelled sets* $\mathbf{A} = (A, \ell_A)$ *and* $\mathbf{B} = (B, \ell_B)$ *is the labelled set* $\mathbf{A} \times \mathbf{B} = (A \times B, \ell_{A \times B})$ *where the labelling function*

$$\ell_{A \times B}(a, b) = \max\{\ell_A(a), \ell_B(b)\}.$$

Every labelled set comes with a natural counting function.

DEFINITION 16.10. *The* counting function $\varphi_{\mathbf{A}} : \mathbb{N}_0 \to \mathbb{N}_0$ *of a labelled set* $\mathbf{A} = (A, \ell)$ *is defined by putting:*

$$\varphi_{\mathbf{A}}(n) = |\{a \in A \mid \ell(a) \leq n\}|.$$

The finite sets $A_n = \{a \in A \mid \ell(a) \leq n\}$ *are called* finite approximations *of* $\mathbf{A}$, *since* A *is obtained as the increasing union* $\bigcup_{n \geq 0} A_n$.

PROPOSITION 16.11. *Two labelled sets* $\mathbf{A} = (A, \ell_A)$ *and* $\mathbf{B} = (B, \ell_B)$ *have the same counting function* $\varphi_{\mathbf{A}} = \varphi_{\mathbf{B}}$ *if and only if they are isomorphic:* $\mathbf{A} \cong \mathbf{B}$.

PROOF. Assume first that $\varphi_{\mathbf{A}} = \varphi_{\mathbf{B}}$. For every $n \in \mathbb{N}_0$, let $X_n = \{a \in A \mid \ell_A(a) = n\}$ and let $Y_n = \{b \in B \mid \ell_B(b) = n\}$. Then $|X_0| = \varphi_{\mathbf{A}}(0) = \varphi_{\mathbf{B}}(0) = |Y_0|$; and for every $n \geq 1$,

$$|X_n| = \varphi_{\mathbf{A}}(n) - \varphi_{\mathbf{A}}(n-1) = \varphi_{\mathbf{B}}(n) - \varphi_{\mathbf{B}}(n-1) = |Y_n|.$$

So, for every $n \in \mathbb{N}_0$ we can pick a bijection $\Psi_n : X_n \to Y_n$. Clearly, the sets in the family $\{X_n\}_n$, as well as the sets in the family $\{Y_n\}_n$, are pairwise disjoint. So, by putting together the functions Ψ_n one obtains a bijection $\Psi : A \to B$ between $A = \bigcup_{n \geq 0} X_n$ and $B = \bigcup_{n \geq 0} Y_n$, and the following equivalences hold:

$$\ell_A(a) = n \iff a \in X_n \iff \Psi(a) = \Psi_n(a) \in Y_n \iff \ell_B(\Psi(a)) = n.$$

Conversely, assume $\mathbf{A} \cong \mathbf{B}$ and let $\Psi : A \to B$ be a bijection where $\ell_B(\Psi(a)) = \ell_A(a)$ for every $a \in A$. Then, with the same notation as above, one has that $Y_n = \{\Psi(a) \mid a \in X_n\}$. It follows that $|X_n| = |Y_n|$ for all $n \geq 0$, and hence

$$\varphi_{\mathbf{A}}(n) = \left| \bigcup_{m=0}^{n} X_m \right| = \sum_{m=0}^{n} |X_m| = \sum_{m=0}^{n} |Y_m| = \left| \bigcup_{m=0}^{n} Y_m \right| = \varphi_{\mathbf{B}}(n).$$

$\square$

PROPOSITION 16.12. *Every non-decreasing function* $f : \mathbb{N}_0 \to \mathbb{N}_0$ *is the counting function* $\varphi_{\mathbf{A}}$ *of a suitable labelled set* $\mathbf{A}$.

PROOF. Let B_0 be a set of cardinality $|B_0| = f(0)$ and, inductively, let B_{n+1} be a set disjoint from $\bigcup_{i=1}^{n} B_n$ and with cardinality $|B_{n+1}| = f(n+1) - f(n)$. The labelled set $\mathbf{A} = (A, \ell)$ where $A = \bigcup_{n \geq 0} B_n$ and $\ell(a) = n \Leftrightarrow a \in B_n$ has the desired counting function $\varphi_{\mathbf{A}}(n) = |\bigcup_{i=0}^{n} B_i| = f(n)$, as it directly follows from the definition. $\square$

2. Alpha-numerosity

Thanks to the Alpha-Theory, one has a natural way of assigning a hypernatural number to each labelled set.

DEFINITION 16.13. *The* Alpha-numerosity $\mathfrak{n}_\alpha(\mathbf{A})$ *of a labelled set* $\mathbf{A}$ *is defined by setting*

$$\mathfrak{n}_\alpha(\mathbf{A}) \;=\; \lim_{n \uparrow \alpha} \varphi_{\mathbf{A}}(n).$$

The above notion of numerosity is coherent with inclusions and isomorphisms. Indeed, the following properties are readily verified:

- $\mathbf{A} \subseteq \mathbf{B} \;\Rightarrow\; \mathfrak{n}_\alpha(\mathbf{A}) \leq \mathfrak{n}_\alpha(\mathbf{B})$;
- $\mathbf{A} \cong \mathbf{B} \;\Rightarrow\; \mathfrak{n}_\alpha(\mathbf{A}) = \mathfrak{n}_\alpha(\mathbf{B})$.

The class of labelled sets with Alpha-numerosity fulfills the properties of a numerosity system (see Definition 15.3).

THEOREM 16.14. *Let* $\mathfrak{L}$ *be the class of all labelled sets, and let* $\mathfrak{N}_\alpha \subseteq \mathbb{N}_0^*$ *be the range of* $\mathfrak{n}_\alpha$. *Then the triple* $(\mathfrak{L}, \mathfrak{N}_\alpha, \mathfrak{n}_\alpha)$ *satisfies the following properties, and hence it is a numerosity system.*[1]

(1) *Finite Set Principle: If* $\mathbf{A}$ *has finite domain* A *then* $\mathfrak{n}(\mathbf{A}) = |A|$;

(2) *Monotonicity Principle: If* $\mathbf{A} \subseteq \mathbf{B}$ *then* $\mathfrak{n}_\alpha(\mathbf{A}) \leq \mathfrak{n}_\alpha(\mathbf{B})$;

(3) *Union Principle: For all labelled sets* $\mathbf{A}$ *and* $\mathbf{B}$ *whose domains are disjoint,* $\mathfrak{n}_\alpha(\mathbf{A} \sqcup \mathbf{B}) = \mathfrak{n}_\alpha(\mathbf{A}) + \mathfrak{n}_\alpha(\mathbf{B})$ *where* $+$ *is the sum operation on* $\mathbb{N}_0^*$;

(4) *Product Principle: For all labelled sets* $\mathbf{A}$ *and* $\mathbf{B}$, $\mathfrak{n}_\alpha(\mathbf{A} \times \mathbf{B}) = \mathfrak{n}_\alpha(\mathbf{A}) \cdot \mathfrak{n}_\alpha(\mathbf{B})$ *where* $\cdot$ *is the product operation on* $\mathbb{N}_0^*$;

(5) *Unit Principle:* $\mathfrak{n}_\alpha(\{\mathbf{P}\} \times \mathbf{A}) = \mathfrak{n}_\alpha(\mathbf{A} \times \{\mathbf{P}\}) = \mathfrak{n}_\alpha(\mathbf{A})$ *for all labelled sets* $\mathbf{A}$ *and all labelled singletons* $\{\mathbf{P}\}$;

(6) *Euclid's Principle: If* $\mathbf{A}$ *is a proper labelled subset of* $\mathbf{B}$ *then* $\mathfrak{n}_\alpha(\mathbf{A}) < \mathfrak{n}_\alpha(\mathbf{B})$.

---

[1] The problem whether $\mathfrak{N}_\alpha = \mathbb{N}_0^*$ will be addressed in Section 8.

PROOF. (1). Since A is finite, we can pick $k = \max\{\ell_A(a) \mid a \in A\}$. Then $\varphi_\mathbf{A}(n) = |A|$ for every $n \geq k$, and so $\mathfrak{n}_\alpha(\mathbf{A}) = \lim_{n \uparrow \alpha} \varphi_\mathbf{A}(n) = |A|$.

(2). Trivial, since $\varphi_\mathbf{A}(n) \leq \varphi_\mathbf{B}(n)$ for all n.

(3). Since the domains A and B are disjoint, for every n we have that $\varphi_{A \sqcup B}(n) = |\{x \in A \cup B \mid \ell_{A \sqcup B}(x) \leq n\}| = |\{x \in A \mid \ell_{A \sqcup B}(x) \leq n\}| + |\{x \in B \mid \ell_{A \sqcup B}(x) \leq n\}| = \varphi_\mathbf{A}(n) + \varphi_\mathbf{B}(n)$. By passing to the Alpha-limits, we obtain $\mathfrak{n}_\alpha(\mathbf{A} \sqcup \mathbf{B}) = \mathfrak{n}_\alpha(\mathbf{A}) + \mathfrak{n}_\alpha(\mathbf{B})$.

(4). Notice that for every n, $\varphi_{\mathbf{A} \times \mathbf{B}}(n) = |\{(x,y) \in A \times B \mid \max\{\ell_\mathbf{A}(x), \ell_\mathbf{B}(y)\} \leq n\}| = |\{x \in A \mid \ell_\mathbf{A}(x) \leq n\}| \cdot |\{y \in B \mid \ell_\mathbf{B}(y) \leq n\}| = \varphi_\mathbf{A}(n) \cdot \varphi_\mathbf{B}(n)$, and the equality $\mathfrak{n}_\alpha(\mathbf{A} \times \mathbf{B}) = \mathfrak{n}_\alpha(\mathbf{A}) \cdot \mathfrak{n}_\alpha(\mathbf{B})$ follows by passing to the Alpha-limits.

(5). If k is the label of the single point P, then for every $n \geq k$ we have that $\varphi_{\mathbf{A} \times \{\mathbf{P}\}}(n) = |\{(x,y) \in A \times \{P\} \mid \max\{\ell_A(x), \ell_{\{P\}}(y)\} \leq n\}| = |\{x \in A \mid \ell_A(x) \leq n\}| = \varphi_\mathbf{A}(n)$. So, by passing to the Alpha-limits, we obtain the desired equality $\mathfrak{n}_\alpha(\mathbf{A} \times \{\mathbf{P}\}) = \mathfrak{n}_\alpha(\mathbf{A})$. Equality $\mathfrak{n}_\alpha(\{\mathbf{P}\} \times \mathbf{A}) = \mathfrak{n}_\alpha(\mathbf{A})$ is proved in the same way.

(6). Pick an element $b \in B \setminus A$. If $k = \ell_B(b)$, then $A_n = \{x \in A \mid \ell_A(x) \leq n\}$ is a proper subset of $B_n = \{x \in B \mid \ell_B(x) \leq n\}$ for all $n \geq k$; so, $\varphi_\mathbf{A}(n) < \varphi_\mathbf{B}(n)$ for all $n \geq k$, and by passing to the Alpha-limits, we get $\mathfrak{n}_\alpha(\mathbf{A}) < \mathfrak{n}_\alpha(\mathbf{B})$. □

3. Finite parts and sets of functions

If $\mathbf{A} = (A, \ell_A)$ is an infinite labelled set then there exist no labelling functions on the powerset $\mathcal{P}(A)$ because that set is uncountable. However, a labelled set structure can be naturally defined on the finite parts.

DEFINITION 16.15. *If $\mathbf{A} = (A, \ell_A)$ is a labelled set, then the labelled set $\mathfrak{Fin}(\mathbf{A}) = (\mathfrak{Fin}(A), \ell_{\mathfrak{Fin}(A)})$ of its finite parts is defined by letting for every $X \in \mathfrak{Fin}(A)$:*

$$\ell_{\mathfrak{Fin}(A)}(X) = \max\{\ell_A(x) \mid x \in X\}.$$

Next, we will see that also a notion of exponentiation makes sense for labelled sets. In analogy to real functions, for functions $f : A \to B$ where $0 \in B$, let us call the *support* of f the following set:

$$\mathrm{supp}(f) = \{a \in A \mid f(a) \neq 0\}.$$

DEFINITION 16.16. *Let $\mathbf{A} = (A, \ell_A)$ and $\mathbf{B} = (B, \ell_B)$ be labelled sets, and assume that $0 \in B$ with its canonical label $\ell_B(0) = 0$. Then the* exponentiation $\mathbf{B}^\mathbf{A} = (\mathfrak{Fun}_0(A, B), \ell_{\exp})$ *is the labelled set where:*

- *The domain $\mathfrak{Fun}_0 = \{f : A \to B \mid supp(f) \in \mathfrak{Fin}\}$ is the set of functions from A to B with finite support;*
- *$\ell_{\exp}(f) = \max\{\ell_{A \times B}(a, f(a)) \mid a \in supp(f)\}$ is the greatest label among elements in the support of f and in the range of f.[2]*

The definition of exponentiation is coherent with that of finite parts.

EXERCISE 16.17. *Let $\mathbf{2}$ be the canonical labelled set with 2 elements, that is, $\mathbf{2} = (\{0, 1\}, \ell)$ where $\ell(0) = 0$ and $\ell(1) = 1$. Then for every labelled set $\mathbf{A}$ there exists an isomorphism $\mathbf{2^A} \cong \mathfrak{Fin}(\mathbf{A})$.*

PROPOSITION 16.18. *For all labelled sets $\mathbf{A}$ and $\mathbf{B}$:*

(1) $\mathfrak{n}_\alpha(\mathfrak{Fin}(\mathbf{A})) = 2^{\mathfrak{n}_\alpha(\mathbf{A})}$; *in particular, $\mathfrak{n}_\alpha(\mathfrak{Fin}(\mathbb{N})) = 2^\alpha$.*

(2) $\mathfrak{n}_\alpha(\mathbf{B^A}) = \mathfrak{n}_\alpha(\mathbf{B})^{\mathfrak{n}_\alpha(\mathbf{A})}$.

PROOF. For $n \in \mathbb{N}_0$, let us consider the following finite approximations:

- $A_n = \{a \in A \mid \ell_A(a) \le n\}$;
- $B_n = \{b \in B \mid \ell_B(b) \le n\}$;
- $(B^A)_n = \{f \in \mathfrak{Fun}_0(A, B) \mid \ell_{\exp}(f) \le n\}$.

(1). For every $n \in \mathbb{N}$, we have that

$$
\begin{aligned}
\varphi_{\mathfrak{Fin}(\mathbf{A})}(n) &= |\{X \in \mathfrak{Fin}(A) \mid \ell_{\mathfrak{Fin}(\mathbf{A})}(X) \le n\}| \\
&= |\{X \in \mathfrak{Fin}(A) \mid \max\{\ell_A(x) \mid x \in X\} \le n\}| \\
&= |\{X \in \mathfrak{Fin}(A) \mid X \subseteq A_n\}| = 2^{|A_n|} = 2^{\varphi_\mathbf{A}(n)}.
\end{aligned}
$$

We reach the desired conclusion by passing to the Alpha-limits on both sides.

(2). Fix n. Given $f \in (B^A)_n$, denote by f_n the restriction of f to A_n; notice that $f_n : A_n \to B_n$. It is readily seen that the correspondence $f \mapsto f_n$ is a bijection between $(A^B)_n$ and the set $\mathfrak{Fun}(A_n, B_n)$ of all functions from A_n to B_n. So, the desired result is obtained by passing the following equalities to the Alpha-limits:

$$
\varphi_{\mathbf{B^A}}(n) = |(B^A)_n| = |\mathfrak{Fun}(A_n, B_n)| = |B_n|^{|A_n|} = \varphi_\mathbf{B}(n)^{\varphi_\mathbf{A}(n)}.
$$

$\square$

---

[2] We agree that the label of the constant function c_0 with empty support is 0.

4. Point sets of natural numbers

In this section, we focus on the "minimal" universe of a counting system that assigns a size also to infinite sets.[3] The obvious request is that the set of natural numbers $\mathbb{N}$ – which is the "canonical" infinite set – be in that universe. So, we come to the following

DEFINITION 16.19. *The* point set universe $\mathfrak{U}(\mathbb{N})$ *over the natural numbers is the smallest universe that contains* $\mathbb{N}$.

The universe $\mathfrak{U}(\mathbb{N})$ can be characterized as the family of all collections of tuples of natural numbers that have a bounded length.

PROPOSITION 16.20. *The point set universe over* $\mathbb{N}$ *is given by the following family:*

$$\mathfrak{U}(\mathbb{N}) \; = \; \bigcup_{n \in \mathbb{N}} \mathcal{P}\left(\bigcup_{k=1}^{n} \mathbb{N}^k\right) \; = \; \left\{ A \; \middle| \; A \subseteq \bigcup_{k=1}^{n} \mathbb{N}^k \textit{ for some } n \right\}.$$

PROOF. Denote by $\mathfrak{U}_n = \mathcal{P}(\bigcup_{k=1}^{n} \mathbb{N}^k)$. Trivially $\mathbb{N} \in \mathcal{P}(\mathbb{N}) = \mathfrak{U}_1$. Let us now verify that $\bigcup_n \mathfrak{U}_n$ satisfies the three properties of a universe as defined above. If $A, B \in \bigcup_n \mathfrak{U}_n$, pick n such that $A, B \subseteq \mathfrak{U}_n$. Then $A \cup B \subseteq \mathfrak{U}_n$ and $A \times B \subseteq \mathfrak{U}_n \times \mathfrak{U}_n = \mathfrak{U}_{2n}$, and so $A \cup B, A \times B \in \bigcup_n \mathfrak{U}_n$. The property (1) of closure under subsets is proved by noticing that $B \subseteq A \in \mathfrak{U}_n \Rightarrow B \in \mathfrak{U}_n$.

Let us now show that any universe $\mathfrak{V}$ that contains $\mathbb{N}$ must include $\bigcup_n \mathfrak{U}_n$. We proceed by induction on n and show that $\mathfrak{U}_n \subseteq \mathfrak{V}$. Since $\mathbb{N} \in \mathfrak{V}$, also all subsets of $\mathbb{N}$ belong to $\mathfrak{V}$, that is, $\mathfrak{U}_1 = \mathcal{P}(\mathbb{N}) \subseteq \mathfrak{V}$. Now assume $\mathfrak{U}_n \subseteq \mathfrak{V}$; then $\mathbb{N}, \mathbb{N}^n \in \mathfrak{V}$, and so also $\mathbb{N}^{n+1} = \mathbb{N} \times \mathbb{N}^n \in \mathfrak{V}$. Now let $B \subseteq \bigcup_{k=1}^{n+1} \mathbb{N}^k$ be given. Notice that $B_1 = B \cap \bigcup_{k=1}^{n} \mathbb{N}^k \in \mathfrak{V}$ because it is a subset of $\bigcup_{k=1}^{n} \mathbb{N}^k$, which belongs to $\mathfrak{V}$. Similarly, $B_2 = B \cap \mathbb{N}^{n+1} \in \mathfrak{V}$ because it is a subset of $\mathbb{N}^{n+1} \in \mathfrak{V}$. Finally, we can conclude that the disjoint union $B = B_1 \sqcup B_2 \in \mathfrak{V}$, thus completing the proof that $\mathfrak{U}_{n+1} = \mathcal{P}(\bigcup_{k=1}^{n+1} \mathbb{N}^k) \subseteq \mathfrak{V}$. □

As it was already pointed out in Example 16.3, every point set in $\mathfrak{U}(\mathbb{N})$ comes with a canonical labelling.

DEFINITION 16.21. *The* canonical label *of a point set* $A \in \mathfrak{U}(\mathbb{N})$ *is the function* ℓ_A *where for every tuple* $(n_1, \ldots, n_k) \in A$ *one puts:*

$$\ell_A(n_1, \ldots, n_k) = \max\{n_1, \ldots, n_k\}.$$

---

[3] Recall that by *universe* we mean a family of sets that is closed under subsets, disjoint unions, and Cartesian products. (See Definition 15.1.)

The canonical labelling is a finite-to-one function, and hence it is an actual labelling function. Indeed, by definition, every $A \in \mathfrak{U}(\mathbb{N})$ is included in $\bigcup_{i=1}^{k} \mathbb{N}^i$ for some k, and so for every n the pre-image

$$\ell_A^{-1}(n) = \{x \in A \mid \ell_A(x) \leq n\} \subseteq I_n \cup I_n^2 \cup \ldots \cup I_n^k,$$

where we denoted $I_n = \{1, \ldots, n\}$. This shows that $\ell_A^{-1}(n) \leq \sum_{i=1}^{k} n^i$ is finite.

- In the sequel, if not specified otherwise, we will identify each point set $A \in \mathfrak{U}(\mathbb{N})$ with the labelled set $\mathbf{A} = (A, \ell_A)$ where ℓ_A is the canonical label.

It directly follows from the definitions that the *counting function* $\varphi_A : A \to \mathbb{N}_0$ of a point set $A \in \mathfrak{U}(\mathbb{N})$ is given by

$$\varphi_A(n) = \left| \bigcup_{k \in \mathbb{N}} \{(a_1, \ldots, a_k) \in A \mid a_1, \ldots, a_k \leq n\} \right|.$$

Every point set $A \in \mathfrak{U}(\mathbb{N})$ has a "canonical" hypernatural size as given by its Alpha-numerosity $\mathfrak{n}_\alpha(A) \in \mathfrak{N}_\alpha \subseteq \mathbb{N}_0^*$.

EXAMPLE 16.22. *For every* k,

$$\mathfrak{n}_\alpha(\mathbb{N}^k) = \lim_{n \uparrow \alpha} \varphi_{\mathbb{N}^k}(n) = \lim_{n \uparrow \alpha} |I_n^k| = \lim_{n \uparrow \alpha} n^k = \boldsymbol{\alpha}^k.$$

The above discussion about point sets of natural numbers can be directly generalized to the *universe* $\mathfrak{U}(\mathbb{Z})$ *of point sets over the integers*. Indeed, also every $B \in \mathfrak{U}(\mathbb{Z})$ can be seen as a labelled set (B, ℓ_B) by considering the *canonical labelling* ℓ_B defined by means of absolute values:

$$\ell_B(x_1, \ldots, x_k) = \max\{|x_1|, \ldots, |x_k|\}.$$

EXAMPLE 16.23. *Let* $\mathbb{Z}^-$ *denote the set of negative integers. For every* $n \in \mathbb{N}$, *the counting functions:*

- $\varphi_{\mathbb{Z}^-}(n) = |\{x \in \mathbb{Z}^- \mid |x| \leq n\}| = n$, *and*
- $\varphi_{\mathbb{Z}}(n) = |\{x \in \mathbb{Z} \mid |x| \leq n\}| = 2n + 1$.

In consequence, we obtain:

- $\mathfrak{n}_\alpha(\mathbb{Z}^-) = \mathfrak{n}_\alpha(\mathbb{N}) = \boldsymbol{\alpha}$, *and*
- $\mathfrak{n}_\alpha(\mathbb{Z}) = 2\boldsymbol{\alpha} + 1$.

5. Numerosity of sets of natural numbers

When the elements of an infinite set of natural numbers are obtained as values of an increasing sequence, then its Alpha-numerosity is explicitly found, as shown in the next proposition.

Let $\lfloor x \rfloor = \max\{k \in \mathbb{Z} \mid k \leq x\}$ denote the *integer part* of x.

PROPOSITION 16.24. *Let* $f : [1, +\infty) \to [1, +\infty)$ *be an increasing real function such that* $A = \{f(n) \mid n \in \mathbb{N}\} \subseteq \mathbb{N}$. *Then*

$$\mathfrak{n}_\alpha(A) \;=\; \lim_{n \uparrow \alpha} \lfloor f^{-1}(n) \rfloor \;=\; \lfloor (f^*)^{-1}(\alpha) \rfloor.$$

PROOF. Since f is increasing, $|A \cap [1, n]| = \{f(1) < \ldots < f(k)\}$ where $k = \max\{h \mid f(h) \leq n\} = \lfloor f^{-1}(n) \rfloor$. So, the counting function $\varphi_A(n) = |A \cap [1, n]| = \lfloor f^{-1}(n) \rfloor$, and $\mathfrak{n}_\alpha(A) = \lim_{n \uparrow \alpha} \lfloor f^{-1}(n) \rfloor$. Moreover, by passing the inequality $\lfloor f^{-1}(n) \rfloor \leq f^{-1}(n) < \lfloor f^{-1}(n) \rfloor + 1$ to the Alpha-limits, we obtain that $\mathfrak{n}_\alpha(A) \leq \lim_{n \uparrow \alpha} f^{-1}(n) < \mathfrak{n}_\alpha(A) + 1$, and so $\mathfrak{n}_\alpha(A)$ is the hyper-integer part of $\lim_{n \uparrow \alpha} f^{-1}(n) = (f^*)^{-1}(\alpha)$ (see Proposition 2.36 (5)). $\qquad\square$

EXAMPLE 16.25. *If* $f : [1, +\infty) \to [1, +\infty)$ *is the square function* $f(x) = x^2$, *then* $\{f(n) \mid n \in \mathbb{N}\} = \{n^2 \mid n \in \mathbb{N}\}$ *is the set of perfect squares. Since the inverse function* $f^{-1}(y) = \sqrt{y}$ *is the square root, we obtain that* $\mathfrak{n}_\alpha(\{1, 4, 9, 16, \ldots\}) = \lfloor \sqrt{\alpha} \rfloor$.

Many more examples are easily obtained similarly as above.

EXAMPLE 16.26. *By applying Proposition 16.24 to the functions* $f_1(x) = 2x^2$, $f_2(x) = 2^x$, *and* $f_3(x) = 2^{(2^x)}$ *respectively, we obtain the following:*

(1) $\mathfrak{n}_\alpha\left(\{2 \cdot 1^2, 2 \cdot 2^2, 2 \cdot 3^2, \ldots\}\right) = \left\lfloor \sqrt{\alpha/2} \right\rfloor$;

(2) $\mathfrak{n}_\alpha\left(\{2, 2^2, 2^3, 2^4, 2^5, \ldots\}\right) = \lfloor \log_2 \alpha \rfloor$;

(3) $\mathfrak{n}_\alpha\left(\left\{2^2, 2^{(2^2)}, 2^{(2^3)}, \ldots\right\}\right) = \lfloor \log_2 \log_2(\alpha) \rfloor$.

Trivially, the positive rational numbers are obtained as ratios of numerosities of finite sets:

$$\frac{n}{m} = \frac{\mathfrak{n}_\alpha(\{1, 2, \ldots, n\})}{\mathfrak{n}_\alpha(\{1, 2, \ldots, m\})}.$$

Similarly, we may represent every positive real number r as the ratio of the numerosity of two (possibly infinite) sets $A, B \subseteq \mathbb{N}$, up to an infinitesimal quantity. For example, we have that

$$\sqrt{2} \sim \frac{\lfloor\sqrt{\alpha}\rfloor}{\lfloor\sqrt{\alpha/2}\rfloor} = \frac{\mathfrak{n}_\alpha(\{1^2, 2^2, 3^2, \ldots\})}{\mathfrak{n}_\alpha(\{2 \cdot 1^2, 2 \cdot 2^2, 2 \cdot 3^2, \ldots\})}.$$

Indeed, if $\nu = \mathfrak{n}_\alpha(\{1^2, 2^2, 3^2, \ldots\})$ and $\mu = \mathfrak{n}_\alpha(\{2 \cdot 1^2, 2 \cdot 2^2, 2 \cdot 3^2, \ldots\})$, from the inequalities $\nu \le \sqrt{\alpha} < \nu+1$ and $\mu \le \sqrt{\alpha/2} < \mu+1$, it follows that

$$\frac{\nu}{\mu+1} < \frac{\sqrt{\alpha}}{\sqrt{\alpha/2}} = \sqrt{2} < \frac{\nu+1}{\mu}.$$

It is easily verified that $\frac{\nu}{\mu+1} \sim \frac{\nu}{\mu} \sim \frac{\nu+1}{\mu}$, and so $\frac{\nu}{\mu} \sim \sqrt{2}$.

Recall that if one assumes a suitable *Qualified Set Axiom*, then both α and $\alpha/2$ are square hypernatural numbers, and so the infinite closeness $\frac{\nu}{\mu} \sim \sqrt{2}$ can be made an actual equality. (See Exercise 2.42.)

EXERCISE 16.27. *Assume* $(QSA)_Q$ *where* $Q = \{m!^{m!} \mid m \in \mathbb{N}\}$. *Then*

(1) $\mathfrak{n}_\alpha(\{h \cdot 1, h \cdot 2^k, h \cdot 3^k, \ldots\}) = \sqrt[k]{\alpha/h}$ *for every* $k, h \in \mathbb{N}$;

(2) $\frac{\mathfrak{n}_\alpha(\{1^2, 2^2, 3^2, \ldots\})}{\mathfrak{n}_\alpha(\{2 \cdot 1^2, 2 \cdot 2^2, 2 \cdot 3^2, \ldots\})} = \sqrt{2}$;

(3) $\mathfrak{n}_\alpha(\{2, 2^2, 2^3, 2^4, 2^5, \ldots\}) < \log_2 \alpha$.

The following general result holds.

THEOREM 16.28. *For every positive real number* r *there exist sets* $A, B \subseteq \mathbb{N}$ *such that*

$$\frac{\mathfrak{n}_\alpha(A)}{\mathfrak{n}_\alpha(B)} \sim r.$$

PROOF. Assume first that $0 < r < 1$, and for $n \in \mathbb{N}$, let $\varphi(n) = \max\{k \in \mathbb{N}_0 \mid \frac{k}{n} \le r\}$. Notice that $\varphi(n) \le \varphi(n+1) \le \varphi(n) + 1$ for every n. Indeed, by the definition,

$$\frac{\varphi(n)}{n} \le r < \frac{\varphi(n)+1}{n} \quad \text{and} \quad \frac{\varphi(n+1)}{n+1} \le r < \frac{\varphi(n+1)+1}{n+1}.$$

Then $\varphi(n) \le r \cdot n < \varphi(n+1) + 1 - r$, and so $\varphi(n) \le \varphi(n+1)$. Moreover,

$$\varphi(n+1) \le r \cdot (n+1) = r \cdot n + r < \varphi(n) + 1 + r < \varphi(n) + 2,$$

and $\varphi(n+1) < \varphi(n) + 2$ implies that $\varphi(n+1) \le \varphi(n) + 1$.

Now let $A \subseteq \mathbb{N}$ be the set where $1 \notin A$, and $n+1 \in A \Leftrightarrow \varphi(n+1) = \varphi(n) + 1$ for every n. It is easily verified that the counting function

φ_A coincides with φ, and so $\mathfrak{n}_\alpha(A) = \lim_{n\uparrow\alpha} \varphi(n)$. By passing to the Alpha-limit, we obtain that

$$\frac{\mathfrak{n}_\alpha(A)}{\alpha} = \lim_{n\uparrow\alpha} \frac{\varphi(n)}{n} \leq r < \lim_{n\uparrow\alpha} \frac{\varphi(n)+1}{n} = \frac{\mathfrak{n}_\alpha(A)}{\alpha} + \frac{1}{\alpha},$$

and so $r \sim \frac{\mathfrak{n}_\alpha(A)}{\alpha} = \frac{\mathfrak{n}_\alpha(A)}{\mathfrak{n}_\alpha(\mathbb{N})}$.

If r is any positive real number, pick numbers $0 < s, t < 1$ such that $r = \frac{s}{t}$. By what proved above, there exist sets $A, B \subseteq \mathbb{N}$ such that $s \sim \frac{\mathfrak{n}_\alpha(A)}{\alpha}$ and $t \sim \frac{\mathfrak{n}_\alpha(B)}{\alpha}$. Finally, we obtain the desired result by noticing that

$$r = \frac{s}{t} \sim \frac{\frac{\mathfrak{n}_\alpha(A)}{\alpha}}{\frac{\mathfrak{n}_\alpha(B)}{\alpha}} = \frac{\mathfrak{n}_\alpha(A)}{\mathfrak{n}_\alpha(B)}.$$

$\square$

6. Properties of Alpha-numerosity

As seen in the previous sections, Alpha-Theory naturally includes a counting system on the universe of point sets of natural numbers, namely $(\mathfrak{U}(\mathbb{N}), \mathfrak{N}_\alpha, \mathfrak{n}_\alpha)$, that retains many of the nice features of finite cardinality. Let us summarize below its main properties.

- Alpha-numerosities of (nonempty) point sets are hyper-natural numbers;
- $\mathfrak{n}_\alpha(F) = |F|$ for every finite point set F;
- If A, B are point sets then
  - $\mathfrak{n}_\alpha(A \cup B) + \mathfrak{n}(A \cap B) = \mathfrak{n}_\alpha(A) + \mathfrak{n}_\alpha(B)$.
  - $\mathfrak{n}_\alpha(A \times B) = \mathfrak{n}_\alpha(A) \cdot \mathfrak{n}_\alpha(B)$.

As $\mathbb{N}$ is the canonical fundamental example of an infinite set, it seems natural to require that its numerosity be "canonical" as well. In the context of Alpha-Theory, we do have a canonical number at hand, namely the number α, and Alpha-numerosity actually fulfills the above request, since

- $\mathfrak{n}_\alpha(\mathbb{N}) = \alpha$.

Equality $\mathfrak{n}_\alpha(\mathbb{N}) = \alpha$ can be seen as the counterpart for numerosities of the equality $|\mathbb{N}| = \aleph_0$ for cardinalities, and of the equality $\mathrm{ot}(\mathbb{N}) = \omega$ for well-ordered types. Notice that we have:

$$\mathfrak{n}(\mathbb{N}_0) = \mathfrak{n}(\mathbb{N} \cup \{0\}) = \mathfrak{n}(\mathbb{N}) + \mathfrak{n}(\{0\}) = \alpha + 1 > \alpha = \mathfrak{n}(\mathbb{N}).$$

The simple inequalities above already suffice to distinguish "numerosities" from cardinal and ordinal numbers, since

$$|\mathbb{N}| = |\mathbb{N}_0| = \aleph_0 \quad \text{and} \quad \mathrm{ot}(\mathbb{N}) = \mathrm{ot}(\mathbb{N}_0) = \omega.$$

In this section we will consider other additional properties of Alpha-numerosity that naturally come to mind. To begin with, notice that we have already considered disjoint unions and Cartesian products and their connections with the sum and product operations, respectively. But what can we say about divisions?

Let us start by considering the following example. Let

$$\mathbb{E} = \{n \in \mathbb{N} \mid n \text{ is even}\}; \quad \mathbb{O} = \{n \in \mathbb{N} \mid n \text{ is odd}\}.$$

Our naive intuition suggests that $\mathfrak{n}(\mathbb{E}) = \mathfrak{n}(\mathbb{O})$. By taking this for granted, from $\mathfrak{n}(\mathbb{E}) + \mathfrak{n}(\mathbb{O}) = \mathfrak{n}(\mathbb{E} \sqcup \mathbb{O}) = \mathfrak{n}(\mathbb{N}) = \alpha$ it follows that

$$\mathfrak{n}(\mathbb{E}) = \mathfrak{n}(\mathbb{O}) = \frac{\alpha}{2},$$

and we conclude that $\alpha \in \mathbb{N}^*$ is an *even* number.

Now that we have divided the natural numbers $\mathbb{N}$ into two equinumerous sets, why not trying to perform similar operations and divide $\mathbb{N}$ into k-many equinumerous sets, for any given $k \in \mathbb{N}$?

$$
\begin{array}{ccccc}
1 & 2 & 3 & \dots & k \\
k+1 & k+2 & k+3 & \dots & 2k \\
2k+1 & 2k+2 & 2k+3 & \dots & 3k \\
\dots & \dots & \dots & \dots & \dots
\end{array}
$$

By requiring all columns to have the same numerosity, one would conclude that each column contains exactly $\frac{\alpha}{k}$-many elements, and hence that the number α is a multiple of k. These considerations lead us to assume the following:

- **Divisibility Property (DP).** *For every $k \in \mathbb{N}$, the number α is a multiple of k and the numerosity of the set of multiples of k:*

$$\mathfrak{n}_\alpha(\{k, 2k, 3k, \dots, nk, \dots\}) = \frac{\alpha}{k}.$$

In the same vein, another question which is amusing to ask is whether the number α is a square. We recall that the ancient Greeks named $1, 4, 9, 16, \dots$ *square numbers* because one can arrange natural numbers to form the following squares:

$$
\begin{array}{cc}
1 & 2 \\
4 & 3
\end{array}
$$

$$
\begin{array}{ccc}
1 & 2 & 5 \\
4 & 3 & 6 \\
9 & 8 & 7
\end{array}
$$

$$
\begin{array}{cccc}
1 & 2 & 5 & 10 \\
4 & 3 & 6 & 11 \\
9 & 8 & 7 & 12 \\
16 & 15 & 14 & 13
\end{array}
$$

$\cdots$

Clearly, in the above *finite* squares the numerosity of every row and every column is exactly the square root of the numerosity of the full square. This suggests to ask for the same property about an infinite square containing *all* natural numbers: *"All rows and columns of the following infinite square have the same numerosity, namely $\sqrt{\alpha}$"*:

$$
\begin{array}{cccccc}
1 & 2 & 5 & 10 & 17 & 26 \quad \cdots \\
4 & 3 & 6 & 11 & 18 \quad \cdots \\
9 & 8 & 7 & 12 \quad \cdots \\
16 & 15 & 14 & 13 \\
25 & 24 \quad \cdots \\
36 \quad \cdots \\
\cdots
\end{array}
$$

In particular, the number $\alpha \in \mathbb{N}^*$ is a square.

A similar construction can be done with cubes of dimension 3, and more generally, for hypercubes of any dimension k. So, we come to the following property:

- **Root Property (RP).** *For every $k \in \mathbb{N}$, the number α is a k-th power and the numerosity of the set of k-th powers:*

$$
\mathfrak{n}_\alpha(\{1^k, 2^k, 3^k, \ldots, n^k, \ldots\}) \;=\; \sqrt[k]{\alpha}.
$$

If one assume the *Qualified Set Axiom* $(QSA)_Q$ for a suitable choice of Q, then both the *Divisibility* and *Root Properties* follow.

PROPOSITION 16.29. *Assume* $(QSA)_Q$ *where* $Q = \{m!^{m!} \mid m \in \mathbb{N}\}$. *Then both (DP) and (RP) hold.*

PROOF. For every k, let A_k be the set $\{k, 2k, \ldots, nk, \ldots\}$ and let B_k be the set $\{1^k, 2^k, \ldots, n^k, \ldots\}$. Then, by the properties seen in Exercise 2.42,

$$\mathfrak{n}_\alpha(A_k) = \lim_{n\uparrow\alpha} \varphi_{A_k}(n) = \frac{\alpha}{k} \quad \text{and} \quad \mathfrak{n}_\alpha(B_k) = \lim_{n\uparrow\alpha} \varphi_{B_k}(n) = \sqrt[k]{\alpha}.$$

$\square$

Now let us turn to powersets. Recall that if F is any finite set with n elements, then the collection of its subsets $\mathcal{P}(F)$ has 2^n-many elements. This fact suggests two reasonable ways to give a meaning to the number 2^α, namely

- $2^\alpha = \mathfrak{n}_\alpha(\mathcal{P}(\mathbb{N}))$, or
- $2^\alpha = \mathfrak{n}_\alpha\left(\bigcup_{n\in\mathbb{N}} \mathcal{P}(\{1, \ldots, n\})\right) = \mathfrak{n}_\alpha(\mathfrak{Fin}(\mathbb{N}))$.

As we already remarked, the first option makes no sense since numerosities are defined only for countable sets, while $\mathcal{P}(\mathbb{N})$ has the cardinality of the *continuum*. However, the second option actually holds, as it was shown in Section 3:

- $\mathfrak{n}_\alpha(\mathfrak{Fin}(\mathbb{N})) = 2^\alpha$.

7. Numerosities of sets of rational numbers

It seems really feasible that the set of negative integers

$$\mathbb{Z}^- = \{k \in \mathbb{Z} \mid k < 0\}$$

has the same size as the set of positive integers $\mathbb{N}$; and indeed we already noticed that this is actually the case when we consider the canonical labelling $\ell_\mathbb{Z}(x) = |x|$ given by the absolute value:

- $\mathfrak{n}_\alpha(\mathbb{Z}^-) = \mathfrak{n}_\alpha(\mathbb{N}) = \alpha$.

Now, what can we expect about the numerosity of the rational numbers $\mathbb{Q}$? Let us begin by considering that

$$\mathbb{Q} = \bigcup_{k\in\mathbb{Z}} [k - 1, k)_\mathbb{Q}.$$

It is a natural assumption that all such intervals of equal length 1 have the same numerosity.

(Q1) $\mathfrak{n}_\alpha((k - 1, k]_\mathbb{Q}) = \mathfrak{n}_\alpha((0, 1]_\mathbb{Q})$ for all $k \in \mathbb{Z}$.

The positive rational numbers $\mathbb{Q}^+ = \{q \in \mathbb{Q} \mid q > 0\}$ are canonically obtained as the union of intervals $\mathbb{Q}^+ = \bigcup_{n \in \mathbb{N}} (n-1, n]_\mathbb{Q}$ indexed by natural numbers n; so, it is natural to also expect the following:

(Q2) $\mathfrak{n}_\alpha(\mathbb{Q}^+) = \mathfrak{n}_\alpha(\mathbb{N}) \cdot \mathfrak{n}_\alpha((0, 1]_\mathbb{Q})$.

Another really natural assumption is that the negative rationals $\mathbb{Q}^- = \{q \in \mathbb{Q} \mid q < 0\}$ be equinumerous to the positive rationals $\mathbb{Q}^+$.

(Q3) $\mathfrak{n}_\alpha(\mathbb{Q}^-) = \mathfrak{n}_\alpha(\mathbb{Q}^+)$.

As for the unitary rational interval $(0, 1]_\mathbb{Q}$, unfortunately it seems there is no simple and intuitive way of relating its numerosity to our canonical infinite number α. In particular, it seems there is no definitive way to decide whether $\mathfrak{n}_\alpha((0, 1]_\mathbb{Q}) \geq \alpha$ or $\mathfrak{n}_\alpha((0, 1]_\mathbb{Q}) \leq \alpha$. So, in the absence of any reason to choose one of the two possibilities, we go for the simplest option.

(Q4) $\mathfrak{n}_\alpha((0, 1]_\mathbb{Q}) = \alpha$.

By putting together the above assumptions (Q1), (Q2), (Q3), (Q4) we obtain that

$$\mathfrak{n}_\alpha(\mathbb{Q}) = \mathfrak{n}_\alpha(\mathbb{Q}^-) + \mathfrak{n}_\alpha(\{0\}) + \mathfrak{n}_\alpha(\mathbb{Q}^+) = 2 \cdot \mathfrak{n}_\alpha(\mathbb{N}) \cdot \mathfrak{n}_\alpha((0, 1]_\mathbb{Q}) + 1 = 2\alpha^2 + 1.$$

Now the question arises as whether one can find a "natural" labelling for $\mathbb{Q}$ in such a way that the considered properties (Q1), (Q2), (Q3), (Q4) are satisfied. The answer is positive, as shown below.

DEFINITION 16.30. *Let* $\mathbb{H}(n) = \{\pm i/n \mid i = 0, \ldots, n^2\}$ *be the n-grid (see Definition 3.39). For every* $q \in \mathbb{Q}$, *set*

$$\ell_\mathbb{Q}(q) = \min\{n! \mid q \in \mathbb{H}(n!)\}.$$

Notice that for $m \leq n$ one has $\mathbb{H}(m!) \subseteq \mathbb{H}(n!)$, but in general $\mathbb{H}(m) \not\subseteq \mathbb{H}(n)$.

THEOREM 16.31. *Assume* $(QSA)_Q$ *where* $Q = \{m! \mid m \in \mathbb{N}\}$. *With the above labelling* $\ell_\mathbb{Q}$:

(1) $\mathfrak{n}_\alpha((k-1, k]_\mathbb{Q}) = \alpha$ *for all* $k \in \mathbb{Z}$;
(2) $\mathfrak{n}_\alpha(\mathbb{Q}^-) = \mathfrak{n}_\alpha(\mathbb{Q}^+) = \alpha^2$.

In consequence, (Q1), (Q2), (Q3), (Q4) *are all satisfied, and so* $\mathfrak{n}_\alpha(\mathbb{Q}) = 2\alpha^2 + 1$.

PROOF. Notice first that for every $X \subseteq \mathbb{Q}$ and for every $n \in \mathbb{N}$, one has

$$\varphi_X(n!) = |\{q \in X \mid \ell_{\mathbb{Q}}(q) \leq n!\}| = \left| \bigcup_{m \leq n} (\mathbb{H}(m!) \cap X) \right| = |\mathbb{H}(n!) \cap X|.$$

(1). Given k, for every $m \geq k$ the intersection

$$(k-1, k]_{\mathbb{Q}} \cap \mathbb{H}(m) = \left\{ \frac{(k-1)m+1}{m}, \frac{(k-1)m+2}{m}, \ldots, \frac{km}{m} \right\}$$

contains exactly m-many elements. So, for every $n \geq k$ we have $\varphi_{(k-1,k]_{\mathbb{Q}}}(n!) = n!$, and by taking the Alpha-limits we obtain the desired result $\mathfrak{n}_\alpha((k-1, k]_{\mathbb{Q}}) = \alpha$.

(2). For every m, both intersections below contain exactly m^2-many elements:

- $\mathbb{H}(m) \cap \mathbb{Q}^- = \left\{ -\frac{m^2}{m}, -\frac{m^2-1}{m}, \ldots, -\frac{1}{m} \right\}$;

- $\mathbb{H}(m) \cap \mathbb{Q}^+ = \left\{ \frac{1}{m}, \frac{2}{m}, \ldots, \frac{m^2}{m} \right\}$.

So, for every n we have $\varphi_{\mathbb{Q}^-}(n!) = \varphi_{\mathbb{Q}^+}(n!) = (n!)^2$, and by taking the Alpha-limits we obtain $\mathfrak{n}_\alpha(\mathbb{Q}^-) = \mathfrak{n}_\alpha(\mathbb{Q}^+) = \alpha^2$. $\square$

EXERCISE 16.32. *Consider the following alternative labelling $\ell_{\mathbb{Q}}'$ for the rationals:*

$$\ell_{\mathbb{Q}}'(x) = \max\{|k|, \ell_{\mathbb{Q}}(q-k)\} \quad \text{where } k = \lfloor q \rfloor.$$

Show that $(\mathbb{Q}, \ell_{\mathbb{Q}}') \cong \mathbb{Z} \times [0, 1)_{\mathbb{Q}}$ and so, by assuming $(QSA)_Q$ where $Q = \{m! \mid m \in \mathbb{N}\}$, the corresponding numerosity would be:

$$\mathfrak{n}_\alpha'(\mathbb{Q}) = \mathfrak{n}(\mathbb{Z}) \cdot \mathfrak{n}([0, 1)_{\mathbb{Q}}) = 2\alpha^2 + \alpha \neq \mathfrak{n}_\alpha(\mathbb{Q}).$$

Show also that $\mathfrak{n}_\alpha'(\mathbb{Q}^-) = \alpha^2 < \alpha^2 + \alpha = \mathfrak{n}_\alpha'(\mathbb{Q}^+)$.

The reader at this point may be dissatisfied with the fact that the numerosity of a set depends on some arbitrary choices. However, this fact appears to be in the very nature of infinite sets. With respect to this, probably the only "absolute" measure of size of infinite sets is the Cantorian cardinality.

In developing the theory of numerosity, we will see that the arbitrariness of properties such as (Q4) is related to the arbitrariness of the properties of the number α. We tell in advance that every model of numerosity is grounded on an ultrafilter; in consequence, many features of the theory will depend on the choice of such an underlying ultrafilter. Although this fact may appear disappointing, it also has some advantages, since it makes the theory flexible. Depending on the applications we have in mind, we will be free to choose the more convenient properties for α and for the numerosities of relevant sets. For instance, property (Q4) might be safely replaced by $\mathfrak{n}([0,1)_{\mathbb{Q}}) = 2^\alpha$, or by other similar assumptions.

8. Zermelo's principle and Alpha-numerosity

It is interesting to observe that *Zermelo's Principle* for Alpha-numerosity is equivalent to *Cauchy's Infinitesimal Principle*.

THEOREM 16.33. *Assume the axioms of Alpha-Theory. Then the following properties are equivalent:*

(1) *Cauchy infinitesimal principle (CIP);*

(2) *The numerosity system $(\mathfrak{L}, \mathfrak{N}_\alpha, \mathfrak{n}_\alpha)$ is Zermelian;*

(3) *The set of Alpha-numerosities coincides with the set of non-negative hyperintegers: $\mathfrak{N}_\alpha = \mathbb{N}_0^*$.*

PROOF. $(1) \Leftrightarrow (3)$. By Theorem 6.12 (6), (CIP) holds if and only if $\mathbb{N}_0^* = \{\lim_{n\uparrow\alpha} f(n) \mid f : \mathbb{N}_0 \to \mathbb{N}_0 \text{ non-decreasing}\}$. On the other hand, $\mathfrak{N}_\alpha = \{\lim_{n\uparrow\alpha} f(n) \mid f : \mathbb{N}_0 \to \mathbb{N}_0 \text{ non-decreasing}\}$, by Proposition 16.12, and the desired equivalence follows.

$(1) \Rightarrow (2)$. Assume that (CIP) holds, and let $\mathfrak{n}_\alpha(\mathbf{A}) < \mathfrak{n}_\alpha(\mathbf{B})$. We will construct a proper labelled subset $\mathbf{A}' \subset \mathbf{B}$ with the property that $\varphi_{\mathbf{A}'}(n) = \varphi_{\mathbf{A}}(n)$ a.e., so that $\mathfrak{n}_\alpha(\mathbf{A}') = \lim_{n\uparrow\alpha} \varphi_{\mathbf{A}'}(n) = \lim_{n\uparrow\alpha} \varphi_{\mathbf{A}}(n) = \mathfrak{n}_\alpha(\mathbf{A})$.

By the hypothesis, $\lim_{n\uparrow\alpha}(\varphi_{\mathbf{B}}(n) - \varphi_{\mathbf{A}}(n)) = \mathfrak{n}_\alpha(\mathbf{B}) - \mathfrak{n}_\alpha(\mathbf{A}) \in \mathbb{N}^*$ is a hypernatural number and so there exists a non-decreasing sequence $\vartheta : \mathbb{N} \to \mathbb{N}$ with

$$\lim_{n\uparrow\alpha} \vartheta(n) = \lim_{n\uparrow\alpha}(\varphi_{\mathbf{B}}(n) - \varphi_{\mathbf{A}}(n)) > 0.$$

Now consider the following qualified set X and enumerate its elements in increasing order:

$$X = \{n \mid \vartheta(n) = \varphi_{\mathbf{B}}(n) - \varphi_{\mathbf{A}}(n) > 0\} = \{x_1 < \ldots < x_k < x_{k+1} < \ldots\}.$$

Notice that $|A_{x_1}| = \varphi_{\mathbf{A}}(x_1) = \varphi_{\mathbf{B}}(x_1) - \vartheta(x_1) > \varphi_{\mathbf{B}}(x_1) = |B_{x_1}|$. Moreover, since $\vartheta(n)$ is non-decreasing, for every k we have:

$$\begin{aligned}
|A_{x_{k+1}} \setminus A_{x_k}| &= \varphi_{\mathbf{A}}(x_{k+1}) - \varphi_{\mathbf{A}}(x_k) = \\
&= \varphi_{\mathbf{B}}(x_{k+1}) - \varphi_{\mathbf{B}}(x_k) - (\vartheta(x_{k+1}) - \vartheta(x_k)) \\
&\geq \varphi_{\mathbf{B}}(x_{k+1}) - \varphi_{\mathbf{B}}(x_k) = |B_{x_{k+1}} \setminus B_{x_k}|.
\end{aligned}$$

As a consequence, we can pick a proper subset $A_1' \subset B_{x_1}$ with cardinality $|A_1'| = \varphi_A(x_1)$; and for every $k \in \mathbb{N}$, a subset

$$A_{k+1}' \subseteq B_{x_{k+1}} \setminus B_{x_k} = B \cap (n_k, n_{k+1}]$$

of cardinality $\varphi_{\mathbf{A}}(x_{k+1}) - \varphi_{\mathbf{A}}(x_k)$. Now let $\mathbf{A}'$ be the labelled subset of $\mathbf{B}$ with domain $A' = \bigcup_{k \in \mathbb{N}} A_k'$. Clearly, $A' \cap [1, x_1] = A_1'$, and so $\varphi_{\mathbf{A}'}(x_1) = \varphi_{\mathbf{A}}(x_1)$. Moreover, for every $k \in \mathbb{N}$,

$$
\begin{aligned}
\varphi_{\mathbf{A}'}(x_{k+1}) &= |A' \cap [1, x_{k+1}]| = |A_1'| + |A_2'| + \ldots + |A_{k+1}'| \\
&= \varphi_{\mathbf{A}}(x_1) + (\varphi_{\mathbf{A}}(x_2) - \varphi_{\mathbf{A}}(x_1)) + \ldots + (\varphi_{\mathbf{A}}(x_{k+1}) - \varphi_{\mathbf{A}}(x_k)) \\
&= \varphi_{\mathbf{A}}(x_{k+1}).
\end{aligned}
$$

We conclude that the counting functions $\varphi_{\mathbf{A}'}(n)$ and $\varphi_{\mathbf{A}}(n)$ agree on the qualified set X, and hence $\mathfrak{n}_\alpha(\mathbf{A}') = \mathfrak{n}_\alpha(\mathbf{A})$ as desired.

$(2) \Rightarrow (3)$. Let us assume that the numerosity system $(\mathfrak{L}, \mathfrak{N}_\alpha, \mathfrak{n}_\alpha)$ satisfies *Zermelo's Principle* (ZP). Given $\nu \in \mathbb{N}^*$, pick $f : \mathbb{N} \to \mathbb{N}$ such that $\lim_{n \uparrow \alpha} f(n) = \nu$. To reach the desired result we need to find a labelled set $\mathbf{C}$ such that $f(n) = \varphi_{\mathbf{C}}(n)$ a.e., so that $\nu = \mathfrak{n}_\alpha(C) \in \mathfrak{N}_\alpha$. To this end, let $F : \mathbb{N} \to \mathbb{N}_0$ be the function where $F(1) = 0$ and $F(n) = \sum_{i=1}^{n-1} f(i)$ for $n \geq 2$; and let $G : \mathbb{N} \to \mathbb{N}_0$ be the function where $G(n) = \sum_{i=1}^{n} f(i)$. As F and G are non-decreasing, by Proposition 16.12 there exist labelled sets $\mathbf{A}$ and $\mathbf{B}$ such that their counting functions $\varphi_{\mathbf{A}}(n) = F(n)$ and $\varphi_{\mathbf{B}}(n) = G(n)$, so that $\varphi_{\mathbf{A}}(n) + f(n) = \varphi_{\mathbf{A}}(n) + G(n) - F(n) = \varphi_{\mathbf{B}}(n)$ for every n. Since $\varphi_{\mathbf{A}}(n) < \varphi_{\mathbf{B}}(n)$ for every n, $\mathfrak{n}_\alpha(\mathbf{A}) < \mathfrak{n}_\alpha(\mathbf{B})$ and by (ZP) there exists a proper labelled subset $\mathbf{A}' \subset \mathbf{B}$ such that $\mathfrak{n}_\alpha(\mathbf{A}') = \mathfrak{n}_\alpha(\mathbf{A})$. Now consider the qualified set $Q = \{n \mid \varphi_{\mathbf{A}}(n) = \varphi_{\mathbf{A}'}(n)\}$ and the labelled subset $\mathbf{C} \subseteq \mathbf{B}$ the with domain $C = B \setminus A'$. Clearly $\mathbf{A}' \sqcup \mathbf{C} = \mathbf{B}$, and so $\varphi_{\mathbf{A}'}(n) + \varphi_{\mathbf{C}}(n) = \varphi_{\mathbf{B}}(n)$ for every n. But then $Q = \{n \mid \varphi_{\mathbf{A}}(n) + \varphi_{\mathbf{C}}(n) = \varphi_{\mathbf{B}}(n)\} = \{n \mid \varphi_{\mathbf{C}}(n) = f(n)\}$, and so $f(n) = \varphi_{\mathbf{C}}(n)$ a.e..

$\square$

As a corollary, we obtain the following.

THEOREM 16.34. *Assume* Cauchy infinitesimal principle *(CIP). Then for every $\nu \in \mathbb{N}^*$ there exists a labelled set $\mathbf{A}$ such that $\mathfrak{n}_\alpha(\mathbf{A}) = \nu$; moreover, $\nu \in \mathbb{N}^*$ is the Alpha-numerosity of a point set $A \in \mathfrak{U}(\mathbb{N})$ if and only if $\nu \leq \boldsymbol{\alpha}^m$ for some $m \in \mathbb{N}$.*

PROOF. The first statement is just the equality $\mathfrak{N}_\alpha = \mathbb{N}_0^*$.

Given a point set $A \in \mathfrak{U}(\mathbb{N})$, let n be such that $A \subseteq \bigcup_{k=1}^{n} \mathbb{N}^k$. As the union $\bigcup_{k=1}^{n} \mathbb{N}^k$ is disjoint, we have that

$$\mathfrak{n}_\alpha(A) \leq \mathfrak{n}_\alpha \left(\bigcup_{k=1}^{n} \mathbb{N}^k \right) = \sum_{k=1}^{n} \mathfrak{n}_\alpha(\mathbb{N}^k) = \sum_{k=1}^{n} \alpha^k < \alpha^{n+1}.$$

Finally, notice that $\alpha^m = \mathfrak{n}_\alpha(\mathbb{N}^m)$. So, given any $\nu < \alpha^m$, one can find a proper subset $B \subset \mathbb{N}^m$ with $\mathfrak{n}_\alpha(\mathbf{B}) = \nu$, and clearly such a $B \in \mathfrak{U}(\mathbb{N})$. $\qquad\square$

9. Asymptotic density and numerosity

As a straight application of Alpha-numerosity, one can introduce a notion of density that generalizes the asymptotic density. Recall the following notion.

DEFINITION 16.35. The *asymptotic density* $d(A)$ of a set A of natural numbers is the following limit (provided it exists):

$$d(A) = \lim_{n \to \infty} \frac{|A \cap [1, n]|}{n}.$$

DEFINITION 16.36. *The* Alpha-density $d_\alpha(A)$ *of a set* $A \subseteq \mathbb{N}$ *is the standard part of the ratio between the Alpha-numerosity of A and the Alpha-numerosity* α *of* $\mathbb{N}$:

$$d_\alpha(A) = \text{st} \left(\frac{\mathfrak{n}_\alpha(A)}{\alpha} \right) = \text{st} \left(\lim_{n \uparrow \alpha} \frac{|A \cap [1, n]|}{n} \right).$$

We remark that the Alpha-density is defined for *all* subsets of $\mathbb{N}$, while examples of sets that do not have an asymptotic density are easily found. This drawback of asymptotic density is usually overcome by considering the *upper density* (or the *lower density*) as obtained by taking the limit superior (the limit inferior, respectively):

$$\overline{d}(A) = \limsup_{n \to \infty} \frac{|A \cap [1, n]|}{n}; \quad \underline{d}(A) = \liminf_{n \to \infty} \frac{|A \cap [1, n]|}{n}.$$

However, even the upper and lower densities fail badly to satisfy the additive property of a measure: $m(A \cup B) + m(A \cap B) = m(A) + m(B)$. An example is given below.

EXAMPLE 16.37. The sets

$$A = \bigcup_{n \in \mathbb{N}} [2n!, (2n+1)!) \quad \text{and} \quad B = \bigcup_{n \in \mathbb{N}} [(2n+1)!, (2n+2)!)$$

form a partition of $\mathbb{N}$ and have upper densities $\overline{d}(A) = \overline{d}(B) = 1$ and lower densities $\underline{d}(A) = \underline{d}(B) = 0$.

Let us now show that the Alpha-density is indeed a (finitely additive) probability measure, and that it generalizes the asymptotic density.

THEOREM 16.38. *The Alpha-density* $d_\alpha : \mathcal{P}(\mathbb{N}) \to [0, 1]$ *is a finitely additive probability measure on* $\mathbb{N}$ *where all finite sets have measure zero. That is:*

(1) $d_\alpha(F) = 0$ *for all finite sets* F, *and* $d_\alpha(\mathbb{N}) = 1$;

(2) $d_\alpha(A \cup B) + d_\alpha(A \cap B) = d_\alpha(A) + d_\alpha(B)$.

Moreover,

(3) $d_\alpha(A)$ *is a limit point of the sequence* $a_n = |A \cap [1, n]|/n$;

(4) *The Alpha-density is an extension of the asymptotic density, that is,* $d_\alpha(A) = d(A)$ *whenever the density* $d(A)$ *exists.*

PROOF. (1). Since $\mathfrak{n}_\alpha(F) = |F| \in \mathbb{N}$ is finite and α is infinite, then $d_\alpha(F) = \text{st}\left(\frac{|F|}{\alpha}\right) = 0$.

(2). It trivially follows from the additivity of $\mathfrak{n}_\alpha$ and the fact that standard parts are coherent with sums: $\text{st}(\xi + \eta) = \text{st}(\xi) + \text{st}(\eta)$.

(3). Recall that the standard part of the Alpha-limit of a real sequence is one of its limit points (see Proposition 3.23).

(4) directly follows from (3). □

CHAPTER 17

A General Numerosity Theory for Labelled Sets

1. Definition and first properties

In this section, we formalize a general definition of numerosity for labelled sets along the lines outlined in Chapter 15, and show its close relationships with the Alpha-Calculus.

DEFINITION 17.1. *A numerosity* for labelled sets is a triple $(\mathfrak{L}, \mathfrak{N}, \mathfrak{n})$ *where:*

- *$\mathfrak{L}$ is a "universe" of labelled sets that includes the universe of point sets $\mathfrak{U}(\mathbb{N})$ over the natural numbers, and is closed under labelled subsets, unions, and Cartesian products;*
- *$\mathfrak{N}$ is an ordered set of "numbers" that are used to count;*
- *$\mathfrak{n} : \mathfrak{L} \twoheadrightarrow \mathfrak{N}$ is a surjective map that assigns to every labelled set $\mathbf{A} \in \mathfrak{L}$ its "numerosity" $\mathfrak{n}(\mathbf{A}) \in \mathfrak{N}$;*

and that satisfies the following properties:

(1) *Union principle: If $\mathfrak{n}(\mathbf{A}) = \mathfrak{n}(\mathbf{A}')$ and $\mathfrak{n}(\mathbf{B}) = \mathfrak{n}(\mathbf{B}')$ and the domains $A \cap B = A' \cap B' = \emptyset$, then $\mathfrak{n}(\mathbf{A} \sqcup \mathbf{B}) = \mathfrak{n}(\mathbf{A}' \sqcup \mathbf{B}')$;*

(2) *Cartesian product principle: If $\mathfrak{n}(\mathbf{A}) = \mathfrak{n}(\mathbf{A}')$ and $\mathfrak{n}(\mathbf{B}) = \mathfrak{n}(\mathbf{B}')$, then $\mathfrak{n}(\mathbf{A} \times \mathbf{B}) = \mathfrak{n}(\mathbf{A}' \times \mathbf{B}')$;*

(3) *Zermelo's principle (ZP): If $\mathfrak{n}(\mathbf{A}) < \mathfrak{n}(\mathbf{B})$ then $\mathfrak{n}(\mathbf{A}) = \mathfrak{n}(\mathbf{A}')$ for some proper labelled subset $\mathbf{A}' \subset \mathbf{B}$;*

(4) *Asymptotic principle (AP): If $\varphi_{\mathbf{A}}(n) \le \varphi_{\mathbf{B}}(n)$ for all n then $\mathfrak{n}(\mathbf{A}) \le \mathfrak{n}(\mathbf{B})$.*

As we already pointed out, the coherence properties given by the *Union Principle* and the *Cartesian Product Principle* make it possible to define a sum and a product operation on numerosities.

DEFINITION 17.2. *For any $\mathbf{A}$ and $\mathbf{B}$, set:*

- $\mathfrak{n}(\mathbf{A}) + \mathfrak{n}(\mathbf{B}) = \mathfrak{n}(\mathbf{A} \sqcup \mathbf{B})$ *whenever $A \cap B = \emptyset$;*
- $\mathfrak{n}(\mathbf{A}) \cdot \mathfrak{n}(\mathbf{B}) = \mathfrak{n}(\mathbf{A} \times \mathbf{B})$.

Let us now derive the first properties, including the fact that numerosity counting systems generalize finite cardinality. To this end, we introduce the canonical finite labelled sets.

- For every $n \in \mathbb{N}$, let us denote by $\widehat{\mathbf{n}} \in \mathfrak{L}$ the point set $\{1, \ldots, n\} \in \mathfrak{U}(\mathbb{N})$ with the canonical labelling given by the identity map.

PROPOSITION 17.3. *Let* $(\mathfrak{L}, \mathfrak{N}, \mathfrak{n})$ *be a numerosity for labelled sets. Then:*

(1) *All labelled singletons have the same numerosity;*

(2) *If a labelled set* $\mathbf{A} \in \mathfrak{L}$ *has finite domain* A, *then* $\mathfrak{n}(\mathbf{A}) = \mathfrak{n}(\widehat{\mathbf{n}})$ *where* $n = |A|$ *;*

(3) *Every numerosity* $\nu = \mathfrak{n}(\mathbf{A})$ *has successor* $\nu + 1 = \mathfrak{n}(\mathbf{A} \cup \{\bullet\})$ *where* $\{\bullet\}$ *is any labelled singleton disjoint from* $\mathbf{A}^1$ *;*

(4) $\mathfrak{N}$ *has a proper initial segment* $\{\mathfrak{n}(\emptyset)\} \cup \{\mathfrak{n}(\widehat{\mathbf{n}}) \mid n \in \mathbb{N}\}$ *that is order-isomorphic to the set of non-negative integers* $\mathbb{N}_0$.

As a consequence of the last property, in the sequel we will directly assume $\mathbb{N}_0 \subset \mathfrak{N}$ by identifying $\mathfrak{n}(\emptyset)$ with 0, and $\mathfrak{n}(\widehat{\mathbf{n}})$ with n for every $n \in \mathbb{N}$.

PROOF. (1). If by contradiction $\mathfrak{n}(\{\bullet\}) < \mathfrak{n}(\{\diamond\})$, by *Zermelo's Principle* there would be a proper labelled subset $\mathbf{A} \subset \{\diamond\}$ with $\mathfrak{n}(\mathbf{A}) = \mathfrak{n}(\{\bullet\})$. As a proper subset of a singleton, it must be $\mathbf{A} = \emptyset$; but then $\mathbf{A} \subset \{\bullet\}$ and we would have $\mathfrak{n}(\mathbf{A}) < \mathfrak{n}(\{\bullet\})$, a contradiction.

(2). We proceed by induction on n. By the previous point, we know already that $\mathfrak{n}(\{\mathbf{a}\}) = \mathfrak{n}(\widehat{\mathbf{1}})$ for all labelled singletons. Given $\mathbf{B}$ be a labelled set whose domain B has cardinality $|B| = n+1$, pick any $b \in B$, and denote by $\{\mathbf{b}\} \subseteq \mathbf{B}$ the labelled subset with domain the singleton $\{b\}$, and by $\mathbf{B}' \subset \mathbf{B}$ the proper labelled subset with domain $B \setminus \{b\}$. By the inductive hypothesis $\mathfrak{n}(\mathbf{B}') = \mathfrak{n}(\widehat{\mathbf{n}})$ and so, by the *Union Principle,* $\mathfrak{n}(\mathbf{B}) = \mathfrak{n}(\mathbf{B}' \cup \{\mathbf{b}\}) = \mathfrak{n}(\widehat{\mathbf{n}} \cup \{\mathbf{n+1}\}) = \mathfrak{n}(\widehat{\mathbf{n+1}})$, where $\{\mathbf{n+1}\} \subseteq \widehat{\mathbf{n+1}}$ is the labelled singleton with domain $\{n+1\}$.

(3). Since $\mathbf{A}$ is a proper labelled subset of $\mathbf{A} \cup \{\bullet\}$, we have the strict inequality $\mathfrak{n}(\mathbf{A}) < \mathfrak{n}(\mathbf{A} \cup \{\bullet\})$. Now assume by contradiction that $\mathfrak{n}(\mathbf{A}) < \mu < \mathfrak{n}(\mathbf{A} \oplus \{\bullet\})$ for some $\mu = \mathfrak{n}(\mathbf{B}) \in \mathfrak{N}$. Take $\mathbf{B}' \subset \mathbf{A} \oplus \{\bullet\}$ with $\mathfrak{n}(\mathbf{B}') = \mathfrak{n}(\mathbf{B}) = \mu$, and take $\mathbf{C} \subset \mathbf{B}'$ with $\mathfrak{n}(\mathbf{C}) = \mathfrak{n}(\mathbf{A})$. Now pick $b \in B' \setminus C$ and denote by $\{\mathbf{b}\} \subseteq \mathbf{B}$ the labelled subset with domain

---

[1] Recall that $\nu + 1$ is the *successor* of ν if $\nu + 1 > \nu$ and there exist no numbers $\mu \in \mathfrak{N}$ such that $\nu < \mu < \nu + 1$.

the singleton $\{b\}$. Notice that $\mathbf{C} \cup \{\mathbf{b}\} \subseteq \mathbf{B}'$. By (1), $\mathfrak{n}(\{\bullet\}) = \mathfrak{n}(\{\mathbf{b}\})$ and so, by the *Union Principle*, we obtain the following contradiction:

$$\mathfrak{n}(\mathbf{A} \cup \{\bullet\}) \;=\; \mathfrak{n}(\mathbf{C} \cup \{\mathbf{b}\}) \;\leq\; \mathfrak{n}(\mathbf{B}') \;<\; \mathfrak{n}(\mathbf{A} \cup \{\bullet\}).$$

(4). By *Zermelo's Principle*, $\mathfrak{n}(\emptyset)$ is the least element of $\mathfrak{N}$, because the empty labelled set contains no proper subsets. The remaining properties directly follow from the previous points (2) and (3). $\qquad\square$

Recall that an *ordered ring* is a ring R endowed with an ordering $\leq$ that is compatible with operations, in the sense that $x \leq y$ implies $x + z \leq y + z$ for every z, and $x \cdot z \leq y \cdot z$ for every $z \geq 0$.

An *ordered semi-ring* has the same properties of an ordered ring, except that inverse elements with respect to sum do not exist. An ordered semi-ring is the positive part of an ordered ring if and only if the following *difference property* holds: "$x \leq y \Leftrightarrow$ there exists a unique z such that $y = x + z$." A typical example of a positive part of an ordered ring is given by the natural numbers $\mathbb{N}$.

THEOREM 17.4. $\langle \mathfrak{N}, +, \cdot, 0, 1, \leq \rangle$ *is the positive part of an ordered semi-ring.*

PROOF. It directly follows from the definitions that two labelled sets are isomorphic if and only if their counting functions coincide. In consequence, by the *asymptotic principle*, the following implications hold:

- If $\mathbf{A} \cong \mathbf{B}$ then $\mathfrak{n}(\mathbf{A}) = \mathfrak{n}(\mathbf{B})$;
- If $\mathbf{A} \cong \mathbf{A}'$ and $\mathbf{B} \cong \mathbf{B}'$ and the domains $A \cap B = A' \cap B' = \emptyset$, then $\mathfrak{n}(\mathbf{A} \sqcup \mathbf{B}) = \mathfrak{n}(\mathbf{A}' \sqcup \mathbf{B}')$;
- If $\mathbf{A} \cong \mathbf{A}'$ and $\mathbf{B} \cong \mathbf{B}'$ then $\mathfrak{n}(\mathbf{A} \times \mathbf{B}) = \mathfrak{n}(\mathbf{A}' \times \mathbf{B}')$.

The sum is commutative and associative, since trivially $\mathbf{A} \sqcup \mathbf{B} = \mathbf{B} \sqcup \mathbf{A}$ and $\mathbf{A} \cup (\mathbf{B} \sqcup \mathbf{C}) = (\mathbf{A} \sqcup \mathbf{B}) \cup \mathbf{C}$ whenever the involved domains are pairwise disjoint. Moreover, the following isomorphisms are easily verified: $\mathbf{A} \times \mathbf{B} \cong \mathbf{B} \times \mathbf{A}$, $\mathbf{A} \times (\mathbf{B} \times \mathbf{C}) \cong (\mathbf{A} \times \mathbf{B}) \times \mathbf{C}$, and $\mathbf{A} \times (\mathbf{B} \sqcup \mathbf{C}) \cong (\mathbf{A} \times \mathbf{B}) \sqcup (\mathbf{A} \times \mathbf{C})$ when $B \cap C = \emptyset$. So, multiplication is commutative and associative, and distributivity holds. The trivial equality $\mathbf{A} = \mathbf{A} \sqcup \emptyset$ and the isomorphism $\mathbf{A} \cong \mathbf{A} \times \widehat{\mathbf{1}}$ show that 0 and 1 are the neutral elements with respect to sum and product, respectively.

As for the ordering, if $\mathfrak{n}(\mathbf{A}) \leq \mathfrak{n}(\mathbf{B})$ pick $\mathbf{A}' \subseteq \mathbf{B}$ with $\mathfrak{n}(\mathbf{A}') = \mathfrak{n}(\mathbf{A})$. Then for every $\mathbf{C}$, we have $\mathbf{A}' \times \mathbf{C} \subseteq \mathbf{B} \times \mathbf{C}$, and so

$$\mathfrak{n}(\mathbf{A}) \cdot \mathfrak{n}(\mathbf{C}) \;=\; \mathfrak{n}(\mathbf{A}') \cdot \mathfrak{n}(\mathbf{C}) \;=\; \mathfrak{n}(\mathbf{A}' \times \mathbf{C}) \;\leq\; \mathfrak{n}(\mathbf{B} \times \mathbf{C}) \;=\; \mathfrak{n}(\mathbf{B}) \cdot \mathfrak{n}(\mathbf{C}).$$

Moreover, if $C \cap A = C \cap B = \emptyset$,

$$\mathfrak{n}(\mathbf{A}) + \mathfrak{n}(\mathbf{C}) \; = \; \mathfrak{n}(\mathbf{A}') + \mathfrak{n}(\mathbf{C}) \; = \; \mathfrak{n}(\mathbf{A}' \sqcup \mathbf{C}) \; \leq \; \mathfrak{n}(\mathbf{B} \sqcup \mathbf{C}) \; = \; \mathfrak{n}(\mathbf{B}) + \mathfrak{n}(\mathbf{C}).$$

This completes the proof that numerosities form an ordered semi-ring. We are left to show the *difference property*. If $\mathbf{D} \subseteq \mathbf{B}$ is the labelled subset with domain $D = B \setminus A'$, then

$$\mathfrak{n}(\mathbf{A}) + \mathfrak{n}(\mathbf{D}) \; = \; \mathfrak{n}(\mathbf{A}') + \mathfrak{n}(\mathbf{D}) \; = \; \mathfrak{n}(\mathbf{A}' \cup \mathbf{D}) \; = \; \mathfrak{n}(\mathbf{B}).$$

Finally, uniqueness of the difference follows from the property that $\mathfrak{n}(\mathbf{A}) + \mathfrak{n}(\mathbf{D}') < \mathfrak{n}(\mathbf{A}) + \mathfrak{n}(\mathbf{D}'')$ implies $\mathfrak{n}(\mathbf{D}') < \mathfrak{n}(\mathbf{D}'')$. □

2. From numerosities to Alpha-Calculus

In Section 2, we presented the notion of Alpha-numerosity for labelled sets that was defined by means of Alpha-limits. The goal of this section is to show that the role of Alpha-limits is in fact essential, because it is the very idea of numerosity that leads to Alpha-Calculus. Following this approach, we will see that the "ideal" infinite natural number α can be defined as the (infinite) numerosity $\mathfrak{n}(\mathbb{N})$ of the set of natural numbers.

As a preliminary result, let us show that every numerosity for labelled sets comes with a family of sets that satisfies the properties of qualified sets of Alpha-Calculus.

- For functions $\varphi, \psi : \mathbb{N} \to \mathbb{N}$, let us denote by

$$E(\varphi, \psi) \; = \; \{n \in \mathbb{N} \mid \varphi(n) = \psi(n)\}.$$

THEOREM 17.5. *Given a numerosity for labelled sets* $\mathfrak{C} = (\mathfrak{L}, \mathfrak{N}, \mathfrak{n})$, *consider the following family of subsets of* $\mathbb{N}$:

$$\mathcal{Q}_{\mathfrak{C}} \; = \; \{E(\varphi_{\mathbf{A}}, \varphi_{\mathbf{B}}) \mid \mathfrak{n}(\mathbf{A}) = \mathfrak{n}(\mathbf{B})\}.$$

Then $\mathcal{Q}_{\mathfrak{C}}$ is a selective ultrafilter on $\mathbb{N}$, that is, it satisfies the following properties:[2]

(1) $\emptyset \notin \mathcal{Q}_{\mathfrak{C}}$ *and* $\mathbb{N} \in \mathcal{Q}_{\mathfrak{C}}$;
(2) $\mathcal{Q}$ *is closed upward:* $B \supseteq A \in \mathcal{Q}_{\mathfrak{C}} \Rightarrow B \in \mathcal{Q}_{\mathfrak{C}}$;
(3) $\mathcal{Q}_{\mathfrak{C}}$ *is closed under intersections:*
     $A_1, \ldots, A_k \in \mathcal{Q}_{\mathfrak{C}} \Rightarrow A_1 \cap \ldots \cap A_k \in \mathcal{Q}_{\mathfrak{C}}$;

---

[2] Properties (1),... ,(6) are the properties of *qualified sets* as itemized in Theorem 2.22; property (7) is the property of *selectiveness* (see Theorems 6.12 and 14.12).

(4) *In every finite partition $A = A_1 \cup \ldots \cup A_k$ of a set $A \in \mathcal{Q}_\mathfrak{c}$ there exists one and only one piece $A_j \in \mathcal{Q}_\mathfrak{c}$;*

(5) $A \notin \mathcal{Q}_\mathfrak{c} \Leftrightarrow A^c \in \mathcal{Q}_\mathfrak{c}$;

(6) *If F is finite, $F \notin \mathcal{Q}_\mathfrak{c}$;*

(7) *For every $f : \mathbb{N} \to \mathbb{N}$ there exists a non-decreasing $g : \mathbb{N} \to \mathbb{N}$ such that $\{n \mid f(n) = g(n)\} \in \mathcal{Q}_\mathfrak{c}$.*

Moreover, $\mathfrak{n}(\mathbf{A}) = \mathfrak{n}(\mathbf{B}) \Leftrightarrow E(\varphi_\mathbf{A}, \varphi_\mathbf{B}) \in \mathcal{Q}_\mathfrak{c}$.

PROOF. The proof that $\mathcal{Q}_\mathfrak{c}$ is an ultrafilter is arranged into three steps. In the first one, we introduce a (seemingly) smaller family $\mathcal{Q}'$, which is easier to deal with. Then, we will prove that such a family satisfies properties $(1), \ldots, (6)$ itemized above. Finally, grounding on the established facts about $\mathcal{Q}'$, we will show that the two families $\mathcal{Q}'$ and $\mathcal{Q}_\mathfrak{c}$ coincide.

For every $X \subseteq \mathbb{N}$, let us denote by $X^+ = \{x + 1 \mid x \in X\}$ the *unit right-translation* of X. It is readily checked that the counting functions of X and X^+ satisfy:

$$\varphi_X(n) = \begin{cases} \varphi_{X^+}(n) & \text{if } n \notin X; \\ \varphi_{X^+}(n) + 1 & \text{if } n \in X, \end{cases}$$

and so, the following inequalities hold for every n:

$$\varphi_{X^+}(n) \leq \varphi_X(n) \leq \varphi_{X^+}(n) + 1 = \varphi_{X^+ \cup \{1\}}(n).$$

By the *Asymptotic principle* (AP), we obtain

$$\mathfrak{n}(X^+) \leq \mathfrak{n}(X) \leq \mathfrak{n}(X^+ \cup \{1\}) = \mathfrak{n}(X^+) + 1,$$

and hence it must be either $\mathfrak{n}(X) = \mathfrak{n}(X^+)$ or $\mathfrak{n}(X) = \mathfrak{n}(X^+) + 1$. Now, let us consider the following family

$$\mathcal{Q}' = \{X \subseteq \mathbb{N} \mid \mathfrak{n}(X) = \mathfrak{n}(X^+) + 1\}.$$

Notice that $n \in X$ if and only if $|X \cap [1, n]| = |(X^+ \cup \{1\}) \cap [1, n]|$, and hence $X = E(\varphi_X, \varphi_{X^+ \cup \{1\}})$; in consequence, $\mathcal{Q}' \subseteq \mathcal{Q}_\mathfrak{c}$. (As the final step of this proof, we will show that in fact $\mathcal{Q}' = \mathcal{Q}_\mathfrak{c}$).

Next, we will show that $\mathcal{Q}'$ satisfies the six properties $(1), \ldots, (6)$. Probably, the shortest way is to derive them from the following fact.

$(\star)$ *In every 3-partition $\mathbb{N} = X_1 \cup X_2 \cup X_3$, exactly one out of the three pieces X_i belongs to $\mathcal{Q}'$.*

Since $\mathbb{N} = \{1\} \cup X_1^+ \cup X_2^+ \cup X_3^+$ is a partition, we have

$$\mathfrak{n}(X_1) + \mathfrak{n}(X_2) + \mathfrak{n}(X_3) = \mathfrak{n}(\mathbb{N}) = \mathfrak{n}(X_1^+) + \mathfrak{n}(X_2^+) + \mathfrak{n}(X_3^+) + 1.$$

We have already noticed above that either $\mathfrak{n}(X_i) = \mathfrak{n}(X_i^+)$ or $\mathfrak{n}(X_i) = \mathfrak{n}(X_i^+) + 1$, and so we conclude that there must be exactly one index i such that $X_i \in \mathcal{Q}'$, and $(\star)$ is proved.

(1) and (5). As trivially $\emptyset = \emptyset^+$, we have that $\emptyset \notin \mathcal{Q}'$. Then, by taking $X_1 = A$, $X_2 = A^c$ and $X_3 = \emptyset$, property $(\star)$ implies that $A \notin \mathcal{Q}' \Leftrightarrow A^c \in \mathcal{Q}'$. In particular, since $\emptyset \notin \mathcal{Q}'$, the complement $\emptyset^c = \mathbb{N} \in \mathcal{Q}'$.

(3). By induction, it is enough to consider the case $k = 2$. Take the 3-partition given by $X_1 = A_1$, $X_2 = A_2 \setminus A_1$, and $X_3 = (A_1 \cup A_2)^c$. Since $A_1 \in \mathcal{Q}'$, it follows that $A_2 \setminus A_1 \notin \mathcal{Q}'$. Now, the sets $A_2 \setminus A_1$, A_2^c and $A_1 \cap A_2$ form a 3-partition where $A_2 \setminus A_1 \notin \mathcal{Q}'$ and $A_2^c \notin \mathcal{Q}'$ because $A_2 \in \mathcal{Q}'$. We conclude that $A_1 \cap A_2 \in \mathcal{Q}'$.

(2). If by contradiction $B \notin \mathcal{Q}'$, then we would have $B^c \in \mathcal{Q}'$, and the intersection $A \cap B^c = \emptyset \in \mathcal{Q}'$, a contradiction.

(4). If it was $A_j \notin \mathcal{Q}'$ for every j, then $A_j^c \in \mathcal{Q}'$ for every j, and we would have $A^c = (A_1 \cup \ldots \cup A_k)^c = A_1^c \cap \ldots \cap A_k^c \in \mathcal{Q}'$, contradicting $A \in \mathcal{Q}'$. Besides, two different pieces of the partition cannot both belong to $\mathcal{Q}'$, otherwise their empty intersection would also belong to $\mathcal{Q}'$.

(6). If $F \subset \mathbb{N}$ is a finite set, then trivially F^+ is a finite set of the same cardinality, and hence $\mathfrak{n}(F) = \mathfrak{n}(F^+)$ and $F \notin \mathcal{Q}'$.

We already noticed that $\mathcal{Q}' \subseteq \mathcal{Q}_\mathfrak{n}$. We will reach the conclusion by showing that in fact $\mathcal{Q}' = \mathcal{Q}_\mathfrak{n}$. To this end, assume by contradiction that $X = E(\varphi_\mathbf{A}, \varphi_\mathbf{B}) \notin \mathcal{Q}'$ where $\mathfrak{n}(\mathbf{A}) = \mathfrak{n}(\mathbf{B})$. Then $\mathfrak{n}(X) = \mathfrak{n}(X^+)$ and the complement $X^c = E(\varphi_X, \varphi_{X^+}) \in \mathcal{Q}' \subseteq \mathcal{Q}_\mathfrak{n}$. Now consider the following labelled sets:[3]

$$\mathbf{C} = \{3\} \times \mathbf{A}\,; \quad \mathbf{D} = \{4\} \times \mathbf{B}\,;$$

$$Y = \{5\} \times X\,; \quad Z = \{6\} \times X^+\,;$$

$$F = \{1\} \times \{1,2,3,4,5\}\,; \quad G = \{2\} \times \{2,3,4,5,6\}.$$

Notice that $\mathfrak{n}(\mathbf{C}) = \mathfrak{n}(\{3\}) \cdot \mathfrak{n}(\mathbf{A}) = \mathfrak{n}(\mathbf{A}) = \mathfrak{n}(\mathbf{B}) = \mathfrak{n}(\{4\}) \cdot \mathfrak{n}(\mathbf{B}) = \mathfrak{n}(\mathbf{D})$, and similarly, $\mathfrak{n}(Y) = \mathfrak{n}(X) = \mathfrak{n}(X^+) = \mathfrak{n}(Z)$. Moreover, by Proposition 17.3, $\mathfrak{n}(F) = \mathfrak{n}(G) = 5$.

The labelled sets $\mathbf{C}, \mathbf{D}, Y, Z, F, G$ are pairwise disjoint, and so we can take the following labelled sets:

$$\mathbf{S} = (\mathbf{C} \times \mathbf{C}) \cup (\mathbf{D} \times \mathbf{D}) \cup (Y \times Y) \cup (Z \times Z) \cup (F \times F) \cup (G \times G)\,;$$

$$\mathbf{T} = (\mathbf{C} \times \mathbf{D}) \cup (\mathbf{D} \times \mathbf{C}) \cup (Y \times Z) \cup (Z \times Y) \cup (F \times G) \cup (G \times F).$$

---

[3] $Y, Z, F, G \subset \mathbb{N} \times \mathbb{N}$ are point sets with the canonical labelling.

By the *Union* and the *Cartesian Product Principle*, we obtain that

$$\begin{aligned}
\mathfrak{n}(\mathbf{S}) &= \mathfrak{n}(\mathbf{C})^2 + \mathfrak{n}(\mathbf{D})^2 + \mathfrak{n}(Y)^2 + \mathfrak{n}(Z)^2 + 25 + 25 \\
&= 2 \cdot \mathfrak{n}(\mathbf{A})^2 + \mathfrak{n}(X)^2 + \mathfrak{n}(X^+)^2 + 50 \\
&= 2 \cdot \mathfrak{n}(\mathbf{C}) \cdot \mathfrak{n}(\mathbf{D}) + 2 \cdot \mathfrak{n}(Z) \cdot \mathfrak{n}(Y) + 25 + 25 = \mathfrak{n}(\mathbf{T}).
\end{aligned}$$

As one can readily verify, if $n \geq 6$ then $\varphi_{\mathbf{C}}(n) = \varphi_{\mathbf{A}}(n)$, $\varphi_{\mathbf{D}}(n) = \varphi_{\mathbf{B}}(n)$, $\varphi_Y(n) = \varphi_X(n)$ and $\varphi_Z(n) = \varphi_{X^+}(n)$. Since $X = E(\varphi_{\mathbf{A}}, \varphi_{\mathbf{B}})$ and $X^c = E(\varphi_X, \varphi_{X^+})$ form a partition of $\mathbb{N}$, it follows that for every $n \geq 6$ either $\varphi_{\mathbf{C}}(n) \neq \varphi_{\mathbf{D}}(n)$ or $\varphi_Y(n) \neq \varphi_Z(n)$. Moreover, if $n \leq 5$, then $\varphi_F(n) = n \neq n-1 = \varphi_G(n)$. So, for every n one has the inequality:

$(\star)$ $\quad (\varphi_{\mathbf{C}}(n) - \varphi_{\mathbf{D}}(n))^2 + (\varphi_Y(n) - \varphi_Z(n))^2 + (\varphi_F(n) - \varphi_G(n))^2 \geq 1.$

Now, for every n we have:

$$\varphi_{\mathbf{S}}(n) = \varphi_{\mathbf{C}}(n)^2 + \varphi_{\mathbf{D}}(n)^2 + \varphi_Y(n)^2 + \varphi_Z(n)^2 + \varphi_F(n)^2 + \varphi_G(n)^2;$$

$$\varphi_{\mathbf{T}}(n) = 2 \cdot \varphi_{\mathbf{C}}(n) \cdot \varphi_{\mathbf{D}}(n) + 2 \cdot \varphi_Y(n) \cdot \varphi_Z(n) + 2 \cdot \varphi_F(n) \cdot \varphi_G(n).$$

By combining with the inequality $(\star)$, we obtain $\varphi_{\mathbf{S}}(n) - \varphi_{\mathbf{T}}(n) \geq 1$, or equivalently, $\varphi_{\mathbf{S}}(n) \geq \varphi_{\mathbf{T} \cup \{1\}}(n)$ for every n. So, by the *asymptotic principle*, $\mathfrak{n}(\mathbf{S}) \geq \mathfrak{n}(\mathbf{T} \cup \{1\}) > \mathfrak{n}(\mathbf{T})$, and this contradicts the equality $\mathfrak{n}(\mathbf{S}) = \mathfrak{n}(\mathbf{T})$ proved above.

Let us now see that $\mathfrak{n}(A) = \mathfrak{n}(B) \Leftrightarrow E(\varphi_A, \varphi_B) \in \mathcal{Q}_{\mathfrak{e}}$. One direction is trivial from the definition of $\mathcal{Q}_{\mathfrak{e}}$. Conversely, assume by contradiction that $X = E(\varphi_{\mathbf{A}}, \varphi_{\mathbf{B}}) \in \mathcal{Q}_{\mathfrak{e}}$ but $\mathfrak{n}(\mathbf{A}) \neq \mathfrak{n}(\mathbf{B})$, say $\mathfrak{n}(\mathbf{A}) < \mathfrak{n}(\mathbf{B})$. Then, by *Zermelo's principle*, we can pick a proper labelled subset $\mathbf{A}' \subset \mathbf{B}$ with $\mathfrak{n}(\mathbf{A}') = \mathfrak{n}(\mathbf{A})$. Notice that $E(\varphi_{\mathbf{A}'}, \varphi_{\mathbf{B}})$ is finite; indeed, if we take any $b_0 \in B \setminus A'$, then $\varphi_{\mathbf{B}}(n) > \varphi_{\mathbf{A}'}(n)$ for all $n \geq \ell_B(b_0)$. Now, $Y = E(\varphi_{\mathbf{A}}, \varphi_{\mathbf{A}'}) \in \mathcal{Q}_{\mathfrak{e}}$ and so, by a property of ultrafilter, also $X \cap Y \in \mathcal{Q}_{\mathfrak{e}}$. This is not possible because $X \cap Y \subseteq E(\varphi_{\mathbf{A}'}, \varphi_{\mathbf{B}})$ is finite.

The selectiveness property (7) is proved by using exactly the same arguments as in the second part of the proof of Theorem 16.33. For completeness, let us repeat them here.

Let $F : \mathbb{N} \to \mathbb{N}_0$ be the function where $F(1) = 0$ and $F(n) = \sum_{i=1}^{n-1} f(i)$ for $n \geq 2$; and let $G : \mathbb{N} \to \mathbb{N}_0$ be the function where $G(n) = \sum_{i=1}^n f(i)$. As F and G are non-decreasing, there exist labelled sets $\mathbf{A}$ and $\mathbf{B}$ such that their counting functions $\varphi_{\mathbf{A}}(n) = F(n)$ and $\varphi_{\mathbf{B}}(n) = G(n)$, so that $\varphi_{\mathbf{A}}(n) + f(n) = \varphi_{\mathbf{B}}(n)$ for every n. Since $\varphi_{\mathbf{A}}(n) < \varphi_{\mathbf{B}}(n)$ for every n, $\mathfrak{n}(\mathbf{A}) < \mathfrak{n}(\mathbf{B})$ and so, by (ZP), there exists a proper labelled subset $\mathbf{A}' \subset \mathbf{B}$ such that $\mathfrak{n}(\mathbf{A}') = \mathfrak{n}_\alpha(\mathbf{A})$, and so $Q = E(\varphi_{\mathbf{A}}, \varphi_{\mathbf{A}'}) = \{n \mid \varphi_{\mathbf{A}}(n) = \varphi_{\mathbf{A}'}(n)\} \in \mathcal{Q}_{\mathfrak{e}}$. Now let $\mathbf{C} \subseteq \mathbf{B}$ the labelled set with domain $B \setminus A'$. Clearly $\mathbf{A}' \cup \mathbf{C} = \mathbf{B}$, and so $\varphi_{\mathbf{A}'}(n) + \varphi_{\mathbf{C}}(n) = \varphi_{\mathbf{B}}(n)$ for every n. But then $\{n \mid \varphi_{\mathbf{C}}(n) = f(n)\} =$

$\{n \mid \varphi_\mathbf{A}(n) + \varphi_\mathbf{C}(n) = \varphi_\mathbf{B}(n)\} = Q \in \mathcal{Q}_\mathfrak{C}$. This completes the proof because $\varphi_\mathbf{C}(n)$ is non-decreasing. □

THEOREM 17.6. *Let* $\mathfrak{C} = (\mathfrak{L}, \mathfrak{N}, \mathfrak{n})$ *be a numerosity for labelled sets. Then there exists a model of Alpha-Calculus where:*

(1) Cauchy's infinitesimal principle *(CIP)* holds;
(2) *The Alpha-numerosity function* $\mathfrak{n}_\alpha : \mathfrak{L} \to \mathbb{N}_0^*$ *(see Definition 16.13) coincides with* $\mathfrak{n}$, *that is,* $\mathfrak{n}_\alpha(\mathbf{A}) = \mathfrak{n}(\mathbf{A})$ *for all* $\mathbf{A} \in \mathfrak{L}$.

PROOF. By the previous Theorem 17.5, the family

$$\mathcal{Q}_\mathfrak{C} = \{E(\varphi_\mathbf{A}, \varphi_\mathbf{B}) \mid \mathfrak{n}(\mathbf{A}) = \mathfrak{n}(\mathbf{B})\}$$

is a selective ultrafilter. By Theorem 11.18, there exists a model of Alpha-Calculus whose family of qualified sets $\mathcal{Q} = \mathcal{Q}_\mathfrak{C}$. Then *Cauchy's Infinitesimal Principle* holds, since $\mathcal{Q}$ is selective (see Theorem 14.12 (2)). So, by Theorem 16.34 and by Proposition 16.12,

$$\mathbb{N}_0^* = \left\{ \lim_{n\uparrow\alpha} f(n) \,\middle|\, f : \mathbb{N} \to \mathbb{N}_0 \text{ non-decreasing} \right\}$$

$$= \left\{ \lim_{n\uparrow\alpha} \varphi_\mathbf{A}(n) \,\middle|\, \mathbf{A} \in \mathfrak{L} \right\} = \{\mathfrak{n}_\alpha(\mathbf{A}) \mid \mathbf{A} \in \mathfrak{L}\}.$$

Finally, for all labelled sets $\mathbf{A}, \mathbf{B} \in \mathfrak{L}$, we have the following equivalences:

$$\mathfrak{n}_\alpha(\mathbf{A}) = \mathfrak{n}_\alpha(\mathbf{B}) \iff \lim_{n\uparrow\alpha} \varphi_\mathbf{A}(n) = \lim_{n\uparrow\alpha} \varphi_\mathbf{B}(n)$$

$$\iff E(\varphi_\mathbf{A}, \varphi_\mathbf{B}) \in \mathcal{Q} = \mathcal{Q}_\mathfrak{C} \iff \mathfrak{n}(\mathbf{A}) = \mathfrak{n}(\mathbf{B}).$$

In consequence, for every $\mathbf{A} \in \mathfrak{L}$ we can identify $\mathfrak{n}_\alpha(\mathbf{A})$ with $\mathfrak{n}(\mathbf{A})$ and have $\mathbb{N}_0^* = \mathfrak{N}$ and $\mathfrak{n}_\alpha = \mathfrak{n}$, as desired. □

Let us summarize below the construction of a the model of Alpha-Calculus plus *Cauchy infinitesimal principle*, as determined by a numerosity counting system $\mathfrak{C} = (\mathfrak{L}, \mathfrak{N}, \mathfrak{n})$ for labelled sets.

- Consider the ring $\mathfrak{Fun}(\mathbb{N}, \mathbb{R})$ of real sequences;
- For $f, g \in \mathfrak{Fun}(\mathbb{N}, \mathbb{R})$, denote by
  - $E(f, g) = \{n \mid f(n) = g(n)\}$,
  - $E(f, g)^+ = \{n + 1 \mid f(n) = g(n)\}$;
- Consider the following equivalence relation on $\mathfrak{Fun}(\mathbb{N}, \mathbb{R})$:

$$f \equiv_\mathfrak{n} g \iff \mathfrak{n}(E(f, g)) = \mathfrak{n}(E(f, g)^+) + 1;$$

- Let $\mathbb{R}^*$ be the quotient $\mathfrak{Fun}(\mathbb{N}, \mathbb{R})/\equiv_n$, where we identify every $r \in \mathbb{R}$ with the equivalence class $[c_r]$ of the corresponding constant sequence;

- For every $f : \mathbb{N} \to \mathbb{R}$, let $\lim_{n \uparrow \alpha} f(n) = [f]$ be the equivalence class of f.

Notice that, by the results proved in this section, there exists a selective ultrafilter $\mathcal{Q}$ such that

$$f \equiv_n g \iff E(f, g) = \{n \mid f(n) = g(n)\} \in \mathcal{Q}.$$

So, the defined $\mathbb{R}^*$ is the ultrapower $\mathbb{R}^{\mathbb{N}}/\mathcal{Q}$, and one obtains a model of Alpha-Calculus, where *Cauchy infinitesimal principle* holds because the family of qualified sets $\mathcal{Q}$ is selective.[4] In particular, recall that $\mathbb{R}^*$ can be characterized as the residue field $\mathfrak{Fun}(\mathbb{N}, \mathbb{R})/\mathfrak{m}$, where $\mathfrak{m} = \{f \in \mathfrak{Fun}(\mathbb{N}, \mathbb{R}) \mid Z(f) \in \mathcal{Q}\}$ is the maximal ideal of those functions whose zero sets belongs to $\mathcal{Q}$.

---

[4] See Theorems 11.17, 11.18, 11.21, 11.22, and 14.12.

3. Remarks and comments

1. A preliminary notion of numerosity for labelled sets was first introduced by V. Benci in [**6**], and then developed and formally axiomatized by V. Benci and M. Di Nasso in [**11**]. Subsequently, several other axiomatizations of the notion of numerosity (or equinumerosity) in various different contexts have been proposed and studied by the authors of this book jointly with M. Forti (see [**12, 14, 38, 24, 49**]). Historical and philosophical issues related to the problem of measuring the size of infinite sets, including a discussion on the notion of numerosity and the "part-whole principle", can be found in [**74, 75**].

2. Relationships between the notion of numerosity and the notion of measure are investigated in [**8, 9**]. A theory of non-Archimedean probability closely related to numerosities is presented in [**17, 18**].

3. In the recent paper [**15**], the theory of labelled sets have been developed to a theory of "Euclidean numbers" where labels are ordinal numbers. In that theory, ordinal numbers are a subclass of numerosities, and their *natural Hessenberg operations* (see, *e.g.*, [**89**, Ch.XIV §28]) are extended to sums and products of numerosities.

Bibliography

[1] S. Albeverio, J.E. Fenstad, R. Høegh-Krohn, T. Lindstrøm, *Nonstandard Methods in Stochastic Analysis and Mathematical Physics*, 1987, reprinted 2009, Dover Books on Mathematics.

[2] R.M. Anderson (1976), A nonstandard representation for Brownian motion and Itô integration, *Bull. Amer. Math. Soc.* **82**, 99–101.

[3] L.O. Arkeryd, N.J. Cutland, and C.W. Henson (1997), *Nonstandard Analysis – Theory and Applications*, NATO ASI Series C **493**, Kluwer A.P..

[4] D. Ballard D. and K. Hrbàček (1992), Standard foundations for nonstandard analysis, *J. Symb. Log.* **57**, 471–478.

[5] T. Bartoszynski and S. Shelah (2008), On the density of Hausdorff ultrafilters, in *"Logic Colloquium 2004"* (A. Andretta, K. Kearnes, and D. Zambella, eds.), Lecture Notes in Logic **29**, A.S.L. - Cambridge University Press, 18–32.

[6] V. Benci (1995), I numeri e gli insiemi etichettati (Italian), *Conferenze del Seminario di Matematica dell'Università di Bari* **261**.

[7] V. Benci (1995), A construction of a nonstandard universe, in *"Advances in Dynamical Systems and Quantum Physics"* (S. Albeverio et als. eds.), World Scientific, 207–237.

[8] V. Benci, E. Bottazzi, and M. Di Nasso (2014), Elementary numerosity and measures, *J. Log. Anal.* **6**, Paper 3.

[9] V. Benci, E. Bottazzi, and M. Di Nasso (2015), Some applications of numerosities in measure theory, *Atti Accad. Naz. Lincei Rend. Lincei Mat. Appl.* **26**, 37–47.

[10] V. Benci V. and M. Di Nasso (2003), Alpha-Theory: an elementary axiomatics for nonstandard analysis, *Expo. Math.* **21**, 355–386.

[11] Benci, V. and Di Nasso M. (2003), Numerosities of labelled sets: a new way of counting, *Adv. Math.* **173**, 50–67.

[12] V. Benci, M. Di Nasso, and M. Forti (2006), An Aristotelian notion of size, *Ann. Pure Appl. Logic* **143**, 43–53.

[13] V. Benci, M. Di Nasso, and M. Forti (2006), The eightfold path to nonstandard analysis, in [**34**], 3–44.

[14] V. Benci, M. Di Nasso, and M. Forti (2007), A Euclidean measure of size for mathematical universes, *Log. Anal.* **197**, 43–62.

[15] V. Benci and M. Forti (2017), The Euclidean numbers, arXiv:1702.04163.

[16] V. Benci, S. Galatolo, and M. Ghimenti (2010), An elementary approach to stochastic differential equations using the infinitesimals, in [**20**], 1–22.

[17] V. Benci, L. Horsten, and S. Wenmackers (2013), Non-Archimedean probability, *Milan J. Math.* **81**, 121–151.

[18] V. Benci, L. Horsten, and S. Wenmackers (2018), Infinitesimal probabilities, *Brit. J. Phil. Sci.* **0** (Advance Access), 1–44.

[19] V. Benci and L. Luperi Baglini (2017), Generalized solutions in PDEs and the Burgers' equation, *J. Differential Equations* **263**, 6916–6952.

[20] V. Bergelson, A. Blass, M. Di Nasso, R.Jin (eds.), *Ultrafilters across Mathematics*, Contemporary Mathematics **530**, Amer. Math. Soc..

[21] A. Blass, *Orderings of ultrafilters*, Ph.D. Thesis, Harvard University, 1970.

[22] A. Blass (2010), Combinatorial cardinal characteristics of the continuum, in *"Handbook of Set Theory"* (Foreman, M. and Kanamori, A., eds.), 395–489.

[23] Blass, A. (2010), Ultrafilters in set theory, in [**20**], 49–71.

[24] A. Blass, M. Di Nasso, and M. Forti (2012), Quasi-selective ultrafilters and asymptotic numerosities, *Adv. Math.* **231**, 1462–1486.

[25] M. Boffa (1972), Forcing et négation de l'axiome de fondement, *Mém. Acad. Sc. Belg.*, tome XL, fasc. 7.

[26] Booth, D. (1970/71), Ultrafilters on a countable set, *Ann. Math. Logic* **2**, 1–24.

[27] E. Bottazzi (2017), *Nonstandard Models in Measure Theory and in Functional Analysis*, PhD thesis, University of Trento.

[28] J. Cleave (1971), Cauchy, convergence and continuity, *Brit. J. Phil. Sci.* **22**, 27–37.

[29] C.C. Chang and H.J. Keisler (1990), Model Theory (3rd edition), North-Holland.

[30] W.W. Comfort and S. Negrepontis (1974), *The Theory of Ultrafilters*, Springer-Verlag.

[31] N.J. Cutland, C. Kessler, E. Kopp, and D.A. Ross (1988), On Cauchy's notion of infinitesimal, *Brit. J. Phil. Sci.* **39**, 375–378.

[32] N.J. Cutland, ed. (1997), *"Nonstandard Analysis and its Applications"*, London Mathematical Society Student Texts **10**, Cambridge University Press.

[33] N.J. Cutland, *Loeb Measures in Practice: Recent Advances*, Lecture Notes in Mathematics, vol. **1751**, Springer, 1997.

[34] N.J. Cutland, M. Di Nasso, and D.A. Ross, eds. (2006), *Nonstandard Methods and Applications in Mathematics*, Lecture Notes in Logic **25**, Association for Symbolic Logic – A K Peters.

[35] M. Daguenet-Teissier (1979), Ultrafiltres à la façon de Ramsey, *Trans. Am. Math. Soc.* **250**, 91–120.

[36] M. Di Nasso (2001), Nonstandard analysis by means of ideal values of sequences, in [**86**], 63–73.

[37] M. Di Nasso and M. Forti (2006), Hausdorff ultrafilters, *Proc. Am. Math. Soc.* **134**, 1809–1818.

[38] M. Di Nasso and M. Forti (2010), Numerosities of point sets over the real line. *Trans. Amer. Math. Soc.*, **362**, 5355–5371.

[39] M. Di Nasso, I. Goldbring, and M. Lupini, *Nonstandard Methods in Ramsey Theory and Combinatorial Number Theory*. 2018, submitted.

[40] F. Diener and K.D. Stroyan (1997). Syntactical Methods in Infinitesimal Analysis, in [**32**], 258–281.

[41] P. Du Bois-Reymond (1877), Über die Paradoxen des Infinitär-Calcüls, *Math. Ann.* **11**, 150–167.

[42] J. Dugundji (1966). *Topology*, Allyn and Bacon, Boston.

[43] P. Ehrlich (2006), The Rise of non-Archimedean Mathematics and the Roots of a Misconception I: The Emergence of non-Archimedean Systems of Magnitudes, *Arch. Hist. Exact Sci.* **60**, 1–121.

[44] P. Ehrlich (2012), The absolute arithmetic continuum and the unification of all numbers great and small, *Bull. Symb. Logic* **18**, 1–45.

[45] F. Enriques, ed. (1923), *Questioni riguardanti le Matematiche Elementari* (Parte Prima), vol. **1**, Zanichelli.

[46] Euclid, *Elements*, Translation from Greek of T.L. Heath (2003), D. Densmore ed. (2nd edition), Green Lion Press.

[47] U. Felgner (1971), Comparison of the axioms of local and universal choice, *Fund. Math.* **71**, 43–62.

[48] G. Fisher (1979), Cauchy's variables and orders of the infinitesimally small, *Brit. J. Phil. Sci.* **30**, 361–365.

[49] M. Forti and G. Morana Roccasalvo (2015), Natural numerosities of sets of tuples, *Trans. Amer. Math. Soc.* **367**, 275–292.

[50] P. Giordano (2010), The ring of Fermat reals, *Adv. Math.* **225**, 2050–2075.

[51] P. Giordano and E. Wu (2016), Calculus in the ring of Fermat reals. Part I: Integral calculus, *Adv. Math.* **289**, 888–927.

[52] R. Goldblatt (1998), *Lectures on the Hyperreals – An Introduction to Nonstandard Analysis*, Graduate Texts in Mathematics **188**, Springer.

[53] J.M. Henle (2003), Second-order non-nonstandard analysis, *Studia Logica* **74**, 399–426.

[54] C.W. Henson (1997), A gentle introduction to nonstandard extensions, in [**3**], 1–49.

[55] W. Hodges (1993), *Model Theory*, Cambridge University Press.

[56] K. Hrbàček (2001), Realism, nonstandard set theory, and large cardinals, *Ann. Pure Appl. Logic* **109**,15–48.

[57] T. Jech (1973), *The Axiom of Choice*, North-Holland. (Re-printed Dover Publications, 2008.)

[58] T. Jech (2002), *Set Theory* (3rd edition), Springer.

[59] V. Kanovei and M. Reeken (2004), *Nonstandard Analysis, Axiomatically*, Springer, Berlin.

[60] H.J. Keisler (1976), *Foundations of Infinitesimal Calculus*, Prindle, Weber & Schmidt, Boston. [Freely downloadable from the author's homepage.]

[61] H.J. Keisler (1984), *An Infinitesimal Approach to Stochastic Analysis*, Mem. Amer. Math. Soc. **48**, AMS.

[62] H.J. Keisler (1986), *Elementary Calculus – An Infinitesimal Approach* (2nd edition), Prindle, Weber & Schmidt, Boston. [Freely downloadable from the author's homepage.]

[63] H.J. Keisler (2010), The ultraproduct construction, in: [**20**], 163–179.

[64] J.L. Kelly (1955) *General Topology*, van Nostrand. (Re-printed: Ishi Press, 2008.)

[65] K. Kunen (1980) *Set Theory: an Introduction to Independent Proofs*, North-Holland.

[66] I. Lakatos (1978), Cauchy and the continuum: the significance of non-standard analysis for the history and philosophy of mathematics, in *"Mathematics, Science and Epistemology"* (J. Worrall, and G. Curries, eds.), Philosophical Papers **2**, Cambridge University Press, 43–60.

[67] S. Lang (2002). *Algebra* (3rd edition), Graduate Texts in Mathematics, Springer.

[68] D. Laugwitz (1980). *The Theory of Infinitesimals – An Introduction to Non-standard Analysis*, Accademia Nazionale dei Lincei.

[69] T. Levi-Civita (1892), Sugli infiniti ed infinitesimi attuali quali elementi analitici, *Atti del R. Istituto Veneto di Scienze Lettere ed Arti* **7**, 1765–1815.

[70] T. Lindstrom (1988), An invitation to nonstandard analysis, in [**33**], 1–105.

[71] P.A. Loeb (1975), Conversion from nonstandard to standard measure spaces and applications in probability theory, *Trans. Amer. Math. Soc.* **211**, 113–122.

[72] P.A. Loeb and M. Wolff, eds. (2015), *Nonstandard Analysis for the Working Mathematician* (2nd edition), Springer.

[73] W.A.J. Luxemburg (1962), *Non-standard Analysis*, Lecture Notes, CalTech - Pasadena.

[74] P. Mancosu (2009), Measuring the size of infinite collections of natural numbers: was Cantor's theory of infinite number inevitable?, *Rev. Symb. Log.* **2**, 612–646.

[75] P. Mancosu (2015), In good company? On Hume's principle and the assignment of numbers to infinite concepts, *Rev. Symb. Log.* **8**, 370–410.

[76] J. Mikusinski J. and R. Sikorski (1973), *Theory of distributions: The sequential approach*, Elsevier.

[77] E. Nelson (1977), Internal Set Theory: a new approach to nonstandard analysis, *Bull. Am. Math. Soc.*, **83**, 1165–1198.

[78] E. Nelson (1987), *Radically elementary probability theory*, Princeton University Press.

[79] E. Palmgren (1995), A constructive approach to nonstandard analysis, *Ann. Pure Appl. Logic* **73**, 297–325.

[80] E. Palmgren (1998), Developments in constructive nonstandard analysis, *Bull. Symb. Logic* **4**, 233–272.

[81] M.M. Richter (1982), *Ideale Punkte, Monaden und Nichtstandard-Methoden*, Vieweg.

[82] A. Robinson (1961), Non-standard analysis, *Proc. Royal Acad. Amsterdam* **64** (= *Indag. Math.* **23**), 432–440.

[83] A. Robinson (1966), *Non-standard Analysis*, North-Holland. (Re-printed by Princeton University Press, 1996.)

[84] A. Robinson and E. Zakon (1969), A set-theoretical characterization of enlargements, in: *"Applications of Model Theory to Algebra, Analysis and Probability"* (W.A.J. Luxemburg, ed.), Holt, Rinehart & Winston, New York, 109–122.

[85] C. Schmieden and D. Laugwitz (1958), Eine erweiterung der infinitesimalrechnung, *Math. Zeitschr.* **69**, 1–39.

[86] P. Schuster, U. Berger, and H. Osswald, eds. (2001), *Reuniting the Antipodes – Constructive and Nonstandard Views of the Continuum*, Synthèse Library **306**, Kluwer A.P..

[87] L. Schwartz (1950/1951). *Théorie des distributions*, Hermann, Paris, 2 vols. (New edition 1966.)

[88] S. Shelah (1998), *Proper and Improper Forcing* (2nd edition), Springer.

[89] W. Sierpiński (1965), *Cardinal and ordinal numbers* (2nd edition), Polska Akademia Nauk Monografie Matematyczne **34**.

[90] K.D. Stroyan and W.A.J. Luxemburg (1976), *Introduction to the Theory of Infinitesimals*, Academic Press.

[91] K.D. Stroyan (1997), *Mathematical Background – Foundations of Infinitesimal Calculus*, Academic Press. [Freely downloadable from the author's homepage.]

[92] T. Todorov and H. Vernaeve (2008), Full algebra of generalized functions and non-standard asymptotic analysis, *Logic and Analysis* **1**, 205–234.

[93] G. Veronese (1890), Il continuo rettilineo e l'assioma V d'Archimede, *Atti della reale Accademia dei Lincei, Memorie* **6**, 603–624.

[94] G. Veronese (1896), Intorno ad alcune osservazioni sui segmenti infiniti e infinitesimi attuali, *Math. Ann.* **47**, 423–432.

Index

A^c is the complement of A, 21

$Z(\varphi) = \{n \mid \varphi(n) = 0\}$ is the zero-set of a sequence φ, 38

$\mathbb{F}$ is a field, 3

$\mathfrak{Fun}(\mathbb{N}, \mathbb{R})$ is the ring of real-valued sequences, 38

$\mathfrak{Fun}_I(A, B)$ is the set of internal functions from A to B, 103

$\mathbb{H}(n)$ is the n-grid of ratio $1/n$, 63

$\mathbb{N}_\infty = \mathbb{N}^* \setminus \mathbb{N}$ is the set of infinite hypernaturals, 102

$\mathcal{P}_\mathcal{I}(A)$ is the family of all internal subsets of A, 103

$\mathbb{Q}(\alpha)$ is the smallest non-Archimedean field, 6

$\mathcal{Q}$ is the family of qualified sets, 24

$\mathbb{R}^*_{\text{fin}}$ is the set of finite hyperreals, 31

$\mathfrak{U}(\mathbb{N})$ is the point set universe over the natural numbers, 283

$\lim_\alpha \varphi$ is the Alpha-limit of the sequence φ, 15

χ_k^c is the characteristic function of $\mathbb{N} \setminus \{k\}$, 39

$d_\alpha(A)$ is the Alpha-density of a set $A \subseteq \mathbb{N}$, 295

η is the infinitesimal number $1/\alpha$, 30

$\mathfrak{gal}(\xi)$ is the galaxy of ξ, 11

$\mathfrak{G}$ is the family of grid functions, 160

$\mathfrak{G}([a, b]$ is the family of grid functions on $[a, b]^*$, 160

$\mathfrak{i}_0$ is the ideal of sequences that eventually vanish, 38

$\lim_{n \uparrow \alpha} \varphi(n)$ is the Alpha-limit of the sequence φ, 15

$\mathfrak{m}_k$ is the principal ideal of sequences that vanish at k, 39

$\mathfrak{mon}(\xi)$ is the monad of ξ, 11

μ_α is the Alpha-measure, 23

$\sim$ is the relation of being infinitely close, 9

$\sim_f$ is the relation of being finitely close, 11

$\varphi_{\restriction X}$ is the restriction of the function φ to the set X, 131

$\xi \ll \zeta$ means that ξ has a smaller order than ζ, 58

$\xi \simeq \zeta$ is the relation of having the same order of magnitude, 58

k-sets, 157

$\mathbb{R}^*$ is the hyperreal field, 15

algebra of sets, 215

almost everywhere (a.e.), 24

Alpha Set Theory AST, 242

Alpha-Calculus, 14
  countably equivalent models, 228
  equivalent models, 225
  model, 223

Alpha-Calculus Theory ACT, 15

Alpha-measure, 23, 216

Alpha-morphism, 42, 218, 223
  reflexive, 143

Alpha-numerosity
  of a labelled set, 280

Alternative Alpha-Theory ALT, 121

anti-derivative, 67

approximating sequence, 64

Archimedean property, 117

Asymptotic principle (AP), 297

atom, 77, 81, 242

axiom
  alpha number, 15, 77, 121, 243
  atom-set, 121
  Boffa's superuniversality, 248